Extracellular Chaperones and Related miRNA as Diagnostic Tools of Chronic Diseases

Extracellular Chaperones and Related miRNA as Diagnostic Tools of Chronic Diseases

Editors

Magdalena Gorska-Ponikowska
Francesco Cappello
Claudia Marino
Francesca Rappa

MDPI • Basel • Beijing • Wuhan • Barcelona • Belgrade • Manchester • Tokyo • Cluj • Tianjin

Editors

Magdalena Gorska-Ponikowska
Department of Medical Chemistry
Medical University of Gdańsk
Gdańsk
Poland

Francesco Cappello
Department of Biomedicine, Neurosciences and Advanced Diagnostics
University of Palermo (UNIPA)
Palermo
Italy

Claudia Marino
Department of Ophthalmology
Institute of Mass Eye and Ear Research
Harvard Medical School
Boston
USA

Francesca Rappa
Department of Biomedicine Neurosciences and Advanced Diagnostics (BIND)
University of Palermo
Palermo
Italy

Editorial Office
MDPI
St. Alban-Anlage 66
4052 Basel, Switzerland

This is a reprint of articles from the Special Issue published online in the open access journal *Applied Sciences* (ISSN 2076-3417) (available at: www.mdpi.com/journal/applsci/special_issues/exosomal_molecular_chaperones).

For citation purposes, cite each article independently as indicated on the article page online and as indicated below:

LastName, A.A.; LastName, B.B.; LastName, C.C. Article Title. *Journal Name* **Year**, *Volume Number*, Page Range.

ISBN 978-3-0365-2148-0 (Hbk)
ISBN 978-3-0365-2147-3 (PDF)

Contents

About the Editors

Magdalena Gorska-Ponikowska

Magdalena Gorska-Ponikowska is a Professor at the Medical University of Gdansk and the Head of the Medical Chemistry Department, Medical University of Gdansk, Poland. Professor Magdalena Gorska-Ponikowska completed her master's degree at the Medical University of Gdansk, Faculty of Pharmacy (2003-2009).

Her master's thesis was entitled "Glucocorticoids and caspase-8 in T lymphocyte apoptosis". The project and study was created as part of the Lifelong Learning Program Erasmus (University of Perugia, Italy). Professor Magdalena Gorska-Ponikowska defended her PhD thesis in 2013. The title of her doctoral dissertation was as follows: "Influence of geldanamycin on the activity and expression of neuronal nitric oxide synthase nNOS in 2-methoxyestradiol-treated osteosarcoma 143B cells". Subsequently, in 2018, she obtained her habilitation in medical sciences.

Professor Magdalena Gorska-Ponikowska is an author or co-author of 49 publications in peer-reviewed international scientific journals and 17 chapters in Polish books (Na Pograniczu Biologii i Chemii, Wiadomosci Chemiczne) and Springer book (Hsp60 in Modifications of Nervous System Homeostasis and Neurodegeneration, DOI: 10.1007/978-3-030- 23154-5 16 in the book: Heat Shock Protein 60 in Human Diseases and Disorders). In addition, she holds the position of Applied Sciences Editor, MDPI, Editor auxiliary, Special Issue in Applied Sciences: Extracellular Chaperones and Related miRNA as Diagnostic Tools of Chronic Diseases as well as Editor of Special Issue in Applied Sciences: Celebrating Applied Sciences Reaches 20,000 Articles Milestone: Feature Papers in Applied Biosciences and Bioengineering Section.

Honorific appointments:

- 2018 - Visiting Professor, Istituto Euro Mediterraneo di Scienza e Tecnologia, Palermo, Italy.

- 2017 - Visiting Professor, Department of Biophysics, University of Stuttgart, Stuttgart, Germany.

- 2017 - Scholarship of Polish Ministry of Science and Higher Education for Eminent Young Scientists No. 0589 / E389 / STYP/ 12/2017.

- 2017 - Laureate of "The Sopot Muse 2017" in the field of science, award of the Mayor of the city of Sopot for research on cancer signaling routes, didactic and organizational achievements.

- 2014 - annual Gdansk Medical University Rector's award for scientific.

Francesco Cappello

Francesco Cappello is Full Professor of Human Anatomy at the University of Palermo (UNIPA), Palermo, Italy since 2016. In 1997 he defended his Degree with honor in Medicine and Surgery at UNIPA, awarded with Best Thesis prize. Subsequently, in 2002, Professor Francesco Cappello became a Specialist in Pathological Anatomy at UNIPA with maximum score. Research interests include primarily cell differentiation, tissue homeostasis and organ regeneration, cell stress and chaperones, as well as nanovesicles.

Honorific appointments:

- Since 2009, Scientific Director of the Euro-Mediterranean Institute of Science and Technology, Palermo, Italy.

- Since 2014, Associate Member of the Faculty at the Neuroscience Graduate Program, University of Texas Medical Branch, Galveston (TX), USA.

- Since 2015, Adjunct Associate Professor at the Department of Biology, Temple University, Philadelphia (PA), USA.

- Since 2019, President of the Italian Society of Experimental Biology.

- Since 2020, Senior Fellow of the Cell Stress Society International.

Claudia Marino

Claudia Marino since 2019 is a Postdoctoral Research Fellow at Schepens Eye Research institute at Mass Eye and Ear, Dept. of Ophthalmology, Harvard Medical School, Boston, MA, USA. In 2013, she defended her Degree with honor in Pharmacy at STEBICEF Department, University of Palermo. Subsequently, in 2017, Dr. Claudia Marino defended her PhD in Neuroscience at the department of Experimental Biomedicine and Clinical Neuroscience, University of Palermo. In 2019, she defended PhD in Neuroscience at the department of Neurology, University of Texas Medical Branch, Galveston, Texas, USA. Research interests include primarily mitochondria, chaperones, Alzheimer's disease, neurodegeneration, protein misfolding, proteinopatheies, and chaperonopathies.

Honorific appointments:

- 2016-2017: President of the Neuroscience Student Organization, University of Texas Medical Branch, Galveston, TX.

- 2015-2016: Vice President of the Neuroscience Graduate Program club, University of Texas Medical Branch, Galveston, TX.

Francesca Rappa

Francesca Rappa from the Department of Biomedicine Neurosciences and Advanced Diagnostics (BIND), University of Palermo, Italy currently is Lecturer of Human Anatomy, University of Palermo, Italy since 2014. In 2001, she defended her Degree in Medicine, Faculty of Medicine, University of Palermo, Italy. Experimental thesis was on the characteristics of cribrous and tubulat areas of prostate carcinoma. In 2006, Professor Francesca Rappa completed a specialization in Surgical Pathology, Faculty of Medicine, University of Palermo, Italy. Final thesis on the biologic and immunohistochemical characteristics of prostate carcinoma with cribrous growth pattern titled: "Hsp60 is the most predictive Heat Shock Protein during large bowel carcinogenesis". In 2014, she defended Doctor of Philosophy in Experimental and Molecular Medicine, Cicle 24th at the Department of Experimental Biomedicine and Clinical Neurosciences (BIONEC), University of Palermo, Italy.

Preface to "Extracellular Chaperones and Related miRNA as Diagnostic Tools of Chronic Diseases"

Chaperones and related miRNAs are actively involved in crucial biological processes such as cell differentiation, tissue homeostasis, and organ remodeling throughout the life span of an individual. An impairment of their function, if not counterbalanced by rescue mechanisms, can lead to the rise and/or perpetuation of human diseases by several—mostly unknown—etiopathophysiological mechanisms.

Magdalena Gorska-Ponikowska, Francesco Cappello, Claudia Marino, Francesca Rappa
Editors

applied sciences

MDPI

Editorial

Editorial for the Special Issue "Extracellular Chaperones and Related miRNA as Diagnostic Tools of Chronic Diseases"

Claudia Marino [1], Magdalena Gorska-Ponikowska [2], Francesca Rappa [3] and Francesco Cappello [3,*]

[1] Schepens Eye Research Institute of Mass Eye and Ear, Department of Ophthalmology, Harvard Medical School, Boston, MA 02114, USA; claudia_marino@meei.harvard.edu
[2] Department of Medical Chemistry, Medical University of Gdansk, 1 Debinki Street, 80-211 Gdansk, Poland; magdalena.gorska-ponikowska@gumed.edu.pl
[3] Department of Biomedicine, Neuroscience and Advanced Diagnostics, University of Palermo, 90127 Palermo, Italy; francesca.rappa@unipa.it
* Correspondence: francesco.cappello@unipa.it

check for updates

Citation: Marino, C.; Gorska-Ponikowska, M.; Rappa, F.; Cappello, F. Editorial for the Special Issue "Extracellular Chaperones and Related miRNA as Diagnostic Tools of Chronic Diseases". *Appl. Sci.* **2021**, *11*, 5517. https://doi.org/10.3390/app11125517

Received: 3 June 2021
Accepted: 8 June 2021
Published: 15 June 2021

Molecular chaperones are a family of proteins that are highly conserved during phylogenesis. They are among the oldest molecules since the appearance of life on Earth, and it is believed that they contributed to the survival of protocells in response to several stressors, both chemical and physical (radiation, temperature, pH, osmolarity, etc.) [1]. Further, these proteins are critically involved in cell proliferation and differentiation, tissue homeostasis, and organ remodeling [1,2].

Since chaperones are very important in preserving cell physiology and tissue homeostasis, their malfunction (for genetic or environmental factors) can result in the pathogenesis of several diseases [3–9]. Many illnesses, either congenital or not, are now referred to as "chaperonopathies" including a few neurodegenerative disorders, cancers, and autoimmune diseases [10]. This novel classification as chaperonopathies could serve as a guide for physicians and biomedical researchers toward the design of novel diagnostic tools or better disease-modifying therapies.

It has only been a few decades since the discovery that molecular chaperones can be actively released by cells and can have a variety of functions in the extracellular environment [11–13]. The most attentive is the interaction between these molecules and the immune system, thus leading to either triggering or modulating the immune response [14,15]. However, other paracrine roles have also been proposed, e.g., in tumorigenesis, favoring tumor cell survival and metastasizing [16]. This latter discovery promoted increasing interest in elucidating not yet fully understood secretory pathways of molecular chaperones [17,18]. Among them, the multivesicular bodies pathway is gaining growing attention as exosomes were "re-discovered" in the last decades.

More recently, a growing interest among the scientific community aims to explore the significance of miRNA secretion by exosomes. Interestingly, a link has been found between these miRNA and molecular chaperones' expression in target cells [19]. This fact increased enormously the interest in the field of chaperonology, as supported by the increased number of experimental protocols that have been proposed to test molecular chaperones as novel diagnostic tools of human diseases [20].

This Special Issue includes several examples of studies conducted in this sense [21–28]. Further, the goal of this issue is to provide novel scientific perspectives and experimental protocols that can be considered in the study of extracellular chaperones and related miRNA.

Funding: M.G.-P. is supported by the ST46 grant funded by Polish Ministry of Science and Higher Education, Medical University of Gdansk, Gdansk, Poland funding. All other authors received no external funding.

Conflicts of Interest: The authors declare no conflict of interest.

References

1. Cappello, F.; Conway de Macario, E.; Macario, A.J.L. HSP60: A Story as Long as Life on the Earth. In *Heat Shock Protein 60 in Human Diseases and Disorders*; Springer: Cham, Switzerland, 2019. [CrossRef]
2. Calabrese, V.; Scapagnini, G.; Ravagna, A.; Giuffrida Stella, A.M.; Butterfield, D.A. Molecular chaperones and their roles in neural cell differentiation. *Dev. Neurosci.* **2002**, *24*, 1–13. [CrossRef]
3. Bellipanni, G.; Cappello, F.; Scalia, F.; Conway de Macario, E.; Macario, A.J.L.; Giordano, A. Zebrafish as a Model for the Study of Chaperonopathies. *J. Cell. Physiol.* **2016**, *231*, 2107–2114. [CrossRef]
4. Rappa, F.; Farina, F.; Zummo, G.; David, S.; Campanella, C.; Carini, F.; Tomasello, G.; Damiani, P.; Cappello, F.; Conway de Macario, E.; et al. HSP-molecular chaperones in cancer biogenesis and tumor therapy: An overview. *Anticancer. Res.* **2012**, *32*, 5139–5150. [PubMed]
5. Scalia, F.; Marino Gammazza, A.; Conway de Macario, E.; Macario, A.J.L.; Cappello, F. Myelin Pathology: Involvement of Molecular Chaperones and the Promise of Chaperonotherapy. *Brain Sci.* **2019**, *9*, 297. [CrossRef] [PubMed]
6. Basset, C.A.; Cappello, F.; Rappa, F.; Lentini, V.L.; Jurjus, A.R.; Conway de Macario, E.; Macario, A.J.L.; Leone, A. Molecular chaperones in tumors of salivary glands. *J. Mol. Histol.* **2020**, *51*, 109–115. [CrossRef] [PubMed]
7. Marino Gammazza, A.; Légaré, S.; Lo Bosco, G.; Fucarino, A.; Angileri, F.; Conway de Macario, E.; Macario, A.J.L.; Cappello, F. Human molecular chaperones share with SARS-CoV-2 antigenic epitopes potentially capable of eliciting autoimmunity against endothelial cells: Possible role of molecular mimicry in COVID-19. *Cell Stress Chaperones* **2020**, *25*, 737–741. [CrossRef]
8. Paladino, L.; Vitale, A.M.; Caruso Bavisotto, C.; Conway de Macario, E.; Cappello, F.; Macario, A.J.L.; Marino Gammazza, A. The Role of Molecular Chaperones in Virus Infection and Implications for Understanding and Treating COVID-19. *J. Clin. Med.* **2020**, *9*, 3518. [CrossRef]
9. Marino, C.; Krishnan, B.; Cappello, F.; Taglialatela, G. Hsp60 Protects against Amyloid β Oligomer Synaptic Toxicity via Modification of Toxic Oligomer Conformation. *ACS Chem. Neurosci.* **2019**, *10*, 2858–2867. [CrossRef]
10. Macario, A.J.L.; Conway de Macario, E. Chaperonopathies and chaperonotherapy. *FEBS Lett.* **2007**, *581*, 3681–3688. [CrossRef]
11. Reddy, V.S.; Madala, S.K.; Trinath, J.; Reddy, G.B. Extracellular small heat shock proteins: Exosomal biogenesis and function. *Cell Stress Chaperones* **2018**, *23*, 441–454. [CrossRef]
12. Tsutsumi, S.; Neckers, L. Extracellular heat shock protein 90: A role for a molecular chaperone in cell motility and cancer metastasis. *Cancer Sci.* **2007**, *98*, 1536–1539. [CrossRef] [PubMed]
13. De Maio, A.; Vazquez, D. Extracellular Heat Shock Proteins: A new location, a new function. *Shock* **2013**, *40*, 239–246. [CrossRef] [PubMed]
14. Zhao, Y.; Zhang, C.; Wei, X.; Li, P.; Cui, Y.; Qin, Y.; Wei, X.; Jin, M.; Kohama, K.; Gao, Y. Heat shock protein 60 stimulates the migration of vascular smooth muscle cells via Toll-like receptor 4 and ERK MAPK activation. *Sci. Rep.* **2015**, *5*, 15352. [CrossRef] [PubMed]
15. Mansilla, M.J.; Costa, C.; Eixarch, H.; Tepavcevic, V.; Castillo, M.; Martin, R.; Lubetzki, C.; Aigrot, M.-S.; Montalban, X.; Espejo, C. Hsp70 Regulates Immune Response in Experimental Autoimmune Encephalomyelitis. *PLoS ONE* **2014**, *9*, e105737. [CrossRef]
16. Albakova, Z.; Siam, M.K.S.; Sacitharan, P.K.; Ziganshin, R.H.; Ryazantsev, D.Y.; Sapozhnikov, A.M. Extracellular heat shock proteins and cancer: New perspectives. *Transl. Oncol.* **2021**, *14*, 100995. [CrossRef]
17. Marino Gammazza, A.; Restivo, V.; Baschi, R.; Caruso Bavisotto, C.; Cefalù, A.B.; Accardi, G.; Conway de Macario, E.; Macario, A.J.L.; Cappello, F.; Monastero, R. Circulating Molecular Chaperones in Subjects with Amnestic Mild Cognitive Impairment and Alzheimer's Disease: Data from the Zabùt Aging Project. *J. Alzheimers Dis.* **2018**, 1–12. [CrossRef]
18. Caruso Bavisotto, C.; Cipolla, C.; Graceffa, G.; Barone, R.; Bucchieri, F.; Bulone, D.; Cabibi, D.; Campanella, C.; Marino Gammazza, A.; Pitruzzella, A.; et al. Immunomorphological Pattern of Molecular Chaperones in Normal and Pathological Thyroid Tissues and Circulating Exosomes: Potential Use in Clinics. *Int. J. Mol. Sci.* **2019**, *20*, 4496. [CrossRef]
19. Place, R.F.; Noonan, E.J. Non-coding RNAs turn up the heat: An emerging layer of novel regulators in the mammalian heat shock response. *Cell Stress Chaperones* **2014**, *19*, 159–172. [CrossRef]
20. Caruso Bavisotto, C.; Graziano, F.; Rappa, F.; Marino Gammazza, A.; Logozzi, M.; Fais, S.; Maugeri, R.; Bucchieri, F.; Conway de Macario, E.; Macario, A.J.L.; et al. Exosomal Chaperones and miRNAs in Gliomagenesis: State-of-Art and Theranostics Perspectives. *Int. J. Mol. Sci.* **2018**, *19*, 2626. [CrossRef]
21. Magdy Beshbishy, A.; Alghamdi, S.; Onyiche, T.; Zahoor, M.; Rivero-Perez, N.; Zaragoza-Bastida, A.; Ghorab, M.; Meshaal, A.; El-Esawi, M.; Hetta, H.; et al. Biogenesis, Biologic Function and Clinical Potential of Exosomes in Different Diseases. *Appl. Sci.* **2020**, *10*, 4428. [CrossRef]
22. Krawczyk, M.; Pospieszynska, A.; Styczewska, M.; Bien, E.; Sawicki, S.; Marino Gammazza, A.; Fucarino, A.; Gorska-Ponikowska, M. Extracellular Chaperones as Novel Biomarkers of Overall Cancer Progression and Efficacy of Anticancer Therapy. *Appl. Sci.* **2020**, *10*, 6009. [CrossRef]
23. Vitale, A.; Santonocito, R.; Vergilio, G.; Marino Gammazza, A.; Campanella, C.; Conway de Macario, E.; Bucchieri, F.; Macario, A.; Caruso Bavisotto, C. Brain Tumor-Derived Extracellular Vesicles as Carriers of Disease Markers: Molecular Chaperones and MicroRNAs. *Appl. Sci.* **2020**, *10*, 6961. [CrossRef]
24. Nuzziello, N.; Consiglio, A.; Viterbo, R.; Licciulli, F.; Liuni, S.; Trojano, M.; Liguori, M. A Pilot Longitudinal Evaluation of MicroRNAs for Monitoring the Cognitive Impairment in Pediatric Multiple Sclerosis. *Appl. Sci.* **2020**, *10*, 8274. [CrossRef]

25. Alberti, G.; Paladino, L.; Vitale, A.; Caruso Bavisotto, C.; Conway de Macario, E.; Campanella, C.; Macario, A.; Marino Gammazza, A. Functions and Therapeutic Potential of Extracellular Hsp60, Hsp70, and Hsp90 in Neuroinflammatory Disorders. *Appl. Sci.* **2021**, *11*, 736. [CrossRef]
26. Scalia, F.; Vitale, A.; Santonocito, R.; Conway de Macario, E.; Macario, A.; Cappello, F. The Neurochaperonopathies: Anomalies of the Chaperone System with Pathogenic Effects in Neurodegenerative and Neuromuscular Disorders. *Appl. Sci.* **2021**, *11*, 898. [CrossRef]
27. David, S.; Vitale, A.; Fucarino, A.; Scalia, F.; Vergilio, G.; Conway de Macario, E.; Macario, A.; Caruso Bavisotto, C.; Pitruzzella, A. The Challenging Riddle about the Janus-Type Solution. *Appl. Sci.* **2021**, *11*, 1175. [CrossRef]
28. Graziano, F.; Iacopino, D.; Cammarata, G.; Scalia, G.; Campanella, C.; Giannone, A.; Porcasi, R.; Florena, A.; Conway de Macario, E.; Macario, A.; et al. The Triad Hsp60-miRNAs-Extracellular Vesicles in Brain Tumors: Assessing Its Components for Understanding Tumorigenesis and Monitoring Patients. *Appl. Sci.* **2021**, *11*, 2867. [CrossRef]

applied sciences

MDPI

Article

The Triad Hsp60-miRNAs-Extracellular Vesicles in Brain Tumors: Assessing Its Components for Understanding Tumorigenesis and Monitoring Patients

Francesca Graziano [1,2,*], Domenico Gerardo Iacopino [1], Giacomo Cammarata [1], Gianluca Scalia [2], Claudia Campanella [3], Antonino Giulio Giannone [4], Rossana Porcasi [4], Ada Maria Florena [4], Everly Conway de Macario [5], Alberto J.L. Macario [5,6], Giovanni Federico Nicoletti [2] and Celeste Caruso Bavisotto [3,6]

1 Department of Biomedicine, Neurosciences and Advanced Diagnostics, Institute of Neurosurgery, University of Palermo, 90127 Palermo, Italy; gerardo.iacopino@gmail.com (D.G.I.); giacamma95@gmail.com (G.C.)
2 Department of Neurosurgery, Highly Specialized Hospital and of National Importance "Garibaldi", 95122 Catania, Italy; gianluca.scalia@outlook.it (G.S.); gfnicoletti@tiscali.it (G.F.N.)
3 Department of Biomedicine, Neurosciences and Advanced Diagnostics, Section of Human Anatomy, University of Palermo, 90127 Palermo, Italy; claudia.campanella@unipa.it (C.C.); celeste.carusobavisotto@unipa.it (C.C.B.)
4 Department of Health Promotion, Mother and Child Care, Internal Medicine and Medical Specialties—Pathologic Anatomy Unit, University of Palermo, 90100 Palermo, Italy; giulio.giannone@unipa.it (A.G.G.); r.porcasi@libero.it (R.P.); adamaria.florena@unipa.it (A.M.F.)
5 Department of Microbiology and Immunology, School of Medicine, University of Maryland at Baltimore—Institute of Marine and Environmental Technology (IMET), Baltimore, MD 21202, USA; econwaydemacario@som.umaryland.edu (E.C.d.M.); ajlmacario@som.umaryland.edu (A.J.L.M.)
6 Euro-Mediterranean Institute of Science and Technology (IEMEST), 90139 Palermo, Italy
* Correspondence: fragraziano9@gmail.com; Tel./Fax: +39-091-6552391

Citation: Graziano, F.; Iacopino, D.G.; Cammarata, G.; Scalia, G.; Campanella, C.; Giannone, A.G.; Porcasi, R.; Florena, A.M.; Conway de Macario, E.; Macario, A.J.L.; et al. The Triad Hsp60-miRNAs-Extracellular Vesicles in Brain Tumors: Assessing Its Components for Understanding Tumorigenesis and Monitoring Patients. *Appl. Sci.* **2021**, *11*, 2867. https://doi.org/10.3390/app11062867

Academic Editor: Francisco Arrebola

Received: 21 February 2021
Accepted: 19 March 2021
Published: 23 March 2021

Abstract: Brain tumors have a poor prognosis and progress must be made for developing efficacious treatments, but for this to occur their biology and interaction with the host must be elucidated beyond current knowledge. What has been learned from other tumors may be applied to study brain tumors, for example, the role of Hsp60, miRNAs, and extracellular vesicles (EVs) in the mechanisms of cell proliferation and dissemination, and resistance to immune attack and anticancer drugs. It has been established that Hsp60 increases in cancer cells, in which it occurs not only in the mitochondria but also in the cytosol and plasma-cell membrane and it is released in EVs into the extracellular space and in circulation. There is evidence suggesting that these EVs interact with cells near and far from their original cell and that this interaction has an impact on the functions of the target cell. It is assumed that this crosstalk between cancer and host cells favors carcinogenesis in various ways. We, therefore, propose to study the triad Hsp60-related miRNAs-EVs in brain tumors and have standardized methods for the purpose. These revealed that EVs with Hsp60 and related miRNAs increase in patients' blood in a manner that reflects disease status. The means are now available to monitor brain tumor patients by measuring the triad and to dissect its effects on target cells in vitro, and in experimental models in vivo.

Keywords: chaperone system; molecular chaperones; chaperonopathies; Hsp60; miRNAs; extracellular vesicles; high-grade glioma; glioblastoma; meningioma; liquid biopsy; tumor biomarkers

1. Introduction

Primary brain tumors are among the top 10 causes of cancer-related deaths [1]. Glioblastoma is a common malignant primary brain tumor, representing approximately 57% of all gliomas and 48% of all primary malignant tumors of the central nervous system (CNS) [2]. Because of the poor clinical outcomes, glioblastoma multiforme (GBM) is among the most challenging human tumors for patients and physician alike [3]. The possibility of long-term survival is remote and much of the focus and treatment decisions are based on cognitive and quality of life issues. The overall survival is usually around 9–12 months and the overall 5-year survival is less than 5% [4–7]. The dismal GBM clinical outcome, owing to highly infiltrative growth, intra-tumor heterogeneity, and high recurrence rates, has made it an urgent subject of cancer research for identification of novel factors associated with its development [8,9]. With the advancement in the knowledge of the molecular pathology of malignant gliomas, it is now evident that epigenetic abnormalities and variations in gene expression are closely related to the occurrence of these tumors and may provide key targets for developing novel means for diagnosis, assessing prognosis, and disease monitoring [8].

Meningiomas are also common primary intracranial tumors and, in most cases, they are histologically benign (WHO grade I). Due to a lack of prospective, randomized trials, standardized treatment guidelines are difficult to formulate. A gross total resection (GTR) remains the gold standard, though a complete removal is not always achievable. A significant subgroup of patients (WHO grades II and III) cannot be submitted to GTR, but less radical resection and postoperative adjuvant radiotherapy and systemic therapies are indicated. Thus, gliomas and meningiomas constitute a great challenge for physicians and a hopeless situation for many patients who are still awaiting the development of efficacious treatments.

In the last several years, cancerology has progressed along several lines, one of which is the understanding of the chaperone, or chaperoning system (CS) and its role in carcinogenesis. The CS of an organism is composed of the entire set of molecular chaperones, some of which are Heat Shock Proteins (Hsps), co-chaperones, chaperone co-factors, and chaperone interactors, and receptors [10]. The canonical function of the CS is maintenance of protein homeostasis and, in this, its main collaborators are the ubiquitin–proteasome system and the chaperone-mediated autophagy [11,12]. In addition, the CS has other functions that involve interaction with the immune system and pertain to carcinogenesis, and inflammatory and autoimmune conditions [13]. Therefore, a full understanding of the biology and pathology of a variety of cancers should include the study of the CS role in their initiation and progression, as well as in their regression when pertinent. It must be borne in mind that the CS has multiple components that interact with one another and with molecules widespread in the body, so a chaperone, for instance, can exit its cell of origin, reach the intercellular space, and enter in the circulating fluids' lymph and blood, in which it can travel to its destination near or far. For instance, the chaperone Hsp60 is increased in some human brain cancers [14–17], and its depletion in in vivo models of GBM WHO grade IV tumors is associated with intracranial tumor regression [18]. Hsp60 levels are under the control of regulatory molecules, such as microRNAs (miRNAs) [19–24], which are noncoding small RNAs that play various roles in oncogenesis acting as oncogenes or tumor suppressors in certain tumors, including meningiomas and gliomas [25–27]. The Hsp60 chaperone in its physiological or pathological migrations travels alone or in extracellular vesicles (EVs), can interact with cells near and far from the cell of origin and may change their functions [28–35]. For these reasons, Hsp60 and other chaperones can be considered biomarkers potentially useful in diagnosis, and in assessing prognosis and response to treatment [36–40].

It is clear from the existing data that the triad Hsp60 and related miRNAs and EVs play critical roles in tumorigenesis and are promising biomarkers for monitoring patients. Based on this information and considering the need for precise diagnostic tools applicable to brain tumors, we have standardized a battery of techniques to measure the three biomarkers in liquid biopsies from patients with glioblastomas and meningiomas. The methods generate a set of complementary results that reveal patient status and clues on disease mechanisms. In this short report, we describe the methods and the kind of information they provide.

2. Materials and Methods

The procedures described are the result of various standardization efforts done over time to define the best parameters for each of them and are explained in detail for the benefit of future users in clinical pathology and research.

2.1. Samples

To test the standardized methods, a total of 34 patients with gliomas (n = 16) or meningioma (n = 18) were recruited within the period December 2016 to October 2018 (Table 1A,B). The project was approved by the Palermo Ethics Committee I (number 11\2018). Written informed consent was obtained from each subject. The study included only patients without a systemic infection or other tumors and capable of providing consent on their own. The GBM patients selected were only those desiring to participate in the study and in which a maximal safe tumor removal was deemed feasible considering neurological status, including neuroradiological imaging evaluation (Table 1A,B). Brain glial tumors with corpus callosum infiltration were excluded. For meningiomas, we included patients with primary diagnosis of meningiomas, without previous radiotherapy. All brain surgeries were performed by the same neurosurgical team, including the Neurosurgical Unit Chairman D.G.I., senior author of the manuscript and his collaborator F.G. Blood and pathological tissue samples were taken from each patient on the day of the surgery. For each patient, blood samples were collected at different time points, just before surgery (Before Surgery, BS), and at one week (After Surgery 1, AS1), and one month (After Surgery, AS2) after surgical tumor resection, and were processed for serum separation. A brain MRI post-gadolinium within 72 h (when feasible) and at one week, and at three months from discharge, was performed to verify tumor recurrence for correlation with the results of our analysis. Patients affected by GBM, after the definitive histopathological analysis, underwent the STUPP protocol therapy as usual [41,42]. Patients affected by brain meningiomas WHO GI or II after complete tumor removal were followed at our Outpatient Clinic. Patients affected by anaplastic meningiomas underwent oncological management and radiotherapy.

Table 1. (**A**) Key data on GMB patients and follow-up. (**B**) Key data on meningioma patients and follow-up.

(A)

Pt [a]	Sex/Age (years)	Brain Site	Histopathology [b]	Follow-Up	
				Molecular BS-AS1-AS2	Clinical/Months
1	M/65	Left temporal-peritrigonal	HGG GBM IV WHO NOS: GFAP+, IDH+, Synaptofisin-; Ki67+ <10%	C	14
2	M/57	Right frontal	HGG GBM IV WHO NOS: p53+, GFAP+, CAM 5.2-; Ki67 > 60%	C	10
3	M/60	Left temporo-parietal	LGG II WHO: GFAP+, Synaptofisin+, Ki67+ < 3%	C	12
4	M/55	Left fronto-temporal	HGG GBM IV WHO NOS: GFAP++, PanCK+, Ki67+ > 30%	C	8
5	F/65	Right fronto-temporo-parietal	HGG GBM IV WHO IDH mutant: Framments HGG: GFAP+, IDH+, p53+ > 3%; Ki67+ > 30%: Framments Anaplastic Astrocitoma: Ki67+ < 5%	C	9
6	F/70	Left fronto-temporal-peritrigonal	HGG III WHO NOS: High differentiated elements: GFAP +, S-100+, synaptofisin +, rounded cells. CD57+, vimentin+, GFAP-, NSE-, synaptofisin- with Ki67 15%, CAM 5.2 -, PanCK -, p53 -, factor XIII -, desmin -, CD68 -, CD45 -	I	5;D
7	F/58	Right frontal	HGG GBM IV NOS: GFAP+; S-100+; Synaptofisin-; NSE-; CD34-; Ki-67+: 30%	C	4; D
8	M/60	Left temporo-parietal	HGG GBM IV NOS: GFAP (+/-), S100 (+), p53 (+/-), synaptofisin -, CD57 -, CD99 -, Cam 5.2-, bcatenin -, CD45-, Ki67 30%	C	12
9	M/67	Left frontal	HGG GBM IV NOS: GFAP+, Synaptofisin-, NSE+/-, Vimentin +/-; Ki-67+: 30% S-100+	C	10
10	F/56	Right temporo-parietal	HGG GBM IV IDH mutant: GFAP+, S100+, synaptofisin-, PanCK+, p53+, IDH+; Ki67 > 35%	C	14
11	M/68	Left frontal	HGG GBM IV NOS: GFAP+, synaptofisin+, PanCK+, Ki67 30%	I	5; R
12	F/58	Left frontal	HGG GBM IV NOS: GFAP+++; synaptofisin -; Ki67 + 25%	I	5
13	M/63	Right temporo-parietal	HGG GBM IV NOS: LARGE CELLS GFAP+, S100+, PanCK+, p53−/-, synaptofisin-, c-myc-, CD45-, CD20-, CD57-, NSE-, CD99+/-, neurofilaments-. SMALL CELLS GFAP-, S100 -/+, PanCK-, p53+/-, synaptofisin-, c-myc-, CD45-, CD20-, CD57-, NSE-/+, CD99+/-, neurofilaments-	C	18
14	M/72	Right parietal	HGG GBM IV IDH wild type: GFAP+, synaptofisin-, Ki-67+ 20%.	C	6; R
15	M/59	Left frontal	HGG GBM IV IDH MUTANT: GFAP+, synaptofisin-, IDH +, CD34+, Ki67 30%	C	10
16	M/64	Left fronto-parietal	HGG GBM IV NOS: GFAP+, CD34-, synaptofisin-, Ki67 30%	C	8; R

Table 1. *Cont.*

(B)					
Pt [a]	Sex/Age (years)	Anatomical Site	Histopathological Classification [b]	Follow-Up	
				Molecular BS-AS1-AS2	Clinical /Months
1	M/60	Right temporal convexity	Angiomatous meningioma G I	C	14
2	F/55	Right convexity frontal	Fibrous meningioma G I	C	10
3	M/60	Left parietal convexity	Atypical meningioma G II	C	12
4	F/54	Fronto-basal/Olfactory groove	Angiomatous meningioma G I	C	8
5	F/60	Parafalcine parietal convexity	Transitional meningioma G I	C	9
6	F/70	Temporal CPA	Meningothelial meningioma G I	C	10
7	M/60	Right convexity frontal	Atypical meningioma G II	C	12
8	M/68	Occipital/Tentorial	Transitional meningioma G I	C	12
9	F/62	Left temporal convexity	Meningothelial meningioma G I	C	10
10	F/50	Occipital-Foramen Magnum	Fibrous meningioma G I	C	14
11	M/72	Left frontal convexity	Atypical meningioma G II	C	
12	F/65	Right temporal-sphenoid wing	Atypical meningioma G II	C	5
13	M/68	Fronto-basal/Planum	Transitional meningioma G I	C	18
14	M/71	Parafalcine frontoparietal convexity	Atypical meningiomas G II	C	12; D
15	F/59	Skull base/olfactory groove	Atypical meningioma G II	C	10
16	M/67	Parafalicine left parietal convexity	Atypical meningioma G II	C	8
17	F/48	Temporal-CPA	Fibrous meningioma G I	C	12
18	M/69	Right frontal convexity	Meningothelial meningioma G I	C	10

(**A**) [a] Abbreviations: GBM, glioblastoma multiforme; Pt, patient; HGG, high-grade glioma; LGG, low-grade glioma; M, male; F, female; C, complete; I, incomplete; D, Death; R, recurrence; NOS, not otherwise specified; GFAP, Glial fibrillary acidic protein; IDH, Isocitrate-dehydrogenases; PanCK, Protein kinase; NSE, Neuron-Specific Enolase; BS time 0- before surgery; AS1 time 1- at 1 week from surgery; AS2 time II- at one month from surgery (see text for details). [b] According to WHO 2016 guidelines. (**B**) [a] Abbreviations. Pt, patient; M, male; F, female; G, grade; CPA, cerebellopontine angle; C, complete; D, death; BS time 0- before surgery; AS1 time 1- at 1 week from surgery; AS2 time II- at one month from surgery (see text for details). [b] According to WHO 2016 guidelines.

2.2. *Histopathology*

A basic tenet of our standardization strategy was the histological characterization of the brain tumors. These were classified following the guidelines of the WHO classification 2016.

2.3. *Tissue Processing and Immunomorphological Analysis*

The samples of pathological tissues were embedded in paraffin and used for immunohistochemistry. Hematoxylin-eosin (H&E) staining was applied for pathologic assessment. Immunohistochemistry was performed on 5 μm thick sections, which were dewaxed in xylene for 30 min at 60 °C and rehydrated at 23 °C by sequential immersion in a graded series of alcohols. For antigen retrieval, the sections were immersed for 8 min in sodium citrate buffer (pH 6) at 95 °C and, afterwards, immersed for 8 min in acetone at −20 °C to prevent the detachment of the sections from the slide. After washing the sections with phosphate buffer saline (pH 7.4), protein detection was performed by the streptavidin–biotin complex method, using a Histostain®-Plus Third Gen IHC Detection Kit (Life Technologies, Frederik, MD, USA; Cat. No. 85–9073). For the detection of Hsp60, mouse anti-Hsp60 monoclonal antibody (Sigma, St. Louis, MO, USA; catalogue No. H4149, dilution 1:200) was used by immunohistochemistry as previously described [43]. After deparaffinization, sections were treated with Peroxidase Blocking Reagent (Cell and Tissue Staining Kit, R&D Systems, Inc., Minneapolis, MN, USA) to inhibit endogenous peroxidase activity and with serum-blocking reagent D (Cell and Tissue Staining Kit) to block non-specific antigenic sites. Then, the sections were treated with avidin-blocking reagent following the kit instructions (Cell and Tissue Staining Kit), and incubated overnight at 4 °C with primary antibody, as pertinent. After washings, the sections were incubated with biotinylated secondary antibody (Cell and Tissue Staining Kit) for 60 min and, subsequently, with high-sensitivity streptavidin-conjugated HRP (HSS-HRP) in the dark for 5 min with the DAB chromogen. Nuclear counterstaining was carried out using hematoxylin (Hematoxylin REF 05-06012/L Bio-Optica, Milano, Italy). Finally, the slides were prepared for observation with coverslips, using a permanent mounting medium (Vecta Mount, Vector, H-5000).

The examination of the sections was performed with an optical microscope (Leica DM 5000 B) connected to a digital camera (Leica DC 300F), at a magnification of 400X and the percentage of positive cells was calculated in a high-power field (HPF) and repeated for 10 HPF. From these data, the total percentages of cells positive for Hsp60 were determined.

2.4. EV Isolation and Characterization

Blood samples were collected following a standard procedure as previously described [43,44]. EVs isolation from plasma was carried out by several steps of differential ultracentrifugation and by ultrafiltration. Briefly, 3 mL of plasma were centrifuged at 11,000× g for 30 min to remove cell debris. The supernatant was diluted with PBS, then filtered through a 0.2 µm filter (Millex GP, Millipore, Darmstadt, Germany), followed by a two-step ultracentrifugation at 110,000× g for 2 h to pellet the EVs. The EVs were then washed in cold PBS and resuspended in 100 µl of PBS for morphological evaluations or in 70 µl of RIPA (radioimmunoprecipitation assay) lysis buffer (0.3M NaCl, 0.1% SDS, 25 mm HEPES pH 7.5, 1.5 mm MgCl2, 0.2 mm EDTA, 1% Triton X-100, 0.5 mm DTT, 0.5% sodium deoxycholate) for Western blotting (WB) [44].

In order to estimate their morphology, EVs were examined with Transmission Electron Microscopy (TEM) (JEOL JEM 1220 TEM at 120 kV); Atomic Force Microscopy (AFM), using a multimode scanning probe microscope driven by a nanoscope V controller (Digital Instruments, Bruker, Kennewick, WA, USA); and Dynamic Light Scattering (DLS), using a Brookhaven Instrument BI200-SM goniometer; as previous described [43,44].

Western Blot was performed to detect the EVs markers Alix (mouse anti-Alix, 1A12 clone, Santa Cruz Biotechnology, Inc., Dallas, TX, USA); Hsp70/Hsc70 (mouse anti-Hsp70/Hsc70, W27 clone, Santa Cruz Biotechnology); and CD81 (mouse anti-CD81, B-11 clone, Santa Cruz Biotechnology) [43,44].

2.5. Western Blot

The EVs Hsp60 was assessed by WB, using equal amounts of protein (50 µg) for each sample, anti-Hsp60 monoclonal antibody (mouse anti-Hsp60, LK1 clone, Sigma, St. Louis, MO, USA), and horseradish peroxidase-conjugated sheep anti-mouse antibody (GE Healthcare Life Science, Milan, Italy). WBs were detected using the Amersham enhanced chemiluminescence substrate (GE Healthcare Life Science, Marlborough, MA, USA), following the manufacturer's instructions. Densitometric analyses of WB were performed using the National Institutes of Health Image J analysis program (version 1.40. National Institutes of Health, Bethesda, MD, USA).

2.6. MicroRNAs Extraction and Real-Time PCR

Total RNA, including small RNA, was isolated from EVs using the miRNeasy Mini Kit® (Qiagen, Hilden, Germany, Cat No: 74104), following the manufacturer's instructions. The online public miRNA bioinformatic program, TargetScan (http://www.targetscan.org/; last accessed on 6 August 2020), was used to predict miRNAs that can potentially bind to 3′UTR of human Hsp60 mRNA [45]. Next, measurement of miRNAs was performed with the miScript II RT Kit (Qiagen, Cat No: 218161) and the miScript SYBR Green PCR Kit (Qiagen). The reactions assessing the miRNAs levels were performed using the Rotor-gene™ 6000 Real-Time PCR Machine (Qiagen).

The target miRNA expression Ct (Cycle threshold) was normalized with the miR-16 Ct and the mean of Ct values of target and mean of Ct values of miR-16 were compared. The calculation was based on ΔΔCt (Livak method), and fold change values of all samples were calculated as compared to reference [46].

2.7. Statistics

All experimental results are presented as the mean ± S.E.M, with at least three independent replications The statistically significant difference between groups was tested by one-way analysis of variance (ANOVA). Values of $p \leq 0.05$ were considered statistically significant.

3. Results

The standardized methods described above were applied to the study of a small set of clinical samples, and the type of results they provide are reported below:

3.1. Patients Follow-Up

A total of 34 patients were studied, 14 with GBM, 1 with LLG (low-grade glioma), 1 with HGG (high-grade glioma) III and 18 with meningioma (Table 1A,B). Follow-up via blood sampling was completed in 31 patients. Three GBM and 1 meningioma patients died within the period of this investigation.

3.2. Immunomorphological Analysis

Hsp60 levels were assessed in healthy cortical brain tissue derived from autopsy, and in tumor biopsies of GBM and meningioma (Figure 1). As approved by the Palermo Ethics Committee 1, the healthy cortical brain tissue samples were taken from the histopathological archives of the University Hospital Forensic Medicine. Since the samples of healthy cortical brain tissue were obtained from the autopsy of subjects who had died of causes unrelated to brain disease, they were considered as the control group. The hematoxylin-eosin stain showed a high proliferation rate and the immunohistochemical reactions showed high levels of Hsp60 in tumor samples compared with controls. A strong, diffuse cytoplasmic positivity for the Hsp60 protein was present in 100% of the tumor specimens examined (Figure 1).

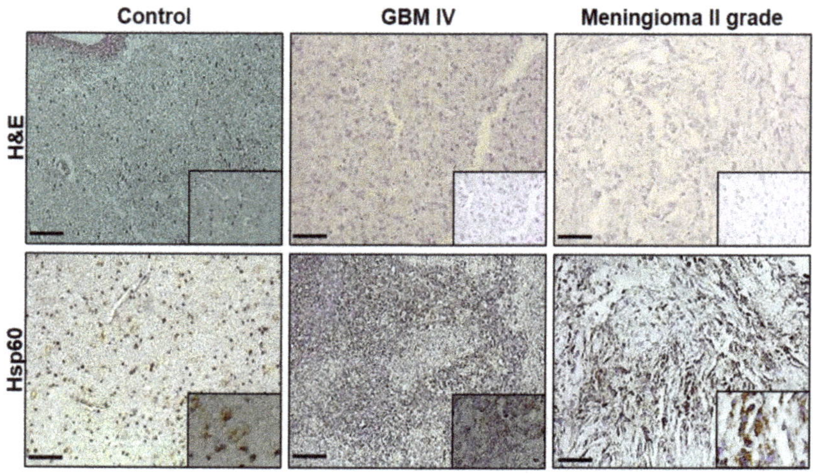

Figure 1. Illustrative examples of the histological and immunohistochemical images provided by the methods used when applied to the study of control and tumor tissues. Top three panels. Hematoxylin-eosin (H&E)-stained tissue sections of Control, and GBM IV, and Meningioma II grade tissues. Bottom three panels. (Hsp60). Immunohistochemical demonstration of Hsp60 in Control, and GBM IV, and Meningioma II grade. Magnification of 200×; insert magnification of 400× Bar = 100 μm.

3.3. Characterization of EVs from GBM and Meningioma II Grade

EVs obtained from plasma of patients with GBM and meningioma G II, before and after surgery, were characterized by TEM, AFM, and DLS to assess size and morphology (Figure 2A–C); and by Western blot to determine the presence of the typical EV markers (protein Alix, Hsc70, and CD81) (Figure 2D). The results all agreed, demonstrating the identity of the EVs.

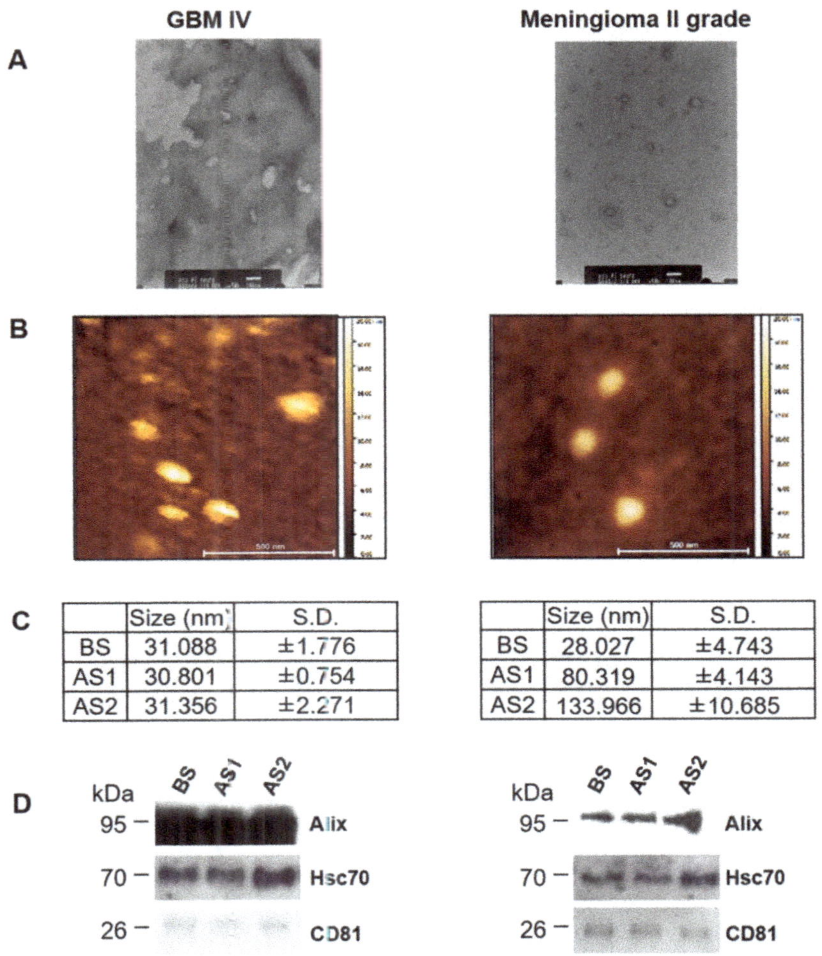

	Size (nm)	S.D.
BS	31.088	±1.776
AS1	30.801	±0.754
AS2	31.356	±2.271

	Size (nm)	S.D.
BS	28.027	±4.743
AS1	80.319	±4.143
AS2	133.966	±10.685

Figure 2. EV characterization. Representative TEM (**A**) and AFM (**B**) images showing the typical characteristics of EVs isolated from the plasma of patients with GBM IV (left) and Meningioma II grade (right). (**C**) Table showing the size, measured by DLS, of the EVs isolated from blood of patients at different times of follow-up: BS, Before Surgery; AS1, 7 days After Surgery; AS2, 30 days After Surgery. (**D**) Evaluation by WB of EVs markers in the EVs from patients with GBM IV (left) and Meningioma II grade (right), at different times of follow-up.

3.4. Hsp60 and Related miRNAs Levels in EVs

The levels of Hsp60 and related miRNAs were assessed in the EVs isolated from the plasma of patients before and at various times after surgery. Hsp60 levels in EVs from GBM patients were successfully measured in all the samples, with the results showing high levels throughout (Figure 3A). The Hsp60 levels in EVs isolated from plasma of patient with atypical meningiomas (WHO G II) were also successfully measured and the results showed that the methods applied can distinguish variations in the chaperonin levels in different patients/situations. For instance, 7 days after surgery, the Hsp60 levels were significantly lower than before surgery, and at 30 days after surgery, a significant increase could be detected (Figure 3A1).

TargetScan prediction revealed that the 3′ UTR of HSPD1 (the Hsp60 gene) contains a putative miR-1 and miR-206 binding site. Therefore, our attention was focused on miR-1 and miR-206 (Figure 3B), which are predicted to regulate Hsp60 expression [45] (confirmed by experimental data [20–24]), and miR-663, considered an in vivo GBM prognostic biomarker [47]. MiR-16 level was used for the normalization of all miRNAs levels (Table 2). The methods applied measured miRNAs successfully and the results indicated that they can detect their quantitative variations. For example, in GBM EVs, the levels of miR-206 and miR-663 did not change during the period tested, whereas the level of miR-1 was increased 30 days after surgery in comparison with the levels at 7 days after surgery (Figure 3C). Furthermore, in meningiomas, miR1, miR-206, and miR-663 showed different levels at the various time points tested. miR-1 levels were low at 7 days after surgery but were again high at 30 days. In contrast, MiR-206 and miR-663 were low at 30 days after surgery (Figure 3C1).

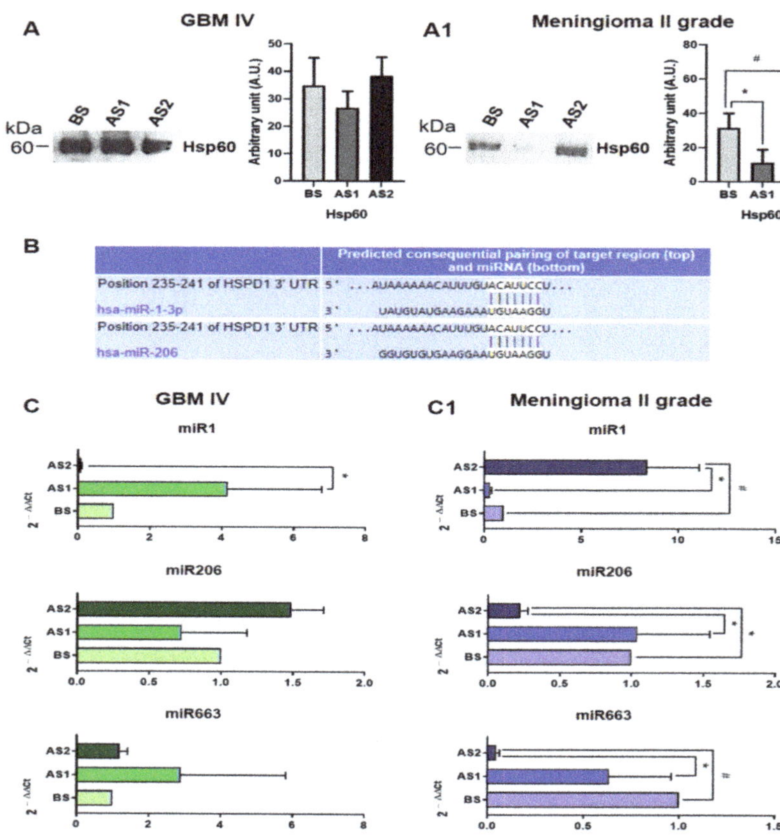

Figure 3. Measurement of Hsp60 and related miRNAs levels in EVs. Western blots and corresponding histograms showing the presence and levels of Hsp60 in EVs from patients with GBM IV (**A**) and in patients with meningioma II grade, at different times of follow-up (**A1**). BS, Before Surgery; AS1, 7 days After Surgery; AS2, 30 days After Surgery. Visible are the high levels of Hsp60 revealed by the standardized procedure in EVs from patients with GBM IV and from patients with meningioma II grade. In the latter, the method revealed differences of Hsp60 levels before and after surgery that were statistically significant (data are presented as the mean ± S.D. * $p < 0.05$; # $p < 0.01$). (**B**) Predicted miR-1 and miR-206 binding sites detected in the HSPD1 3′ UTR region by TargetScan and the underlined base pairs indicate the target region we adopted. (**C**) Examples of results of our measurements with real-time PCR of the levels of miR1, miR-206, and miR-663 in EVs isolated from blood of GBM IV and from meningioma II grade patients (**C1**) The data in the horizontal histograms were normalized with the reference genes, according to the Livak method (2−ΔΔCT). Data are presented as the mean ± S.D. * $p < 0.05$; # $p < 0.01$.

Table 2. Primers used for real-time PCR.

Name	Sequence
Hs_miR-1_2	5′-UGGAAUGUAAAGAAGUAUGUAU
Hs_miR-206_1	5′-UGGAAUGUAAGGAAGUGUGUGG
Hs_miR-663b_2	5′-GGUGGCCCGGCCGUGCCUGAGG
Hs_miR-16	5′-AGCAGCACGUAAAUAUUGGCG

4. Discussion

Brain tumors have as a rule a poor prognosis and treatments are mostly of limited efficaciousness. Therefore, there is a desperate need for disease biomarkers and relevant methods that can offer insight into pathogenesis and be useful in theranostics [37,48]. To address this issue, in this work we focused on three elements pertinent to brain tumor biology, i.e., Hsp60, microRNAs that regulate Hsp60 expression [19–24], and EVs, and we standardized a battery of techniques for measuring these elements. The standardized methods provided quantitative and qualitative information on the three elements. For example, the methods were useful for determining morphology, size distribution, and the presence of specific EVs markers (Alix, Hsc70, and CD81) in the EVs isolated from plasma. The purified vesicles showed typical characteristics of small (<100 nm) and medium (<200 nm) size EVs [49]. The EVs isolated from the plasma of GBM patients were more homogeneous in size distribution during follow-up compared to EVs from atypical meningioma. This variation in size of EVs from plasma of patients with atypical meningioma could be due to disease progression, and is in line with the high variability in biochemical and biophysical properties of tumor cells-derived EVs [50–53]. The standardized methodology revealed quantitative variations of Hsp60 during time, for example, before and after ablative surgery, allowing surveyance of response to treatment.

A major limitation in brain tumor diagnosis stems from the impossibility of molecular profiling using tissue biopsies. A viable alternative is liquid biopsy, which can be obtained via a minimally invasive method such as drawing venous blood. This provides enough material for isolating EVs from plasma to measure their contents in Hsp60 and miRNAs. However, purification and characterization of EVs must be carefully done for the information to be of use to clinicians and surgeons, and to scientists willing to penetrate the secrets of the malignancy of some brain tumors. There are pitfalls that must be avoided that pertain to blood collection; plasma extraction; and specimen manipulation, storage, and testing, and these include type of anticoagulant, sample processing time and temperature, and number of freeze–thaw cycles, just to name a few. Various strict criteria should be met while assessing the physical and biochemical characteristics of EVs for the information obtained to be reliable and reproducible. Currently, the method we recommend for isolating EVs from plasma is ultracentrifugation, including density gradient-based ultracentrifugation. This approach, based on the density of the EVs, ensures a high level of purity, higher than the size-based methods, such as size-exclusion chromatography. Moreover, differential ultracentrifugation can be complemented with ultrafiltration steps to increase the yield, but this imposes a pre-established cutoff. These procedures allow the isolation of defined subpopulations of EVs, such as exosomes, excluding the EVs of larger size. However, the latter larger EVs can also provide useful biomarkers. The EVs population may be highly heterogeneous, depending on the type and/or the state of the cell from which the vesicles derive. Therefore, validation and standardization of the EVs isolation and characterization methods must be carefully done, before applying them to the management of brain tumors patients [49]. Furthermore, the EVs cargo includes a wide range of biomarkers that can be different between EVs subtypes, a diversity not yet fully characterized for all tumors and conditions [49].

In the present work, we propose a set of standardized methods that produce a plethora of complementary results, providing a quantitative picture of three key disease players, Hsp60 and related miRNAs and EVs that can easily be sampled by liquid biopsy (Figure 4). The methods are now available to study more samples from more patients to reveal the quantitative profiles of the triad components comparing results at various times after surgery with pre-operatory data. In this way, standard curves will be obtained for general application when the adequate number of cases has been tested.

A limitation of our study is that it focused on only one CS component, namely, Hsp60. Most likely other CS components such as chaperones of the Hsp70 and Hsp90 families and other chaperonins, e.g., CCT, also play a role in brain carcinogenesis. Our methodology can be applied to measure them with the pertinent adaptations, which should easily be implemented since they will consist mostly of the use of specific antibodies and primers.

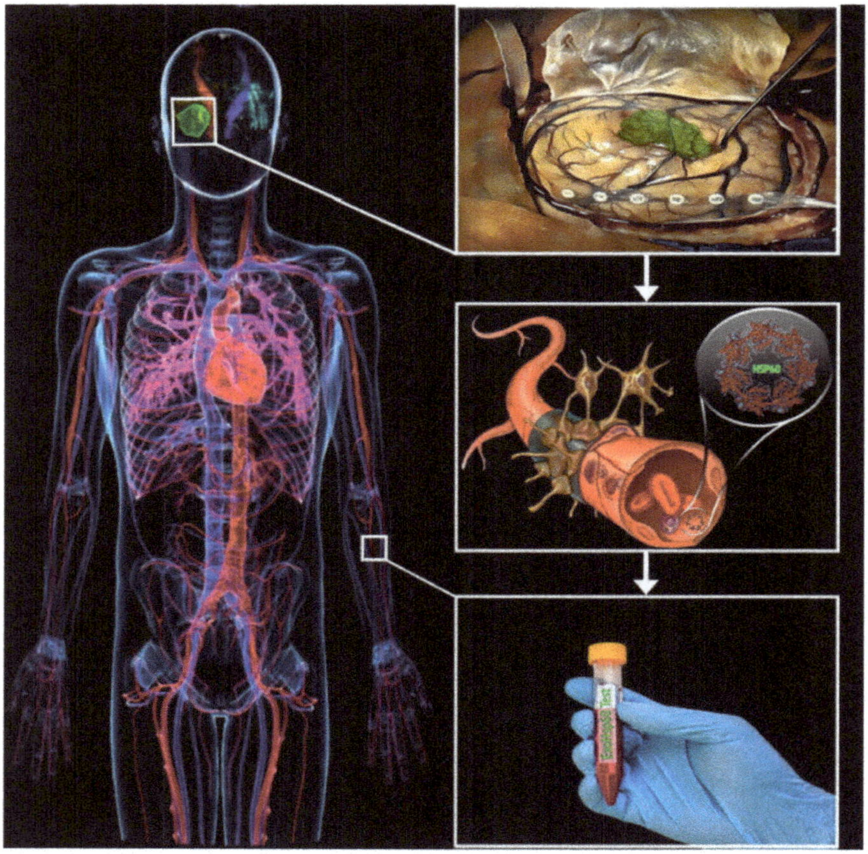

Figure 4. Drawing representing the hypothetical dynamics of Hsp60 from the tumor to the peripheral blood that could be investigated with the methodology described. Hsp60 and related miRNAs can be quantified with the methods used in this study in liquid biopsies containing EVs released by the tumor. This approach is doable in routine settings and would provide a wealth of information of practical and scientific interest that could help in finding ways to improve the management of patients with brain tumors.

Author Contributions: Conceptualization, methodology, software, validation, formal analysis investigation resources, data curation, writing—original draft preparation, writing—review and editing visualization supervision, project administration, funding acquisition: F.G., and C.C.B.; Project administration, funding acquisition, writing—review and editing, visualization supervision: D.G.I., C.C., G.F.N., E.C.d.M., and A.J.L.M.; Software, validation, formal analysis investigation resources: G.C., G.S., A.G.G., R.P., and A.M.F. All authors have read and agreed to the published version of the manuscript.

Funding: Part of this work was funded by the Italian National Operational Programme (PON) for Research and Competitiveness 2007–2013; grant awarded by the Italian Ministry of University and Research to the project titled "Cyber Brain—Polo di innovazione" (Project code: PONa3_00210, European Regional Development Fund). A.J.L.M. and E.C. de M. were partially supported by IMET. This is IMET contribution number IMET 21-005.

Institutional Review Board Statement: The study was conducted according to the guidelines of the Declaration of Helsinki, and approved by Palermo Ethics Committee I (number 11\2018).

Informed Consent Statement: Informed consent was obtained from all subjects involved in the study.

Acknowledgments: A.J.L.M. and E.C. de M. were partially supported by IMET. This is IMET contribution number IMET 21-005.

Conflicts of Interest: The authors report no conflict of interest.

References

1. Ostrom, Q.T.; Cioffi, G.; Gittleman, H.; Patil, N.; Waite, K.; Kruchko, C.; Barnholtz-Sloan, J.S. CBTRUS statistical report: Primary brain and other central nervous system tumors diagnosed in the united states in 2012–2016. *Neurooncology* **2019**, *21*, v1–v100. [CrossRef] [PubMed]
2. Tan, A.C.; Ashley, D.M.; López, G.Y.; Malinzak, M.; Friedman, H.S.; Khasraw, M. Management of glioblastoma: State of the art and future directions. *CA A Cancer J. Clin.* **2020**, *70*, 299–312. [CrossRef] [PubMed]
3. Alifieris, C.; Trafalis, D.T. Glioblastoma multiforme: Pathogenesis and treatment. *Pharmacol. Ther.* **2015**, *152*, 63–82. [CrossRef] [PubMed]
4. Ashby, L.S.; Ryken, T.C. Management of malignant glioma: Steady progress with multimodal approaches. *Neurosurg. Focus* **2006**, *20*, E3. Available online: https://pubmed.ncbi.nlm.nih.gov/16709034/ (accessed on 4 January 2021). [CrossRef] [PubMed]
5. Ryken, T.C.; Frankel, B.; Julien, T.; Olson, J.J. Surgical management of newly diagnosed glioblastoma in adults: Role of cytoreductive surgery. *J. Neuro-Oncology* **2008**, *89*, 271–286. [CrossRef]
6. Brancato, V.; Nuzzo, S.; Tramontano, L.; Condorelli, G.; Salvatore, M.; Cavaliere, C. Predicting survival in glioblastoma patients using diffusion MR imaging metrics—A systematic review. *Cancers* **2020**, *12*, 2858. [CrossRef]
7. Montemurro, N.; Anania, Y.; Cagnazzo, F.; Perrini, P. Survival outcomes in patients with recurrent glioblastoma treated with Laser Interstitial Thermal Therapy (LITT): A systematic review. *Clin. Neurol. Neurosurg.* **2020**, *195*, 105942. [CrossRef] [PubMed]
8. Zang, L.; Kondengaden, S.M.; Che, F.; Wang, L.; Heng, X. Potential epigenetic-based therapeutic targets for glioma. *Front. Mol. Neurosci.* **2018**, *11*, 408. [CrossRef]
9. Liu, L.; Wang, G.; Wang, L.; Yu, C.; Li, M.; Song, S.; Hao, L.; Ma, L.; Zhang, Z. Computational identification and characterization of glioma candidate biomarkers through multi-omics integrative profiling. *Biol. Direct* **2020**, *15*, 1–14. [CrossRef]
10. Macario, A.J.L.; Conway de Macario, E. Chaperone proteins and chaperonopathies. In *Handbook of Stress; Stress Physiology, Biochemistry, and Pathology*; Fink, G., Ed.; Elsevier/Academic Press: Cambridge, MA, USA, 2019; Volume 3, Chapter 12; pp. 135–152. [CrossRef]
11. Kocaturk, N.M.; Gozuacik, D. Crosstalk between mammalian autophagy and the ubiquitin-proteasome system. *Front. Cell Dev. Biol.* **2018**, *6*, 128. [CrossRef]
12. Tekirdag, K.; Cuervo, A.M. Chaperone-mediated autophagy and endosomal microautophagy: Jointed by a chaperone. *J. Biol. Chem.* **2018**, *293*, 5414–5424. [CrossRef] [PubMed]
13. Macario, A.J.L.; Conway de Macario, E. Molecular mechanisms in chaperonopathies: Clues to understanding the histopathological abnormalities and developing novel therapies. *J. Pathol.* **2020**, *250*, 9–18. [CrossRef]
14. Rappa, F.; Unti, E.; Baiamonte, P.; Cappello, F.; Scibetta, N. Different immunohistochemical levels of Hsp60 and Hsp70 in a subset of brain tumors and putative role of Hsp60 in neuroepithelial tumorigenesis. *Eur. J. of Histochem.* **2013**, *57*, 20. [CrossRef]
15. Alexiou, G.A.; Vartholomatos, G.; Stefanaki, K.; Patereli, A.; Dova, L.; Karamoutsios, A.; Lallas, G.; Sfakianos, G.; Moschovi, M.; Prodromou, N. Expression of heat shock proteins in medulloblastoma. *J. Neurosurgery Pediatr.* **2013**, *12*, 452–457. [CrossRef] [PubMed]
16. Caruso Bavisotto, C.; Marino Gammazza, A.; Lo Cascio, F.; Mocciaro, E.; Vitale, A.M.; Vergilio, G.; Pace, A.; Cappello, F.; Campanella, C.; Palumbo Piccionello, A. Curcumin affects HSP60 folding activity and levels in neuroblastoma cells. *Int. J. Mol. Sci.* **2020**, *21*, 661. [CrossRef]
17. Alexiou, G.A.; Karamoutsios, A.; Lallas, G.; Ragos, V.; Goussia, A.; Kyritsis, A.P.; Voulgaris, S.; Vartholomatos, G. Expression of heat shock proteins in brain tumors. *Turk. Neurosurg.* **2014**, *24*, 745–749. [CrossRef]
18. Tang, H.; Li, J.; Liu, X.; Wang, G.; Luo, M.; Deng, H. Down-regulation of HSP60 suppresses the proliferation of glioblastoma cells via the ROS/AMPK/mTOR pathway. *Sci. Rep.* **2016**, *6*, 28388. [CrossRef] [PubMed]
19. Shan, Z.-X.; Lin, Q.-X.; Deng, C.-Y.; Zhu, J.-N.; Mai, L.-P.; Liu, J.-L.; Fu, Y.-H.; Liu, X.-Y.; Li, Y.-X.; Zhang, Y.-Y.; et al. miR-1/miR-206 regulate Hsp60 expression contributing to glucose-mediated apoptosis in cardiomyocytes. *FEBS Lett.* **2010**, *584*, 3592–3600. [CrossRef]

20. Pan, Z.; Sun, X.; Ren, J.; Li, X.; Gao, X.; Lu, C.; Zhang, Y.; Sun, H.; Wang, Y.; Wang, H.; et al. miR-1 exacerbates cardiac ischemia-reperfusion injury in mouse models. *PLoS ONE* **2012**, *7*, e50515. [CrossRef]
21. Neumann, E.; Brandenburger, T.; Santana-Varela, S.; Deenen, R.; Köhrer, K.; Bauer, I.; Hermanns, H.; Wood, J.N.; Zhao, J.; Werdehausen, R. MicroRNA-1-associated effects of neuron-specific brain-derived neurotrophic factor gene deletion in dorsal root ganglia. *Mol. Cell. Neurosci.* **2016**, *75*, 36–43. [CrossRef] [PubMed]
22. Choghaei, E.; Khamisipour, G.; Falahati, M.; Naeimi, B.; Mossahebi-Mohammadi, M.; Tahmasebi, R.; Hasanpour, M.; Shamsian, S.; Hashemi, Z.S. Knockdown of microRNA-29a changes the expression of heat shock proteins in breast carcinoma MCF-7 cells. *Oncol. Res. Featur. Preclin. Clin. Cancer Ther.* **2016**, *23*, 69–78. [CrossRef]
23. Hu, Y.; Chen, X.; Li, X.; Li, Z.; Diao, H.; Liu, L.; Zhang, J.; Ju, J.; Wen, L.; Liu, X.; et al. MicroRNA-1 downregulation induced by carvedilol protects cardiomyocytes against apoptosis by targeting heat shock protein 60. *Mol. Med. Rep.* **2019**, *19*, 3527–3536. [CrossRef]
24. Wang, L.L.; Dong, J.J.; An, B.Z.; Liang, J.; Cai, K.R.; Jin, Z.S.; Jin, H.S.; Hu, J.P. HAS-MIR-17 increases the malignancy of gastric lymphoma by HSP60/TNFR2 pathway. *J. Biol. Regul. Homeost. Agents* **2020**, *34*, 1317–1324. [CrossRef]
25. Cesarini, V.; Silvestris, D.A.; Tassinari, V.; Tomaselli, S.; Alon, S.; Eisenberg, E.; Locatelli, F.; Gallo, A. ADAR2/miR-589-3p axis controls glioblastoma cell migration/invasion. *Nucleic Acids Res.* **2018**, *46*, 2045–2059. [CrossRef]
26. Banelli, B.; Forlani, A.; Allemanni, G.; Morabito, A.; Pistillo, M.P.; Romani, M. MicroRNA in glioblastoma: An overview. *Int. J. Genom.* **2017**, *2017*, 7639084. [CrossRef] [PubMed]
27. Ludwig, N.; Kim, Y.-J.; Mueller, S.C.; Backes, C.; Werner, T.V.; Galata, V.; Sartorius, E.; Bohle, R.M.; Keller, A.; Meese, E. Posttranscriptional deregulation of signaling pathways in meningioma subtypes by differential expression of miRNAs. *Neurooncology* **2015**, *17*, 1250–1260. [CrossRef]
28. Kol, A.; Lichtman, A.H.; Finberg, R.W.; Libby, P.; Kurt-Jones, E.A. Cutting Edge: Heat Shock Protein (HSP) 60 activates the innate immune response: CD14 is an essential receptor for HSP60 activation of mononuclear cells. *J. Immunol.* **2000**, *164*, 13–17. [CrossRef] [PubMed]
29. Flohé, S.B.; Brüggemann, J.; Lendemans, S.; Nikulina, M.; Meierhoff, G.; Flohé, S.; Kolb, H. Human heat shock protein 60 induces maturation of dendritic cells versus a Th1-promoting phenotype. *J. Immunol.* **2003**, *170*, 2340–2348. [CrossRef] [PubMed]
30. Gupta, S.; Knowlton, A.A. HSP60 trafficking in adult cardiac myocytes: Role of the exosomal pathway. *Am. J. Physiol. Heart Circ. Physiol.* **2007**, *292*, H3052–H3056. [CrossRef]
31. Hayoun, D.; Kapp, T.; Edri-Brami, M.; Ventura, T.; Cohen, M.; Avidan, A.; Lichtenstein, R.G. HSP60 is transported through the secretory pathway of 3-MCA-induced fibrosarcoma tumour cells and undergoes N-glycosylation. *FEBS J.* **2012**, *279*, 2083–2095. [CrossRef]
32. Campanella, C.; Bucchieri, F.; Merendino, A.M.; Fucarino, A.; Burgio, G.; Corona, D.F.V.; Barbieri, G.; David, S.; Farina, F.; Zummo, G.; et al. The odyssey of Hsp60 from tumor cells to other destinations includes plasma membrane-associated stages and golgi and exosomal protein-trafficking modalities. *PLoS ONE* **2012**, *7*, e42008. [CrossRef] [PubMed]
33. Cruz Junho, C.V.; Trentin-Sonoda, M.; Alvim, J.M.; Gaisler-Silva, F.; Carneiro-Ramos, M.S. Ca2+/Calmodulin-dependent kinase II delta B is essential for cardiomyocyte hypertrophy and complement gene expression after LPS and HSP60 stimulation in vitro. *Braz. J. Med Biol. Res.* **2019**, *52*, e8732. [CrossRef] [PubMed]
34. Cui, X.; Liu, Y.; Wang, S.; Zhao, N.; Qin, J.; Li, Y.; Fan, C.; Shan, Z.; Teng, W. Circulating exosomes activate dendritic cells and induce unbalanced CD4+ t cell differentiation in Hashimoto thyroiditis. *J. Clin. Endocrinol. Metab.* **2019**, *104*, 4607–4618. [CrossRef] [PubMed]
35. Zabegina, L.; Nazarova, I.; Knyazeva, M.; Nikiforova, N.; Slyusarenko, M.; Titov, S.; Vasilyev, D.; Sleptzov, I.; Malek, A. miRNA let-7 from TPO(+) extracellular vesicles is a potential marker for a differential diagnosis of follicular thyroid nodules. *Cells* **2020**, *9*, 1917. [CrossRef]
36. Graziano, F.; Caruso Bavisotto, C.; Marino Gammazza, A.; Rappa, F.; Conway de Macario, E.; Macario, A.J.L.; Cappello, F.; Campanella, C.; Maugeri, R.; Iacopino, D.G. Chaperonology: The third eye on brain gliomas. *Brain Sci.* **2018**, *8*, 110. [CrossRef] [PubMed]
37. Caruso Bavisotto, C.; Graziano, F.; Rappa, F.; Marino Gammazza, A.; Logozzi, M.; Fais, S.; Maugeri, R.; Bucchieri, F.; Conway de Macario, E.; Macario, A.J.L.; et al. Exosomal chaperones and miRNAs in gliomagenesis: State-of-art and theranostics perspectives. *Int. J. Mol. Sci.* **2018**, *19*, 2626. [CrossRef]
38. Campanella, C.; Caruso Bavisotto, C.; Marino Gammazza, A.; Nikolic, D.; Rappa, F.; David, S.; Cappello, F.; Bucchieri, F.; Fais, S. Exosomal heat shock proteins as new players in tumour cell-to-cell communication. *J. Circ. Biomarkers* **2014**, *3*, 4. [CrossRef]
39. Caruso Bavisotto, C.; Cappello, F.; Macario, A.J.L.; Conway de Macario, E.; Logozzi, M.; Fais, S.; Campanella, C. Exosomal HSP60: A potentially useful biomarker for diagnosis, assessing prognosis, and monitoring response to treatment. *Expert Rev. Mol. Diagn.* **2017**, *17*, 815–822. [CrossRef]
40. Caruso Bavisotto, C.; Marino Gammazza, A.; Rappa, F.; Fucarino, A.; Pitruzzella, A.; David, S.; Campanella, C. Exosomes: Can doctors still ignore their existence? *EuroMediterr. Biomed. J.* **2013**, *8*. [CrossRef]
41. Stupp, R.; Mason, W.P.; van den Bent, M.J.; Weller, M.; Fisher, B.; Taphoorn, M.J.B.; Belanger, K.; Brandes, A.A.; Marosi, C.; Bogdahn, U.; et al. Radiotherapy plus concomitant and adjuvant temozolomide for glioblastoma. *N. Engl. J. Med.* **2005**, *352*, 987–996. [CrossRef] [PubMed]

42. Stupp, R.; Hegi, M.E.; Mason, W.P.; Bent, M.J.V.D.; Taphoorn, M.J.B.; Janzer, R.C.; Ludwin, S.K.; Allgeier, A.; Fisher, B.; Belanger, K.; et al. Effects of radiotherapy with concomitant and adjuvant temozolomide versus radiotherapy alone on survival in glioblastoma in a randomised phase III study: 5-year analysis of the EORTC-NCIC trial. *Lancet Oncol.* **2009**, *10*, 459–466. [CrossRef]
43. Caruso Bavisotto, C.; Cipolla, C.; Graceffa, G.; Barone, R.; Bucchieri, F.; Bulone, D.; Cabibi, D.; Campanella, C.; Marino Gammazza, A.; Pitruzzella, A.; et al. Immunomorphological pattern of molecular chaperones in normal and pathological thyroid tissues and circulating exosomes: Potential use in clinics. *Int. J. Mol. Sci.* **2019**, *20*, 4496. [CrossRef]
44. Campanella, C.; Rappa, F.; Sciumè, C.; Marino Gammazza, A.; Barone, R.; Bucchieri, F.; David, S.; Curcurù, G.; Caruso Bavisotto, C.; Pitruzzella, A.; et al. Heat shock protein 60 levels in tissue and circulating exosomes in human large bowel cancer before and after ablative surgery. *Cancer* **2015**, *121*, 3230–3239. [CrossRef] [PubMed]
45. TargetScan. TargetScanHuman 7.2 Predicted Targeting of Human HSPD1. Available online: http://www.targetscan.org/cgi-bin/targetscan/vert_72/view_gene.cgi?rs=ENST00000388968.3&taxid=9606&members=&showcnc=0&shownc=0&showncf1=&showncf2=&subset=1#miR-1-3p/206 (accessed on 6 August 2020).
46. Schmittgen, T.D.; Livak, K.J. Analyzing real-time PCR data by the comparative CT method. *Nat. Protoc.* **2008**, *3*, 1101–1108. [CrossRef]
47. Shi, Y.; Chen, C.; Yu, S.-Z.; Liu, Q.; Rao, J.; Zhang, H.-R.; Xiao, H.-L.; Fu, T.-W.; Long, H.; He, Z.-C.; et al. miR-663 suppresses oncogenic function of CXCR4 in glioblastoma. *Clin. Cancer Res.* **2015**, *21*, 4004–4013. [CrossRef]
48. Lammers, T.; Aime, S.; Hennink, W.E.; Storm, G.; Kiessling, F. Theranostic Nanomedicine. *Acc. Chem. Res.* **2011**, *44*, 1029–1038. [CrossRef]
49. Théry, C.; Witwer, K.W.; Aikawa, E.; Alcaraz, M.J.; Anderson, J.D.; Andriantsitohaina, R.; Antoniou, A.; Arab, T.; Archer, F.; Atkin-Smith, G.K.; et al. Minimal information for studies of extracellular vesicles 2018 (MISEV2018): A position statement of the International Society for Extracellular Vesicles and update of the MISEV2014 guidelines. *J. Extracell. Vesicles* **2018**, *7*, 1535750. [CrossRef] [PubMed]
50. Shen, W.; Guo, K.; Adkins, G.B.; Jiang, Q.; Liu, Y.; Sedano, S.; Duan, Y.; Yan, W.; Wang, S.E.; Bergersen, K.; et al. A Single Extracellular Vesicle (EV) flow cytometry approach to reveal EV heterogeneity. *Angew. Chem. Int. Ed.* **2018**, *57*, 15675–15680. [CrossRef]
51. Willms, E.; Cabañas, C.; Mäger, I.; Wood, M.J.A.; Vader, P. Extracellular vesicle heterogeneity: Subpopulations, isolation techniques, and diverse functions in cancer progression. *Front. Immunol.* **2018**, *9*, 738. [CrossRef] [PubMed]
52. Caruso Bavisotto, C.; Scalia, F.; Marino Gammazza, A.; Carlisi, D.; Bucchieri, F.; Conway de Macario, E.; Macario, A.J.L.; Cappello, F.; Campanella, C. Extracellular vesicle-mediated cell–cell communication in the nervous system: Focus on neurological diseases. *Int. J. Mol. Sci.* **2019**, *20*, 434. [CrossRef] [PubMed]
53. Colombo, M.; Moita, C.; Van Niel, G.; Kowal, J.; Vigneron, J.; Benaroch, P.; Manel, N.; Moita, L.F.; Théry, C.; Raposo, G. Analysis of ESCRT functions in exosome biogenesis, composition and secretion highlights the heterogeneity of extracellular vesicles. *J. Cell Sci.* **2013**, *126*, 5553–5565. [CrossRef] [PubMed]

applied sciences

MDPI

Review

The Challenging Riddle about the Janus-Type Role of Hsp60 and Related Extracellular Vesicles and miRNAs in Carcinogenesis and the Promises of Its Solution

Sabrina David [1,†], Alessandra Maria Vitale [1,2,†], Alberto Fucarino [1,2,†], Federica Scalia [1,2], Giuseppe Vergilio [1], Everly Conway de Macario [3], Alberto J. L. Macario [2,3,*], Celeste Caruso Bavisotto [1,2,*] and Alessandro Pitruzzella [1,2,4]

1 Department Biomedicine, Neurosciences and Advanced Diagnostics, Section of Human Anatomy, University of Palermo, 90127 Palermo, Italy; sabrina.david@unipa.it (S.D.); alessandramaria.vitale@unipa.it (A.M.V.); alberto.fucarino@unipa.it (A.F.); federica.scalia@unipa.it (F.S.); giuseppe.vergilio@unipa.it (G.V.); alessandro.pitruzzella@unipa.it (A.P.)
2 Euro-Mediterranean Institute of Science and Technology (IEMEST), 90139 Palermo, Italy
3 Department of Microbiology and Immunology, School of Medicine, University of Maryland at Baltimore-Institute of Marine and Environmental Technology (IMET), Baltimore, MD 21202, USA; econwaydemacario@som.umaryland.edu
4 Consorzio Universitario Caltanissetta, University of Palermo, 93100 Caltanissetta, Italy
* Correspondence: ajlmacario@som.umaryland.edu (A.J.L.M.); celeste.carusobavisotto@unipa.it (C.C.B.); Tel.: +39-091-23865700 (C.C.B.)
† These authors contributed equally to this work.

Citation: David, S.; Vitale, A.M.; Fucarino, A.; Scalia, F.; Vergilio, G.; Conway de Macario, E.; Macario, A.J.L.; Caruso Bavisotto, C.; Pitruzzella, A. The Challenging Riddle about the Janus-Type Role of Hsp60 and Related Extracellular Vesicles and miRNAs in Carcinogenesis and the Promises of Its Solution. *Appl. Sci.* **2021**, *11*, 1175. https://doi.org/10.3390/app11031175

Academic Editor: Francisco Arrebola
Received: 21 December 2020
Accepted: 22 January 2021
Published: 27 January 2021

Abstract: Hsp60 is one of the most ancient and evolutionarily conserved members of the chaperoning system. It typically resides within mitochondria, in which it contributes to maintaining the organelle's proteome integrity and homeostasis. In the last few years, it has been shown that Hsp60 also occurs in other locations, intracellularly and extracellularly, including cytosol, plasma-cell membrane, and extracellular vesicles (EVs). Consequently, non-canonical functions and interacting partners of Hsp60 have been identified and it has been realized that it is a hub molecule in diverse networks and pathways and that it is implicated, directly or indirectly, in the development of various pathological conditions, the Hsp60 chaperonopathies. In this review, we will focus on the multi-faceted role of this chaperonin in human cancers, showing the contribution of intra- and extracellular Hsp60 in cancer development and progression, as well as the impact of miRNA-mediated regulation of Hsp60 in carcinogenesis. There are still various aspects of this intricate biological scenario that are poorly understood but ongoing research is steadily providing new insights and we will direct attention to them. For instance, we will highlight the possible applications of the Hsp60 involvement in carcinogenesis not only in diagnosis, but also in the development of specific anti-cancer therapies centered on the use of the chaperonin as therapeutic target or agent and depending on its role, pro- or anti-tumor.

Keywords: Hsp60; chaperonopathies; carcinogenesis; extracellular vesicle (EV); miRNA; chaperonotherapy

1. Stress Responses and the Chaperoning System

Since their first appearance on Earth, living organisms have been oppressed by physical and chemical stressors, such as radiations, extreme temperatures and pH, and hypoxia. These challenges required adaptation, including the development of anti-stress mechanisms. Today, it is possible to observe the success of these protection strategies, since living beings are observable in a wide range of ecosystems, even in those seemingly incompatible with life, such as sulfurous lakes, deep depth of the oceans, and permafrost [1–8].

One of the most important cellular anti-stress machineries is the chaperoning (or chaperone) system, which is highly conserved in the three phylogenetic domains, Bacteria,

Archaea, and Eucarya [9]. It is composed of molecular chaperones and co-chaperones, and their co-factors, interactors, and receptors, which form functional networks working together to ensure protein homeostasis, under normal and stressful conditions [10,11]. The genes of many molecular chaperones are constitutively expressed, but others, named Heat shock protein (Hsp), are transcriptionally upregulated by exposure to stressors (e.g., heat, hyperthermia, hypoxia, heavy metals, ethanol, infections, radiation, and UV) and, thereby, protect against protein misfolding and aggregation and maintain proteins in their functional native state [12–15]. Hsps are commonly classified according to their molecular weight, and, even if not all the chaperones can be considered Hsps, the two terms are commonly used as synonymous [16–18].

Hsp60, One of the Most Ancient Anti-Stress Molecules

Hsp60 belongs to one of the oldest and evolutionarily most conserved protein families of the chaperoning system [19–21]. These proteins are present in all living species, including plants, where they were first discovered [22–24], and, considering their unique molecular characteristics, they were named "chaperonins" to distinguish them from other chaperones [25].

The canonical classification divides chaperonins into two main groups. Group I chaperonins are found in bacteria, as well as inside eukaryotic organelles of endosymbiotic origin (mitochondria in animal cells, and chloroplasts in plant cells), and work together with a co-chaperonin which helps the closing of the folding cage. Group II chaperonins are found in the eukaryotic cytosol and in Archaea, and do not require a co-chaperonin since they have a built-in lid [26]. More recently, a third group (Group III chaperonins) has been discovered and it is now under characterization [27,28].

In humans, the Group I chaperonin is Hsp60 (or Cpn60, or HSPD1), and the Group II chaperonin is CCT (chaperonin-containing TCP-1) or TRiC (T-complex protein Ring Complex). The former typically resides inside mitochondria but also occurs in various other locations intra- and extracellularly, while CCT is in the cytosol.

Hsp60 and CCT form macromolecular double-ring complexes with a central internal cavity in which polypeptides in need of assistance for folding or refolding are encapsulated and assisted to achieve their functional final conformation (native state) via an ATP-dependent mechanism [29–31].

A considerable part of our knowledge about how human chaperonins assemble and work derives from the study of their bacterial homologues. The bacterial counterpart of human Hsp60 is the chaperonin GroEL [32], which, together with the co-chaperonin GroES (the bacterial counterpart of human the Hsp10 or Cpn10), forms a tubular, double-ring complex with a central cavity similar to that described above for the eukaryotic Hsp60 and CCT complexes, inside which protein folding occurs [33,34]. Human Hsp60 has been found in various conformations (monomer, single heptameric ring, double-ring tetradecamer), depending on their concentration, levels of ATP, and presence of Hsp10 and substrate [21,35–37]. Early studies showed that human Hsp60 can assist in productive protein folding without forming a macro-double ring-complex [38,39]. However, more recent studies, based on transmission electron microscopy and X-ray crystallographic investigations, have provided strong evidence that it likely uses both double- and single-ring intermediates during its ATPase cycle [40–42].

Hsp60 has been found in extramitochondrial sites, such as cytoplasm and plasma-cell membrane, as well as in extracellular sites, inside extracellular vesicles (EVs), in circulation, and in body fluids [43–49]. Consequently, in addition to its canonical chaperoning function, Hsp60 also performs various other non-canonical activities, "moonlighting functions" unrelated to protein quality control. (Figure 1) [21,49–51]. For instance, it has been observed that increased amounts of Hsp60 on the surface of cancer cells act as a signal to stimulate the immune system, leading to the activation and maturation of dendritic cells and the generation of an antitumor T-cell response [52–54]. A non-canonical function of human Hsp60 that is still under scrutiny is the regulation of cell apoptosis [55]. Some studies suggested a pro-apoptotic role, involving pro-caspase 3 proteolytic activation [56,57], whereas other investigations support an anti-apoptotic role, involving various mechanisms: the sequestration of Bax-containing complexes [58,59], the maintenance of mitochondria integrity and ATP generation [60], and the triggering of the IKK/NF-κB survival pathway [61]. This dual role in cell apoptosis/survival regulation has been observed both in normal and tumor cells [59,62], thus it may affect cancer progression either positively or negatively.

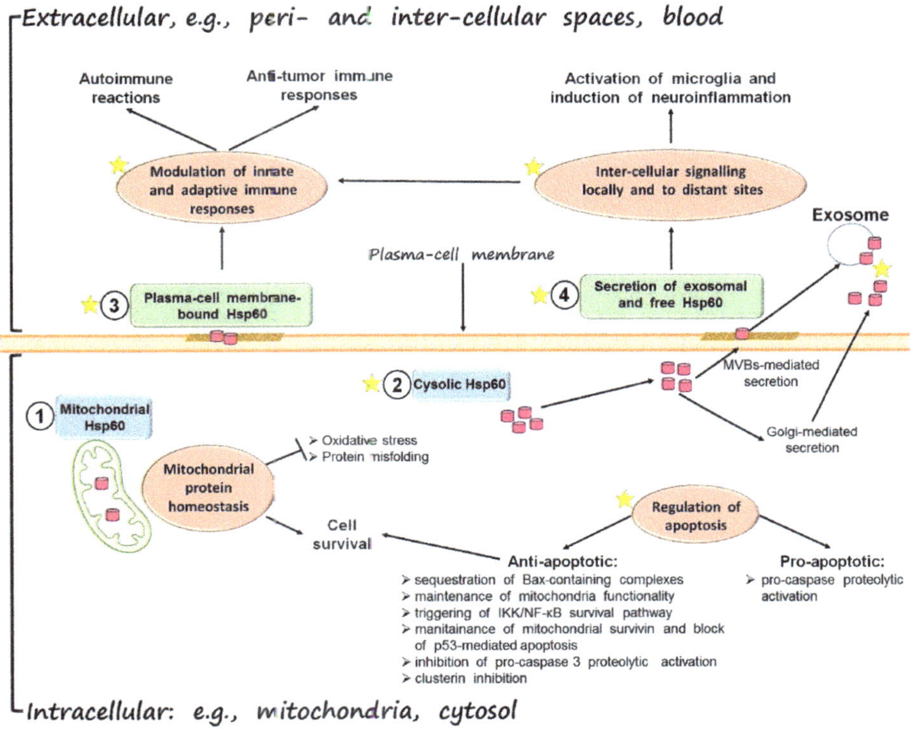

Figure 1. Hsp60 plays multiple roles intra- and extracellularly. Its canonical functions pertain to maintenance of protein homeostasis inside mitochondria, whereas its non-canonical roles are unrelated to protein quality control and are played in a variety of locations (shown with a yellow star). The chaperonin (small dark-pink cylinder) is typically in the mitochondrion matrix (1) in which it assists in the folding of intra-mitochondrial proteins, for instance those of the electron transport system, and therefore it is a vital molecule for maintaining cell viability and organismal physiology under normal conditions and in the face of stress. Consequently, Hsp60 chaperonopathies are usually serious conditions. The non-canonical functions of Hsp60 displayed in a variety of locations beyond the mitochondria, for example in the cytosol (2), are also vital. One example is regulation of apoptosis, which can be in either direction pro- or anti-apoptotic with implications for carcinogenesis. The anti-apoptotic effect of Hsp60 helps cancer cells to become immortal. In the plasma-cell membrane (3), Hsp60 can be recognized by immune cells and antibodies and generate immune reactions that damage the cell, which is a welcome event if the target cell is a cancerous one, but it is a pathogenic reaction when the cell is a normal one, a vascular epithelial cell, for instance, generating autoimmune conditions. Hsp60 exits cells (4) via different mechanisms and can, thus, reach molecules in the extracellular space and other cells nearby and far away, the latter via blood free or on particles, such as platelets, red cells, and microvesicles, like exosomes. In this manner, Hsp60 plays a role in intercellular communication, as illustrated by the microglia activation that occurs in some neurodegenerative diseases via the Hsp60-TLR-4-NF-κB signaling pathway.

2. Hsp60 Chaperonopathies

Hsp60 is multifaceted and plays diverse physiological roles but when abnormal in structure and/or function it can become pathogenic and cause diseases, the Hsp60 chaperonopathies [63,64].

In genetic chaperonopathies, there are pathogenic variants, e.g., missense mutations, in the Hsp60 gene, whereas in the acquired chaperonopathies the gene is normal but the Hsp60 protein is altered structurally and/or functionally. Genetic chaperonopathies are infrequent and typically have an early clinical onset, while the acquired ones are more common, mostly occur in adults, and are often associated with other pathological conditions, especially age-related diseases [65,66]. Another classification of practical value sorts the chaperonopathies according to quantitative parameters pertaining to concentration and functionality into by defect, by excess, and by mistake or collaborationism (Table 1) [65,66].

Table 1. Classification of chaperonopathies according to pathogenic mechanism.

Chaperonopathies	Mechanism	Example	References
Excess	Quantitative variation, in which a gene is dysregulated or overexpressed. Qualitative variation, with a gain of function.	e.g., Alzheimer's disease, Parkinson's disease, Huntington's disease.	[65]
Defect	Quantitative variation with gene downregulation. Qualitative variation, due to structural defect (genetic or acquired)	e.g., Charcot–Marie–Tooth disease, Spastic paraplegia, Hypo-myelinating leukodystrophy	[65]
Mistake or collaborationism	The chaperone is normal but the pathway in which it is involved may promote cell pathology.	e.g., certain tumor types, autoimmune conditions, prion disease	[65]

Acquired chaperonopathies can be caused by aberrant post-translational modifications (PTMs) that have an impact on the structure/function of the chaperone molecule. The Hsp60 amino acid sequence contains various critical sites that can be affected by PTMs [67]. Possible PTMs are phosphorylation, O-GlcNAcylation, nitration, acetylation, S-nitrosylation, citrullination, methylation, oxidation, biotinylation, and ubiquitination [51]. Hsp60 PTMs can have beneficial or deleterious effects. For instance, Hsp60 hyperacetylation in the course of anti-osteosarcoma treatment lead to death of the malignant cells [68]. Hyperacetylation disrupted the Hsp60/p53 complex, restored replicative senescence, and diminished or stopped tumor growth [69]. Phosphorylated Hsp60 on the surface of breast cancer cells induced $\alpha3\beta1$ integrin activation, resulting in enhanced motility and adhesion of these cells [70]. Tyrosine phosphorylation of Hsp60 helps malignant cells to escape immune surveillance by NK and CD8 T cells [71].

Chaperonopathies by mistake include all those pathological conditions in which a chaperone is normal as far it can be determined by current methodologies but contributes to the initiation and/or progression of disease, as it has been observed in various types of cancer, and autoimmune, inflammatory, and neurologic disorders [66].

Other examples of chaperonopathies by mistake are those autoimmune conditions in which human Hsp60 acts as auto-antigen. This situation has been described for various autoimmune conditions such as Behçet's disease [72,73], diabetes mellitus [74], systemic lupus erythematosus and vasculitis-associated systemic autoimmune disorders [75–78], atherosclerosis [79–82], and rheumatoid arthritis [83]. The pathogenic autoimmune mechanism for some of these disorders is triggered by the presence of Hsp60 on the plasma-cell membrane where it becomes a target accessible to autoantibodies and, thereby, leads to apoptosis [75,76]. However, in other situations Hsp60 has shown cytoprotective activity, which reiterates the concept that this chaperonin can play apparently opposing roles (Table 2).

In view of the above considerations, it becomes evident that (1) the kind of activity the Hsp60 chaperonin undertakes depends on its context, i.e., it is determined and/or modulated by the composition of its surroundings, namely, by the receptors and interactors within its reach; and (2) Hsp60 has not only canonical functions, which pertain to protein quality control, but also displays other functions unrelated to protein homeostasis that are also important in health and disease. Therefore, advances in the treatment of Hsp60 chaperonopathies ought to include the development of modulators of the chaperonin in situ. Two types of chaperonin inhibitors are currently being investigated: type I that block the binding and hydrolysis of ATP, and type II that bind cysteine covalently [84]. Natural and synthetic compounds potentially useful have already been identified with various degrees of anti-cancer, anti-inflammatory, or anti-autoimmune potency [63,84].

3. Hsp60 in Carcinogenesis

An indication of Hsp60 involvement in carcinogenesis is its altered expression and localization observed in certain human cancers [47,49,85]. However, the exact role of Hsp60 in cancer remains undefined and seems to change depending on the molecular and cytological context (Table 2). Hsp60 was found increased in various malignancies: cervical and ovarian [86–90], breast [91,92], colorectal [93], lung [94,95], prostate [96], gastric [97,98], and thyroid cancers [99,100]; and in leukemias [101], and glioblastoma multiforme [102–104]. The data in general suggest that Hsp60 is actively involved in carcinogenesis as a pro-tumorigenic factor, because its increase and location changes positively correlated with tumor development and malignancy. Moreover, in some cancers, the Hsp60 increase was also associated with a heightened resistance to anti-cancer drugs and other treatments [89,105], and with metastasization and angiogenesis [106,107]. It has to be emphasized that the quantitative variations of Hsp60 in cancer cells, particularly its increase, may reflect the heightened need of the chaperone system by the malignant cells with their rapid and intensive metabolism and proliferation. Thus, the Hsp60 quantitative patterns observed would be the consequence of the disease but not a distinct etiological factor. We argue that even so, Hsp60 aids the tumor in what we call a chaperonopathy by mistake or collaborationism because the tumor depends on its help. What is the value of this concept? It puts Hsp60 in the stage's center and presents it as a target for developing treatment strategies aimed at inhibiting-blocking this collaborator with the enemy from the inside. For example, Hsp60 downregulation by chemical compounds suppressed cancer-cell proliferation and tumor progression and enhanced the beneficial effects of anti-cancer treatments [62,69,90,108–112]. It was observed that Hsp60 knockdown inhibited tumor progression by altering mitochondrial homeostasis and inactivating the mTOR pathway, in ovarian cancer and glioblastoma [90,109]. In colorectal cancer, Hsp60 inhibition promoted the tumor-suppressive activity of insulin-like growth factor binding protein 7 (IGFBP7) [108]. Treatment of a neuroblastoma cell line with curcumin caused cell death by diminishing the cellular level of Hsp60 [111]. It was suggested that this cytotoxic effect was induced through the downregulation of survivin, whose expression was shown to be positively correlated with the expression of CCAR2 and Hsp60 in neuroblastoma tissues and cell lines [113]. Thus, it is becoming clear that the positive correlation between Hsp60 overexpression and increase in cancer cell proliferation and survival depends on the interaction of the chaperonin with proteins involved in cell cycle and apoptosis. In cancer cells lines, Hsp60 played a cytoprotective and pro-survival role by stabilizing the mitochondrial level of survivin and blocking p53-mediated apoptosis [62,113], or by inhibiting the intracellular isoform of clusterin [114]. In other cases, the pro-tumorigenic role of Hsp60 involved blocking the caspase-dependent apoptosis through the negative regulation of mitochondrial permeability transition [102], or the inhibition of pro-caspase 3 proteolytic activation [115,116].

However, others have reported that in some types of cancer, Hsp60 was decreased, downregulated not increased or, if increased, caused tumor suppression not enhancement [117,118]. For instance, Hsp60 was decreased in hepatocellular carcinoma (HCC) tissue compared to peritumor tissue, and this pattern positively correlated with high serum AFP (alpha-fetoprotein) level and poor overall survival. Conversely, increased Hsp60 inhibited invasion and migration of HCC cells both in vitro and in vivo, correlating with a better prognosis [118].

All the above observations (Table 2) make clear that Hsp60 has different roles in carcinogenesis that deserve investigation to elucidate the molecular mechanisms involved, particularly those determining whether the chaperonin acts against or for the cancer cell.

Some data in the literature support the assumption that Hsp60 acts as a pro-tumor protein, considering its role in the modulation of anti-apoptotic factors. It is known that during tumorigenesis, cells undergo complex transcriptional events that lead to a dysregulation of numerous factors. In this scenario, transformed cells show the typical phenotype of the proteotoxic stress, in which the consequent Hsp60 overexpression is a key event because of its central role in the regulation of protein homeostasis. Conversely, the observation that, in several types of tumors, such as lung cancer, Hsp60 levels are reduced, further demonstrate that its functioning is not only related to protein folding and the maintenance of protein homeostasis, but is complex and dependent on the cell and tissue type, its molecular interactors and its localization inside or outside the mitochondria and the cell. Thus, the multifaceted and at times contradictory functions of Hsp60 in cancer are still poorly understood and deserve active investigation, considering the importance of the chaperonin for survival of cells, normal or malignant.

Table 2. Hsp60 in human cancers.

Cancer	Hsp60 Level, Location and/or Status	Effect	Reference
Lung carcinoma	Presence of Hsp60 on cancer-cell plasma-cell membrane and on membrane of cancer cell-derived exosomes	Possible involvement in cell-to-cell communication and anti-tumor immune response stimulation	[47,48]
	Decrease of intracellular Hsp60 level and increase of Hsp60 acetylation level after doxorubicin treatment	Hsp60/p53 complex dissociation and restoration of cancer-cells replicative senescence	[69]
	Decrease of intracellular Hsp60 level after CubipyOXA treatment	Dissociation of the Hsp60/pro-Caspase-3 complex and cancer-cell apoptosis	[116]
	Increased Hsp60 level	Positive correlation with cancer progression and poor prognosis	[94,96]
Oral cancer	Presence of Hsp60 on cell surface	Interaction with gamma-delta T cells and transduction of anti-cancer immune response	[52]
Osteosarcoma	Hyperacetylation and loss of mitochondrial Hsp60 after Geldanamycin treatment	Decreased viability and augmented cancer-cell death	[68]
Breast cancer	Increase of phosphorylated surface Hsp60	$\alpha3\beta1$ integrin activation and enhancement of cancer cells motility and adhesion	[70]
	Increased cytosolic Hsp60	Enhanced cancer-cell proliferation and reduced apoptosis; positive correlation with worse disease-free survival and poor prognosis	[91,92]
Bronchial carcinoma	Decreased Hsp60 level	Positive correlation with bronchial cancer development and progression	[85,119]
Cervical cancer	Increased Hsp60 level	Positive correlation with cancer progression and malignancy	[86–110]

Table 2. *Cont.*

Cancer	Hsp60 Level, Location and/or Status	Effect	Reference
Ovarian cancer	Increased Hsp60 level	Positive correlation with cancer progression and severity (poor prognosis and resistance to anti-cancer treatment)	[89,90,105]
Colorectal cancer	Increased Hsp60 level	Positive correlation with cancer progression and malignancy	[93,120]
	IGFBP7-dependent down-regulation of intracellular and extracellular Hsp60 level	Involvement in tumor suppressive activity of IGFBP7	[108]
Prostate cancer	Increased Hsp60 level	Positive correlation with tumor progression and hormone resistance	[96]
Gastric cancers	Increased Hsp60 level	Positive correlation with cancer progression, invasiveness, and poor overall survival.	[97,98]
Leukemia	Presence of Hsp60 on the cell surface	Activation/maturation of dendritic cells and generation of potent anti-tumor T-cell response	[53]
	Increased Hsp60 level	Positive correlation with lower complete remission rate and shorter survival	[101]
Glioblastoma multiforme	Increased Hsp60 level	Cytoprotective and pro-survival role	[102]
	Decreased Hsp60 level	Reduced cancer cell proliferation and tumor growth	[109,112]
Hepatocellular carcinoma	Decreased Hsp60 level	Positive correlation with cancer progression and poor prognosis	[118]
	Increased exosomal release of Hsp60 after anti-cancer treatment	Activation of anti-tumor immune response	[121]
Pancreatic cancer	Increased Hsp60 level	Positive correlation with cancer-cell proliferation and tumor growth and progression	[111]
Bladder carcinoma	Increased Hsp60 level	Positive correlation with resistance to anti-cancer treatment	[105]
	Decreased Hsp60 level	Positive correlation with higher tumor stage and cancer recurrence	[122]
Renal cell carcinoma	Decreased Hsp60	Disruption of mitochondria homeostasis and positive correlation with cancer progression	[117]
Large bowel cancer	Increased intracellular and exosomal Hsp60 level	Positive correlation with tumor development and progression	[123]
Thyroid cancers	Increased intracellular and exosomal Hsp60 level	Positive correlation with tumor progression	[99,100]

3.1. Hsp60 in Extracellular Vesicles in Carcinogenesis

Extracellular vesicles (EVs) are membranous particles with a diameter of 30–150 nm, which are found in blood, urine, cerebrospinal fluid, breast milk, and saliva, and are released by normal and tumor cells. The International Society for Extracellular Vesicles (ISEV) classify EVs considering their biogenesis pathways and specific markers received from the cells in which they originate, into exosomes, microvesicles, oncosomes, and apoptotic bodies; and considering size into "small EVs" (sEVs) and "medium/large EVs" (m/lEVs) [124]. However, considerable morphological and biochemical heterogeneity exists among the EVs and, to complicate matters even more, many publications do not provide detailed descriptions and/or use non-standardized terminology. Because of this, we will use here EVs and exosomes as synonyms. EVs are involved in physiological and pathological processes as mediator of cell-to-cell communication by carrying proteins, lipids, and nucleic acids (DNA, mRNA, and miRNAs), that affect recipient cells and modify their functions [125–131]. Their content may vary depending on the cell type that produces them, but in general EVs carry a range of diverse proteins, such as tumor susceptibility gene 101 (TSG101); integrins; and tetraspanins such as CD9, CD53, CD63, CD81, and CD82 [132,133]. EVs are thought to play a role in the remodelling of the pericellular microenvironment that is crucial in maintaining tumor growth and recurrence [123,134]. MicroRNAs transported by EVs are being considered the main players in the modulation of target cell functions [135,136]. Therefore, tumor-derived EVs are attracting attention from scientists interested in molecules that might be used as biomarkers for diagnosis and patient follow up. In addition, EVs are candidates for delivering anti-cancer compounds to specific target tissues because their content can be modified; they show some tissue-specificity; and their immunogenicity is low when they are isolated from the same patient, thus presenting slow risk of generating anti-EV immunity [137–139].

Our research group provided the first evidence that tumor cells actively release Hsp60 via exosomes, through a secretion mechanism that requires the translocation of the chaperonin molecule to the plasma-cell membrane and its association with lipid rafts (Figure 2) [47,48]. The exosomal release of Hsp60 by human tumor cells can be enhanced by anti-cancer treatment. For instance, histone deacetylase inhibitors SAHA causes cell cycle arrest and death in a human lung-derived carcinoma cell line; this cytotoxic effect was associated with generation of oxidative stress, mitochondrial damage, and diminution of the intracellular level of Hsp60, which became nitrated and was released via exosomes [110]. It has been suggested that the released Hsp60 interacts with the immune system, generating an anti-tumor response that potentiates the effect of SAHA. In this manner, exosomal Hsp60 could modulate the tumor microenvironment and, from the practical standpoint it could be consider for use as a diagnostic and prognostic biomarker [123]. Indeed, it was found that the plasma levels of exosomes with Hsp60 diminished after anti-cancer treatment that caused tumor mass reduction in bowel cancer and thyroid papillary carcinoma [100,123]. Additionally, treating hepatocellular carcinoma with anti-cancer drugs caused the release of a larger amount of exosomes containing Hsps, including Hsp60, which was followed by an enhanced anti-tumor immune response mediated by natural killer cells [121].

3.2. *Hsp60 and miRNAs Correlations and Implications for Carcinogenesis*

MicroRNAs (miRNAs) are a class of small non-coding RNAs that regulate gene expression post-transcriptionally by translational inhibition and/or mRNAs destabilization [140]. MiRNAs regulate most protein coding genes, and thus they virtually control all biological processes [141,142]. MiRNAs dysregulation following amplification or deletion of miRNA genes, abnormal transcriptional control of miRNAs, dysregulated epigenetic changes, or defects in the miRNA biogenesis process, are associated with pathological conditions, including cancer [143,144]. Dysregulated, abnormal miRNAs can affect all hallmarks of cancer (e.g., continued proliferative signalling, evasion of growth suppressors, cell death resistance, invasiveness and metastasization, and angiogenesis) by acting either as onco-miRs or tumor suppressors, with inhibition of tumor suppressive mRNAs or oncogenic mRNAs, respectively [145,146]. For this reason, miRNAs are considered for use not only as diagnostic and prognostic biomarkers, but also as potential targets or agents in anti-cancer treatment [147].

MiR-9 and miR-221 have been classified as onco-miRs because an increase in their levels paralleled increased risk for tumorigenesis and resistance to chemotherapeutics in primary cancers [148–151]. In breast cancer, miR-9 and miR-221 increased levels correlated with poor outcome and promoted tumor progression and aggressiveness by favoring epithelial-mesenchymal transition (EMT) and breast cancer stem cell phenotypes [152]. Therefore, they have been suggested as potential biomarkers for breast cancer progression and targets for treatment.

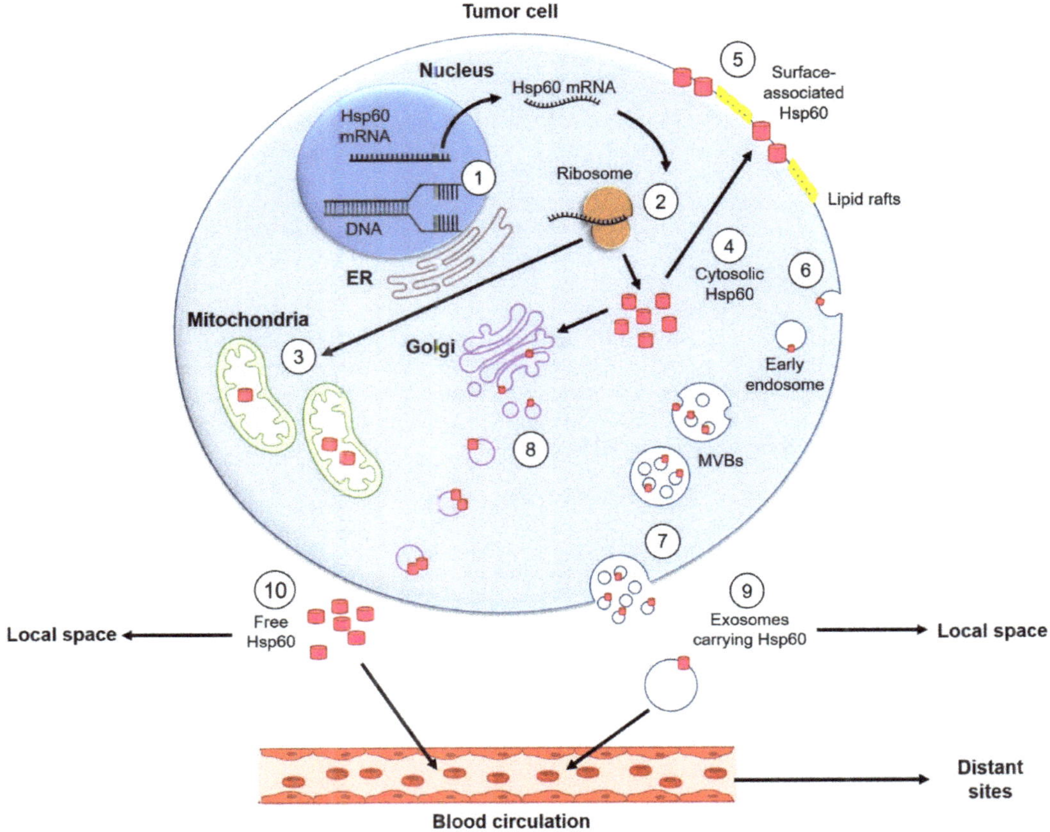

Figure 2. Diagram representing the Hsp60 secreting mechanism. Hsp60 (pink cylinder) is encoded in a single nuclear gene (1). After translation (2), Hsp60 is translocated to the mitochondrion matrix, in which it performs its canonical chaperoning function (3). In tumor cells, Hsp60 accumulates in the cytosol (4), and/or reaches the plasma-cell membrane, near lipid rafts (yellow) (5). The membrane-associated Hsp60 is internalized through a mechanism of endocytosis into early endosomes (6), and then secreted via exosomes after the fusion of multivesicular bodies (MVBs) with the plasma-cell membrane (7). Hsp60 is also exported extracellularly via the classic Golgi-mediated secretion pathway (8). Exosomal Hsp60 (9) and free Hsp60 (10) can reach other cells, e.g., tumor cells and immune cells, nearby or in distant sites via circulation.

miR-30 was found decreased in prostate cancer and acted as tumor suppressor by targeting the EMT-associated gene ERG (Ets-related gene) [153]. Conversely, when increased, miR-30 suppressed EMT and inhibited cell migration and invasion, suggesting it could be used as a therapeutic agent (Table 3) [153].

Hsps, including Hsp60, are also targets for regulation by miRNAs and this has been investigated in various cancers [154–157]. miR-1 and miR-206 increase in rat cardiomyocytes contributed to glucose-mediated apoptosis by diminishing Hsp60 and IGF-1 expression and inhibiting the IGF-1/IGF-1R/PI3 K/Akt pathway [158]. Conversely, treatment of cardiomyocytes with carvedilol, a non-selective β-adrenergic receptor antagonist, inhibited miR-1, which resulted in increased levels of Hsp60 and apoptosis prevention (Table 3) [159].

MiR-29a was found augmented in the serum of breast cancer patients [155]. Downregulation of miR-29a in a breast cancer cell line promoted apoptosis by causing an increase in the level of Hsp60 and a decrease of Hsp27, Hsp40, Hsp70, and Hsp90, suggesting that downregulation of this miRNA is a promising strategy to sensitize cancer cells to chemotherapy [155]. It is likely that the anti-cancer effects observed with the increase of Hsp60 were related to the known pro-apoptotic role of the chaperonin [56,57], or to its ability to stimulate an anti-tumor immune response when exposed on cell surface or released extracellularly [53,54,121]. The level of miR-644a l was found low in hepatocellular carcinoma tissues and cell lines, and negatively correlated with tumor diameter and TNM (Tumor-Node-Metastases) stages [156]. In vitro, an increase in miR-644a promoted cancer cell apoptosis by inhibiting HSF1, Hsp90, Hsp60, Bcl-2, and Bcl-xL proteins while increasing the levels of BID, BAD, BIM, SMAC, Apaf-1, and cleaved caspases-3 and -9, which are all mediators of cell apoptosis [156]. In view of all these observations, miR-644a was suggested as potential prognostic biomarker and therapeutic target in HCC (Table 3) [156]. It is likely that Hsp60 typically displays a pro-tumorigenic activity that can be suppressed by its miRNA-mediated inhibition [157–159]. A further confirmation of the role of miRNAs in the regulation of Hsp60 level in cancer emerged from a study of the pro-tumorigenic activity of miR-17a in gastric lymphoma [160]. The expression of miR-17 was found significantly higher in gastric lymphoma than in normal tissues, which promoted tumor development, progression, and metastasization by regulating the Hsp60/TNFR2 pathway [160]. Therefore, the latter pathway emerges as a potential target for the diagnosis and treatment of gastric lymphoma (Table 3).

Table 3. Examples of miRNAs involved in Hsp60 regulation.

Tissue	miRNAs and miRNAs Status	Effect	Reference
Rat cardiomyocytes	miR-1 and miR-206 high-glucose-dependent up-regulation	Increased cell apoptosis induced by Hsp60 and IGF-1 down-regulation and IGF-1/IGF-1R/PI3 K/Akt pathway inhibition	[158,159]
Breast cancer	miR-29a in vitro down-regulation	Increased cancer cell apoptosis and sensitization to anti-cancer treatment induced by Hsp60 up-regulation	[155]
Gastric Lymphoma	miR-17 higher level	Increased malignancy via regulation of Hsp60/TNFR2 pathway	[160]
Hepatocellular carcinoma	miR-644a in vitro up-regulation	Increased cancer-cell apoptosis induced by Hsp60 inhibition	[156]

4. Concluding Remarks

The involvement of molecular chaperones, including Hsp60, in carcinogenesis has been suggested by various findings [107,161,162]. Molecular chaperones have been found increased in tumor tissues and closely associated with tumor growth and aggressiveness. Along the same lines, decreased levels and expression of chaperones have been found associated with reduced cancer cell proliferation, motility, survival, and metastasization, and with decreased neoangiogenesis and resistance to anti-tumor immune response and treatment [163–167]. The quantitative variations of Hsp60 during carcinogenesis, especially its increase, may be simply the reflection of increased metabolic and proliferative activities of the cancer cell, which would require more chaperonin molecules than a normal cell at its physiological level of metabolism and proliferation. Nevertheless, the role of Hsp60 can still be considered pro-tumoral even if it is not a distinct carcinogenic factor. It would be an example of chaperonopathy by mistake or collaborationism, meaning that the human molecule helps the tumor to grow, proliferate, metastasize, and resist stressors such as anti-cancer drugs. It is crucial to visualize this situation because it then appears very clearly that therapeutic strategies aiming at inhibiting-blocking the chaperonin (negative chaperonotherapy) in tumor cells may offer efficacious ways to defeat cancer. Therefore, negative chaperonopathy, aiming to inhibit or block pro-tumorigenic molecular chaperones deserves investigation [63,168,169].

The role of Hsp60 in carcinogenesis is not yet fully understood, and its elucidation is complicated because it seems to vary depending on the tumor type, tissue affected, stage, and other unidentified factors. While some reports indicate that Hsp60 displays a protumor role others show the contrary [85,90,98,117–119,122]. While the exact role of Hsp60 in every tumor in which its levels and location changes is being studied, the quantitative variations and tissue distribution of the chaperonin may be used as biomarkers potentially useful for diagnosis, prognostication, and disease monitoring [120,123,170,171]. Additionally, investigations on Hsp60 chaperonotherapy in its various forms should be performed and aiming at standardizing the use the chaperonin as a therapeutic target or agent [63]. Some possibilities are to (i) identify compounds and/or methods that will induce a antitumor immune response by promoting the presence of Hsp60 on the surface of tumor-cells [172]; (ii) develop Hsp60 vaccines exploiting its potential for inducing the production of cytokines by interaction with monocytes and macrophages, and by manipulating its ability to bind and activate TLR-2 in CD4+CD25+ regulatory T cells and TLR-4 in B cells [173]; and (iii) identify compounds able to modulate Hsp60 expression, in order to reduce or enhance its activity in those tumors in which it displays a pro-or anti-tumorigenic role, respectively. Understanding the mode of action of Hsp60 and its regulation by miRNAs in different tissue and tumor types will help in the development of novel therapeutic strategies for cancer therapy [68,112,174]. Moreover, the utilization of advanced drug delivery systems, as exosome-like vesicles promise to be, to deliver molecules such as miRNAs known to regulate Hsp60, and internalize them in the target cells, should be actively investigated to standardize the particles and their contents as well as their isolation from cancer patients.

Author Contributions: Conceptualization, S.D., A.M.V., A.F., E.C.d.M. and C.C.B.; original draft preparation, S.D., A.M.V., A.F., F.S., G.V., E.C.d.M., A.P. and C.C.B.; supervision and final writing, A.J.L.M. and C.C.B. All authors have read and agreed to the published version of the manuscript.

Funding: This work was funded in part by the Italian National Operational Programme (PON) «Imprese e Competitività» 2014–2020 FESR, grant awarded by the Italian Ministry of Economic Development to the project titled «Gestione di un servizio integrato multicentrico di diagnostica e terapia personalizzata in oncologia» (Project code: F/090012/01-02/X36). A.J.L.M. and E.C.d.M. were partially supported by IMET and IEMEST. This is IMET contribution number IMET 21-003.

Institutional Review Board Statement: Not applicable.

Informed Consent Statement: Not applicable.

Data Availability Statement: Not applicable.

Conflicts of Interest: The authors declare no conflict of interest.

References

1. Brock, T.D.; Freeze, H. Thermus aquaticus gen. n. and sp. n., a nonsporulating extreme thermophile. *J. Bacteriol.* **1969**, *98*, 289–297. [CrossRef]
2. Moseley, B.E.B. Photobiology and Radiobiology of Micrococcus (Deinococcus) radiodurans. In *Photochemical and Photobiological Reviews*; Springer: Boston, MA, USA, 1983; pp. 223–274.
3. Herbert, R.A. A perspective on the biotechnological potential of extremophiles. *Trends Biotechnol.* **1992**, *10*, 395–402. [CrossRef]
4. Minton, K.W. DNA repair in the extremely radioresistant bacterium *Deinococcus radiodurans*. *Mol. Microbiol.* **1994**, *13*, 9–15. [CrossRef] [PubMed]
5. Minton, K.W.; Daly, M.J. A model for repair of radiation-induced DNA double-strand breaks in the extreme radiophile *Deinococcus radiodurans*. *BioEssays* **1995**, *17*, 457–464. [CrossRef] [PubMed]
6. Gilichinsky, D.; Vishnivetskaya, T.; Petrova, M.; Spirina, E.; Mamykin, V.; Rivkina, E. Bacteria in permafrost. In *Psychrophiles: From Biodiversity to Biotechnology*; Springer: Berlin/Heidelberg, Germany, 2008; pp. 83–102, ISBN 9783540743347.
7. Sorokin, D.Y.; Kuenen, J.G.; Muyzer, G. The microbial sulfur cycle at extremely haloalkaline conditions of soda lakes. *Front. Microbiol.* **2011**, *2*, 44. [CrossRef]

8. Dhakar, K.; Pandey, A. Wide pH range tolerance in extremophiles: Towards understanding an important phenomenon for future biotechnology. *Appl. Microbiol. Biotechnol.* **2016**, *100*, 2499–2510. [CrossRef] [PubMed]
9. Macario, A.J.L.; Conway de Macario, E. Stress genes: An introductory overview. *Stress* **1997**, *1*, 123–134. [CrossRef]
10. Hartl, F.U.; Bracher, A.; Hayer-Hartl, M. Molecular chaperones in protein folding and proteostasis. *Nature* **2011**, *475*, 324–332. [CrossRef]
11. Macario, A.J.L.; Conway de Macario, E. Chapter 12—Chaperone proteins and chaperonopathies. In *Handbook of Stress, Physiology, Biochemistry, and Pathology*; Fink, G., Ed.; Elsevier/Academic Press: Cambridge, MS, USA, 2019; Volume 3, pp. 135–152. Available online: https://doi.org/10.1016/B978-0-12-813146-6.00012-6 (accessed on 18 December 2020).
12. Lindquist, S.; Craig, E.A. The heat-shock proteins. *Annu. Rev. Genet.* **1988**, *22*, 631–677. [CrossRef]
13. Becker, J.; Craig, E.A. Heat-shock proteins as molecular chaperones. *Eur. J. Biochem.* **1994**, *219*, 11–23. [CrossRef]
14. Nollen, E.A.A.; Morimoto, R.I. Chaperoning signaling pathways: Molecular chaperones as stress-sensing "heat shock" proteins. *J. Cell Sci.* **2002**, *115*, 2809–2816.
15. Jacob, P.; Hirt, H.; Bendahmane, A. The heat-shock protein/chaperone network and multiple stress resistance. *Plant Biotechnol. J.* **2017**, *15*, 405–414. [CrossRef] [PubMed]
16. Hendrick, J.P.; Hartl, F.U. Molecular chaperone functions of heat-shock proteins. *Annu. Rev. Biochem.* **1993**, *62*, 349–384. [CrossRef] [PubMed]
17. Welch, W.J. Heat shock proteins functioning as molecular chaperones: Their roles in normal and stressed cells. *Philos. Trans. R. Soc. B Biol. Sci.* **1993**, *339*, 327–333.
18. Kampinga, H.H.; Hageman, J.; Vos, M.J.; Kubota, H.; Tanguay, R.M.; Bruford, E.A.; Cheetham, M.E.; Chen, B.; Hightower, L.E. Guidelines for the nomenclature of the human heat shock proteins. *Cell Stress Chaperones* **2009**, *14*, 105–111. [CrossRef]
19. Gupta, R.S. Evolution of the chaperonin families (HSP60, HSP 10 and TCP-1) of proteins and the origin of eukaryotic cells. *Mol. Microbiol.* **1995**, *15*, 1–11. [CrossRef]
20. Rowland, S.E.; Robb, F.T. *Structure, Function and Evolution of the Hsp60 Chaperonins*; Springer: Singapore, 2017; pp. 3–20.
21. Vilasi, S.; Bulone, D.; Caruso Bavisotto, C.; Campanella, C.; Marino Gammazza, A.; San Biagio, P.L.; Cappello, F.; Conway de Macario, E.; Macario, A.J.L. Chaperonin of Group I: Oligomeric spectrum and biochemical and biological implications. *Front. Mol. Biosci.* **2018**, *4*, 99. [CrossRef]
22. Georgopoulos, C.P.; Hohn, B. Identification of a host protein necessary for bacteriophage morphogenesis (the groE gene product). *Proc. Natl. Acad. Sci. USA* **1978**, *75*, 131–135. [CrossRef]
23. Barraclough, R.; Ellis, R.J. Protein synthesis in chloroplasts IX. Assembly of newly-synthesized large subunits into ribulose bishopshate carboxylase in isolated intact pea chloroplasts. *Biochim. Biophys. Acta* **1980**, *608*, 19–31. [CrossRef]
24. Hemmingsen, S.M.; Ellis, R.J. Purification and Properties of Ribulosebisphosphate Carboxylase Large Subunit Binding Protein. *Plant Physiol.* **1986**, *80*, 269–276. [CrossRef]
25. Hemmingsen, S.M.; Woolford, C.; Van Der Vies, S.M.; Tilly, K.; Dennis, D.T.; Georgopoulos, C.P.; Hendrix, R.W.; Ellis, R.J. Homologous plant and bacterial proteins chaperone oligomeric protein assembly. *Nature* **1988**, *333*, 330–334. [CrossRef] [PubMed]
26. Horwich, A.L.; Fenton, W.A.; Chapman, E.; Farr, G.W. Two families of chaperonin: Physiology and mechanism. *Annu. Rev. Cell Dev. Biol.* **2007**, *23*, 115–145. [CrossRef] [PubMed]
27. Techtmann, S.M.; Robb, F.T. Archaeal-like chaperonins in bacteria. *Proc. Natl. Acad. Sci. USA* **2010**, *107*, 20269–20274. [CrossRef] [PubMed]
28. An, Y.J.; Rowland, S.E.; Na, J.H.; Spigolon, D.; Hong, S.K.; Yoon, Y.J.; Lee, J.H.; Robb, F.T.; Cha, S.S. Structural and mechanistic characterization of an archaeal-like chaperonin from a thermophilic bacterium. *Nat. Commun.* **2017**, *8*, 827. [CrossRef] [PubMed]
29. Spiess, C.; Meyer, A.S.; Reissmann, S.; Frydman, J. Mechanism of the eukaryotic chaperonin: Protein folding in the chamber of secrets. *Trends Cell Biol.* **2004**, *14*, 598–604. [CrossRef] [PubMed]
30. Leitner, A.; Joachimiak, L.A.; Bracher, A.; Mönkemeyer, L.; Walzthoeni, T.; Chen, B.; Pechmann, S.; Holmes, S.; Cong, Y.; Ma, B.; et al. The molecular architecture of the eukaryotic chaperonin TRiC/CCT. *Structure* **2012**, *20*, 814–825. [CrossRef] [PubMed]
31. Willison, K.R. The structure and evolution of eukaryotic chaperonin-containing TCP-1 and its mechanism that folds actin into a protein spring. *Biochem. J.* **2018**, *475*, 3009–3034. [CrossRef] [PubMed]
32. McMullin, T.W.; Hallberg, R.L. A highly evolutionarily conserved mitochondrial protein is structurally related to the protein encoded by the Escherichia coli groEL gene. *Mol. Cell. Biol.* **1988**, *8*, 371–380. [CrossRef]
33. Sigler, P.B.; Xu, Z.; Rye, H.S.; Burston, S.G.; Fenton, W.A.; Horwich, A.L. Structure and function in GroEL-mediated protein folding. *Annu. Rev. Biochem.* **1998**, *67*, 581–608. [CrossRef]
34. Xu, Z.; Sigler, P.B. GroEL/GroES: Structure and function of a two-stroke folding machine. *J. Struct. Biol.* **1998**, *124*, 129–141. [CrossRef]
35. Levy-Rimler, G.; Viitanen, P.; Weiss, C.; Sharkia, R.; Greenberg, A.; Niv, A.; Lustig, A.; Delarea, Y.; Azem, A. The effect of nucleotides and mitochondrial chaperonin 10 on the structure and chaperone activity of mitochondrial chaperonin 60. *Eur. J. Biochem.* **2001**, *268*, 3465–3472. [CrossRef] [PubMed]
36. Vilasi, S.; Carrotta, R.; Mangione, M.R.; Campanella, C.; Librizzi, F.; Randazzo, L.; Martorana, V.; Marino Gammazza, A.; Ortore, M.G.; Vilasi, A.; et al. Human Hsp60 with its mitochondrial import signal occurs in solution as heptamers and tetradecamers remarkably stable over a wide range of concentrations. *PLoS ONE* **2014**, *9*, e97657. [CrossRef] [PubMed]

37. Enriquez, A.S.; Rojo, H.M.; Bhatt, J.M.; Molugu, S.K.; Hildenbrand, Z.L.; Bernal, R.A. The human mitochondrial Hsp60 in the APO conformation forms a stable tetradecameric complex. *Cell Cycle* **2017**, *16*, 1309–1319. [CrossRef] [PubMed]
38. Viitanen, P.V.; Lorimer, G.; Bergmeier, W.; Weiss, C.; Kessel, M.; Goloubinoff, P. Purification of mammalian mitochondrial chaperonin 60 through in vitro reconstitution of active oligomers. *Methods Enzymol.* **1998**, *290*, 203–217.
39. Nielsen, K.L.; Cowan, N.J. A single ring is sufficient for productive chaperonin-mediated folding in vivo. *Mol. Cell* **1998**, *2*, 93–99. [CrossRef]
40. Nisemblat, S.; Parnas, A.; Yaniv, O.; Azem, A.; Frolow, F. Crystallization and structure determination of a symmetrical "football" complex of the mammalian mitochondrial Hsp60-Hsp10 chaperonins. *Acta Crystallogr. Sect. F Struct. Biol. Commun.* **2014**, *70*, 116–119. [CrossRef] [PubMed]
41. Nisemblat, S.; Yaniv, O.; Parnas, A.; Frolow, F.; Azem, A. Crystal structure of the human mitochondrial chaperonin symmetrical football complex. *Proc. Natl. Acad. Sci. USA* **2015**, *112*, 6044–6049. [CrossRef]
42. Gomez-Llorente, Y.; Jebara, F.; Patra, M.; Malik, R.; Nisemblat, S.; Chomsky-Hecht, O.; Parnas, A.; Azem, A.; Hirsch, J.A.; Ubarretxena-Belandia, I. Structural basis for active single and double ring complexes in human mitochondrial Hsp60-Hsp10 chaperonin. *Nat. Commun.* **2020**, *11*, 1916. [CrossRef]
43. Soltys, B.J.; Gupta, R.S. Immunoelectron microscopic localization of the 60-kDa heat shock chaperonin protein (Hsp60) in mammalian cells. *Exp. Cell Res.* **1996**, *222*, 16–27. [CrossRef]
44. Soltys, B.J.; Gupta, R.S. Cell surface localization of the 60 kDa heat shock chaperonin protein (hsp60) in mammalian cells. *Cell Biol. Int.* **1997**, *21*, 315–320. [CrossRef]
45. Soltys, B.J.; Gupta, R.S. Mitochondrial-matrix proteins at unexpected locations: Are they exported? *Trends Biochem. Sci.* **1999**, *24*, 174–177. [CrossRef]
46. Cechetto, J.D.; Soltys, B.J.; Gupta, R.S. Localization of mitochondrial 60-kD heat shock chaperonin protein (Hsp60) in pituitary growth hormone secretory granules and pancreatic zymogen granules. *J. Histochem. Cytochem.* **2000**, *48*, 45–56. [CrossRef] [PubMed]
47. Campanella, C.; Bucchieri, F.; Merendino, A.M.; Fucarino, A.; Burgio, G.; Corona, D.F.V.; Barbieri, G.; David, S.; Farina, F.; Zummo, G.; et al. The odyssey of Hsp60 from tumor cells to other destinations includes plasma membrane-associated stages and Golgi and exosomal protein-trafficking modalities. *PLoS ONE* **2012**, *7*, e42008. [CrossRef] [PubMed]
48. Merendino, A.M.; Bucchieri, F.; Campanella, C.; Marcianò, V.; Ribbene, A.; David, S.; Zummo, G.; Burgio, G.; Corona, D.F.V.; Conway de Macario, E.; et al. Hsp60 is actively secreted by human tumor cells. *PLoS ONE* **2010**, *5*, e9247. [CrossRef] [PubMed]
49. Cappello, F.; Conway de Macario, E.; Marasà, L.; Zummo, G.; Macario, A.J.L. Hsp60 expression, new locations, functions and perspectives for cancer diagnosis and therapy. *Cancer Biol. Ther.* **2008**, *7*, 801–809. [CrossRef] [PubMed]
50. Henderson, B.; Fares, M.A.; Lund, P.A. Chaperonin 60: A paradoxical, evolutionarily conserved protein family with multiple moonlighting functions. *Biol. Rev.* **2013**, *88*, 955–987. [CrossRef]
51. Caruso Bavisotto, C.; Alberti, G.; Vitale, A.M.; Paladino, L.; Campanella, C.; Rappa, F.; Gorska, M.; Conway de Macario, E.; Cappello, F.; Macario, A.J.L.; et al. Hsp60 Post-translational modifications: Functional and pathological consequences. *Front. Mol. Biosci.* **2020**, *7*, 95. [CrossRef]
52. Laad, A.D.; Thomas, M.L.; Fakih, A.R.; Chiplunkar, S.V. Human gamma delta T cells recognize heat shock protein-60 on oral tumor cells. *Int. J. Cancer* **1999**, *80*, 709–714. [CrossRef]
53. Feng, H.; Zeng, Y.; Graner, M.W.; Katsanis, E. Stressed apoptotic tumor cells stimulate dendritic cells and induce specific cytotoxic T cells. *Blood* **2002**, *100*, 4108–4115. [CrossRef]
54. Osterloh, A.; Meier-Stiegen, F.; Veit, A.; Fleischer, B.; Von Bonin, A.; Breloer, M. Lipopolysaccharide-free heat shock protein 60 activates T cells. *J. Biol. Chem.* **2004**, *279*, 47906–47911. [CrossRef]
55. Chandra, D.; Choy, G.; Tang, D.G. Cytosolic accumulation of HSP60 during apoptosis with or without apparent mitochondrial release: Evidence that its pro-apoptotic or pro-survival functions involve differential interactions with caspase-3. *J. Biol. Chem.* **2007**, *282*, 31289–31301. [CrossRef] [PubMed]
56. Samali, A.; Cai, J.; Zhivotovsky, B.; Jones, D.P.; Orrenius, S. Presence of a pre-apoptotic complex of pro-caspase-3, Hsp60 and Hsp10 in the mitochondrial fraction of Jurkat cells. *EMBO J.* **1999**, *18*, 2040–2048. [CrossRef] [PubMed]
57. Xanthoudakis, S.; Sophie, R.; Rasper, D.; Hennessey, T.; Aubin, Y.; Cassady, R.; Tawa, P.; Ruel, R.; Rosen, A.; Nicholson, D.W. Hsp60 accelerates the maturation of pro-caspase-3 by upstream activator proteases during apoptosis. *EMBO J.* **1999**, *18*, 2049–2056. [CrossRef] [PubMed]
58. Shan, Y.X.; Liu, T.J.; Su, H.F.; Samsamshariat, A.; Mestril, R.; Wang, P.H. Hsp10 and Hsp60 modulate Bcl-2 family and mitochondria apoptosis signaling induced by doxorubicin in cardiac muscle cells. *J. Mol. Cell. Cardiol.* **2003**, *35*, 1135–1143. [CrossRef]
59. Gupta, S.; Knowlton, A.A. HSP60, Bax, apoptosis and the heart. *J. Cell. Mol. Med.* **2005**, *9*, 51–58. [CrossRef] [PubMed]
60. Lin, K.M.; Lin, B.; Lian, I.Y.; Mestril, R.; Scheffler, I.E.; Dillmann, W.H. Combined and individual mitochondrial HSP60 and HSP10 expression in cardiac myocytes protects mitochondrial function and prevents apoptotic cell deaths induced by simulated ischemia-reoxygenation. *Circulation* **2001**, *103*, 1787–1792. [CrossRef]
61. Chun, J.N.; Choi, B.; Lee, K.W.; Lee, D.J.; Kang, D.H.; Lee, J.Y.; Song, I.S.; Kim, H.I.; Lee, S.H.; Kim, H.S.; et al. Cytosolic Hsp60 Is Involved in the NF-kB-Dependent Survival of Cancer Cells via IKK Regulation. *PLoS ONE* **2010**, *5*, e9422. [CrossRef]
62. Ghosh, J.C.; Dohi, T.; Byoung, H.K.; Altieri, D.C. Hsp60 regulation of tumor cell apoptosis. *J. Biol. Chem.* **2008**, *283*, 5188–5194. [CrossRef]

63. Cappello, F.; Marino Gammazza, A.; Palumbo Piccionello, A.; Campanella, C.; Pace, A.; Conway de Macario, E.; Macario, A.J.L. Hsp60 chaperonopathies and chaperonotherapy: Targets and agents. *Expert Opin. Ther. Targets* **2014**, *18*, 185–208. [CrossRef]
64. Macario, A.J.L.; Conway de Macario, E. Sick chaperones, cellular stress, and disease. *N. Engl. J. Med.* **2005**, *353*, 1489–1501. [CrossRef]
65. Macario, A.J.L.; Conway de Macario, E. Chaperonopathies by defect, excess, or mistake. *Ann. N. Y. Acad. Sci.* **2007**, *1113*, 178–191. [CrossRef] [PubMed]
66. Macario, A.J.L.; Conway de Macario, E.; Cappello, F. *The Chaperonopathies. Diseases with Defective Molecular Chaperones*; Springer: Dordrecht, The Netherlands; Heidelberg, Germany; New York, NY, USA; London, UK, 2013; Available online: https://www.springer.com/us/book/9789400746664 (accessed on 18 December 2020).
67. Baron, B. Role of the Post-translational Modifications of HSP60 in Disease. In *Heat Shock Protein 60 in Human Diseases and Disorder*; Springer: Cham, Switzerland, 2019; Volume 18, pp. 69–94.
68. Gorska, M.; Marino Gammazza, A.; Zmijewski, M.A.; Campanella, C.; Cappello, F.; Wasiewicz, T.; Kuban-Jankowska, A.; Daca, A.; Sielicka, A.; Popowska, U.; et al. Geldanamycin-Induced Osteosarcoma Cell Death Is Associated with Hyperacetylation and Loss of Mitochondrial Pool of Heat Shock Protein 60 (Hsp60). *PLoS ONE* **2013**, *8*, e71135. [CrossRef] [PubMed]
69. Marino Gammazza, A.; Campanella, C.; Barone, R.; Caruso Bavisotto, C.; Gorska, M.; Wozniak, M.; Carini, F.; Cappello, F.; D'Anneo, A.; Lauricella, M.; et al. Doxorubicin anti-tumor mechanisms include Hsp60 post-translational modifications leading to the Hsp60/p53 complex dissociation and instauration of replicative senescence. *Cancer Lett.* **2017**, *385*, 75–86. [CrossRef] [PubMed]
70. Barazi, H.O.; Zhou, L.; Templeton, N.S.; Krutzsch, H.C.; Roberts, D.D. Identification of heat shock protein 60 as a molecular mediator of alpha 3 beta 1 integrin activation. *Cancer Res.* **2002**, *62*, 1541–1548.
71. Leung, W.-H.; Vong, Q.P.; Lin, W.; Bouck, D.; Wendt, S.; Sullivan, E.; Li, Y.; Bari, R.; Chen, T.; Leung, W. PRL-3 Mediates the Protein Maturation of ULBP2 by Regulating the Tyrosine Phosphorylation of HSP60. *J. Immunol.* **2015**, *194*, 2930–2941. [CrossRef]
72. Direskeneli, H.; Ekşioğlu-Demiralp, E.; Yavuz, Ş.; Ergun, T.; Shinnick, T.; Lehner, T.; Akoğlu, T. T cell responses to 60/65 kDa heat shock protein derived peptides in Turkish patients with Behcet's disease. *J. Rheumatol.* **2000**, *27*, 708–713.
73. Shimizu, J.; Izumi, T.; Suzuki, N. Aberrant activation of heat shock protein 60/65 reactive T cells in patients with Behcet's disease. *Autoimmune Dis.* **2012**, *1*, 105205. [CrossRef]
74. Horváth, L.; Cervenak, L.; Oroszlán, M.; Prohászka, Z.; Uray, K.; Hudecz, F.; Baranyi, É.; Madácsy, L.; Singh, M.; Romics, L.; et al. Antibodies against different epitopes of heat-shock protein 60 in children with type 1 diabetes mellitus. *Immunol. Lett.* **2002**, *80*, 155–162. [CrossRef]
75. Dieudé, M.; Senécal, J.L.; Raymond, Y. Induction of endothelial cell apoptosis by heat-shock protein 60-reactive antibodies from anti-endothelial cell autoantibody-positive systemic lupus erythematosus patients. *Arthritis Rheum.* **2004**, *50*, 3221–3231. [CrossRef]
76. Jamin, C.; Dugué, C.; Alard, J.É.; Jousse, S.; Saraux, A.; Guillevin, L.; Piette, J.C.; Youinou, P. Induction of endothelial cell apoptosis by the binding of anti-endothelial cell antibodies to Hsp60 in vasculitis-associated systemic autoimmune diseases. *Arthritis Rheum.* **2005**, *52*, 4028–4038. [CrossRef]
77. Alard, J.E.; Dueymes, M.; Youinou, P.; Jamin, C. Modulation of endothelial cell damages by anti-Hsp60 autoantibodies in systemic autoimmune diseases. *Autoimmun. Rev.* **2007**, *6*, 438–443. [CrossRef] [PubMed]
78. Alard, J.E.; Dueymes, M.; Youinou, P.; Jamin, C. HSP60 and anti-HSP60 antibodies in vasculitis: They are two of a kind. *Clin. Rev. Allergy Immunol.* **2008**, *35*, 66–71. [CrossRef] [PubMed]
79. Zhu, J.; Quyyumi, A.A.; Rott, D.; Csako, G.; Wu, H.; Halcox, J.; Epstein, S.E. Antibodies to human heat-shock protein 60 are associated with the presence and severity of coronary artery disease: Evidence for an autoimmune component of atherogenesis. *Circulation* **2001**, *103*, 1071–1075. [CrossRef] [PubMed]
80. Benagiano, M.; D'Elios, M.M.; Amedei, A.; Azzurri, A.; van der Zee, R.; Ciervo, A.; Rombolà, G.; Romagnani, S.; Cassone, A.; Del Prete, G. Human 60-kDa Heat Shock Protein Is a Target Autoantigen of T Cells Derived from Atherosclerotic Plaques. *J. Immunol.* **2005**, *174*, 6509–6517. [CrossRef]
81. Mandal, K.; Foteinos, G.; Jahangiri, M.; Xu, Q. Role of antiheat shock protein 60 autoantibodies in atherosclerosis. *Lupus* **2005**, *14*, 742–746. [CrossRef]
82. Bodolay, E.; Prohászka, Z.; Paragh, G.; Csipő, I.; Nagy, G.; Laczik, R.; Demeter, N.; Zöld, E.; Nakken, B.; Szegedi, G.; et al. Increased levels of anti-heat-shock protein 60 (anti-Hsp60) indicate endothelial dysfunction, atherosclerosis and cardiovascular diseases in patients with mixed connective tissue disease. *Immunol. Res.* **2014**, *60*, 50–59. [CrossRef]
83. Mantej, J.; Polasik, K.; Piotrowska, E.; Tukaj, S. Autoantibodies to heat shock proteins 60, 70, and 90 in patients with rheumatoid arthritis. *Cell Stress Chaperones* **2019**, *24*, 283–287. [CrossRef]
84. Meng, Q.; Li, B.X.; Xiao, X. Toward developing chemical modulators of Hsp60 as potential therapeutics. *Front. Mol. Biosci.* **2018**, *5*, 35. [CrossRef]
85. Cappello, F.; Di Stefano, A.; D'Anna, S.E.; Donner, C.F.; Zummo, G. Immunopositivity of heat shock protein 60 as a biomarker of bronchial carcinogenesis. *Lancet Oncol.* **2005**, *6*, 816. [CrossRef]
86. Cappello, F.; Bellafiore, M.; Palma, A.; Marciano, V.; Martorana, G.; Belfiore, P.; Martorana, A.; Farina, F.; Zummo, G.; Bucchieri, F. Expression of 60-kD Heat Shock Protein Increases during Carcinogenesis in the Uterine Exocervix. *Pathobiology* **2002**, *70*, 83–88. [CrossRef]

87. Castle, P.E.; Ashfaq, R.; Ansari, F.; Muller, C.Y. Immunohistochemical evaluation of heat shock proteins in normal and preinvasive lesions of the cervix. *Cancer Lett.* **2005**, *229*, 245–252. [CrossRef] [PubMed]
88. Hwang, Y.J.; Lee, S.P.; Kim, S.Y.; Choi, Y.H.; Kim, M.J.; Lee, C.H.; Lee, J.Y.; Kim, D.Y. Expression of heat shock protein 60 kDa is upregulated in cervical cancer. *Yonsei Med. J.* **2009**, *50*, 399–406. [CrossRef] [PubMed]
89. Hjerpe, E.; Egyhazi, S.; Carlson, J.; Stolt, M.F.; Schedvins, K.; Johansson, H.; Shoshan, M.; Åvall-Lundqvist, E. HSP60 predicts survival in advanced serous ovarian cancer. *Int. J. Gynecol. Cancer* **2013**, *23*, 448–455. [CrossRef] [PubMed]
90. Guo, J.; Li, X.; Zhang, W.; Chen, Y.; Zhu, S.; Chen, L.; Xu, R.; Lv, Y.; Wu, D.; Guo, M.; et al. HSP60-regulated Mitochondrial Proteostasis and Protein Translation Promote Tumor Growth of Ovarian Cancer. *Sci. Rep.* **2019**, *9*, 12628. [CrossRef] [PubMed]
91. Bodoor, K.; Abu-Sheikha, A.; Matalka, I.; Alzou'bi, H.; Batiha, O.; Abu-Awad, A.; Jalboush, S.A.; Fayyad, L.M.; Qadiri, E.; Jarun, Y.; et al. Immunohistochemical analysis of heat shock proteins in triple negative breast cancer: HSP60 expression is a marker of poor prognosis. *Eur. J. Gynaecol. Oncol.* **2018**, *39*, 926–934.
92. Kim, S.K.; Kim, K.; Ryu, J.W.; Ryu, T.Y.; Lim, J.H.; Oh, J.H.; Min, J.K.; Jung, C.R.; Hamamoto, R.; Son, M.Y.; et al. The novel prognostic marker, EHMT2, is involved in cell proliferation via HSPD1 regulation in breast cancer. *Int. J. Oncol.* **2019**, *54*, 65–76. [CrossRef]
93. Cappello, F.; Bellafiore, M.; Palma, A.; David, S.; Marcianò, V.; Bartolotta, T.; Sciumè, C.; Modica, G.; Farina, F.; Zummo, G.; et al. 60 KDa chaperonin (HSP60) is over-expressed during colorectal carcinogenesis. *Eur. J. Histochem.* **2003**, *47*, 105–109. [CrossRef]
94. Xu, X.; Wang, W.; Shao, W.; Yin, W.; Chen, H.; Qiu, Y.; Mo, M.; Zhao, J.; Deng, Q.; He, J. Heat shock protein-60 expression was significantly correlated with the prognosis of lung adenocarcinoma. *J. Surg. Oncol.* **2011**, *104*, 598–603. [CrossRef]
95. Ağababaoğlu, I.; Önen, A.; Demir, A.B.; Aktaş, S.; Altun, Z.; Ersöz, H.; Şanlii, A.; Özdemir, N.; Akkoçlu, A. Chaperonin (HSP60) and annexin-2 are candidate biomarkers for non-small cell lung carcinoma. *Medicine* **2017**, *96*, e5903. [CrossRef]
96. Castilla, C.; Congregado, B.; Conde, J.M.; Medina, R.; Torrubia, F.J.; Japn, M.A.; Sáez, C. Immunohistochemical expression of Hsp60 correlates with tumor progression and hormone resistance in prostate cancer. *Urology* **2010**, *76*, 1017.e1–1017.e6. [CrossRef]
97. Giaginis, C.; Daskalopoulou, S.S.; Vgenopoulou, S.; Sfiniadakis, I.; Kouraklis, G.; Theocharis, S.E. Heat Shock Protein-27, -60 and -90 expression in gastric cancer: Association with clinicopathological variables and patient survival. *BMC Gastroenterol.* **2009**, *9*, 14. [CrossRef]
98. Li, X.S.; Xu, Q.; Fu, X.Y.; Luo, W.S. Heat shock protein 60 overexpression is associated with the progression and prognosis in gastric cancer. *PLoS ONE* **2014**, *9*, e107507. [CrossRef] [PubMed]
99. Pitruzzella, A.; Paladino, L.; Vitale, A.M.; Martorana, S.; Cipolla, C.; Graceffa, G.; Cabibi, D.; David, S.; Fucarino, A.; Bucchieri, F.; et al. Quantitative immunomorphological analysis of heat shock proteins in thyroid follicular adenoma and carcinoma tissues reveals their potential for differential diagnosis and points to a role in carcinogenesis. *Appl. Sci.* **2019**, *9*, 4324. [CrossRef]
100. Caruso Bavisotto, C.; Cipolla, C.; Graceffa, G.; Barone, R.; Bucchieri, F.; Bulone, D.; Cabibi, D.; Campanella, C.; Marino Gammazza, A.; Pitruzzella, A.; et al. Immunomorphological pattern of molecular chaperones in normal and pathological thyroid tissues and circulating exosomes: Potential use in clinics. *Int. J. Mol. Sci.* **2019**, *20*, 4496. [CrossRef]
101. Thomas, X.; Campos, L.; Mounier, C.; Cornillon, J.; Flandrin, P.; Le, Q.H.; Piselli, S.; Guyotat, D. Expression of heat-shock proteins is associated with major adverse prognostic factors in acute myeloid leukemia. *Leuk. Res.* **2005**, *29*, 1049–1058. [CrossRef]
102. Ghosh, J.C.; Siegelin, M.D.; Dohi, T.; Altieri, D.C. Heat shock protein 60 regulation of the mitochondrial permeability transition pore in tumor cells. *Cancer Res.* **2010**, *70*, 8988–8993. [CrossRef] [PubMed]
103. Caruso Bavisotto, C.; Graziano, F.; Rappa, F.; Marino Gammazza, A.; Logozzi, M.; Fais, S.; Maugeri, R.; Bucchieri, F.; Conway de Macario, E.; Macario, A.J.L.; et al. Exosomal chaperones and miRNAs in gliomagenesis: State-of-art and theranostics perspectives. *Int. J. Mol. Sci.* **2018**, *19*, 2626. [CrossRef]
104. Graziano, F.; Caruso Bavisotto, C.; Marino Gammazza, A.; Rappa, F.; Conway de Macario, E.; Macario, A.J.L.; Cappello, F.; Campanella, C.; Maugeri, R.; Iacopino, D.G. Chaperonology: The third eye on brain gliomas. *Brain Sci.* **2018**, *8*, 110. [CrossRef] [PubMed]
105. Abu-Hadid, M.; Wilkes, J.D.; Elakawi, Z.; Pendyala, L.; Perez, R.P. Relationship between heat shock protein 60 (HSP6O) mRNA expression and resistance to platinum analogues in human ovarian and bladder carcinoma cell lines. *Cancer Lett.* **1997**, *119*, 63–70. [CrossRef]
106. Tsai, Y.P.; Yang, M.H.; Huang, C.H.; Chang, S.Y.; Chen, P.M.; Liu, C.J.; Teng, S.C.; Wu, K.J. Interaction between HSP60 and β-catenin promotes metastasis. *Carcinogenesis* **2009**, *30*, 1049–1057. [CrossRef]
107. Wu, J.; Liu, T.; Rios, Z.; Mei, Q.; Lin, X.; Cao, S. Heat Shock Proteins and Cancer. *Trends Pharmacol. Sci.* **2017**, *38*, 226–256. [CrossRef]
108. Ruan, W.; Wang, Y.; Ma, Y.; Xing, X.; Lin, J.; Cui, J.; Lai, M. HSP60, a protein downregulated by IGFBP7 in colorectal carcinoma. *J. Exp. Clin. Cancer Res.* **2010**, *29*, 41. [CrossRef] [PubMed]
109. Tang, H.; Li, J.; Liu, X.; Wang, G.; Luo, M.; Deng, H. Down-regulation of HSP60 Suppresses the Proliferation of Glioblastoma Cells via the ROS/AMPK/mTOR Pathway. *Sci. Rep.* **2016**, *6*, 28388. [CrossRef] [PubMed]
110. Campanella, C.; D'Anneo, A.; Marino Gammazza, A.; Caruso Bavisotto, C.; Barone, R.; Emanuele, S.; Lo Cascio, F.; Mocciaro, E.; Fais, S.; Conway de Macario, E.; et al. The histone deacetylase inhibitor SAHA induces HSP60 nitration and its extracellular release by exosomal vesicles in human lung-derived carcinoma cells. *Oncotarget* **2016**, *7*, 28849–28867. [CrossRef] [PubMed]

111. Zhou, C.; Sun, H.; Zheng, C.; Gao, J.; Fu, Q.; Hu, N.; Shao, X.; Zhou, Y.; Xiong, J.; Nie, K.; et al. Oncogenic HSP60 regulates mitochondrial oxidative phosphorylation to support Erk1/2 activation during pancreatic cancer cell growth. *Cell Death Dis.* **2018**, *9*, 161. [CrossRef] [PubMed]
112. Caruso Bavisotto, C.; Marino Gammazza, A.; Lo Cascio, F.; Mocciaro, E.; Vitale, A.M.; Vergilio, G.; Pace, A.; Cappello, F.; Campanella, C.; Palumbo Piccionello, A. Curcumin affects Hsp60 folding activity and levels in neuroblastoma cells. *Int. J. Mol. Sci.* **2020**, *21*, 661. [CrossRef] [PubMed]
113. Kim, W.; Ryu, J.; Kim, J.E. CCAR2/DBC1 and Hsp60 positively regulate expression of survivin in neuroblastoma cells. *Int. J. Mol. Sci.* **2019**, *20*, 131. [CrossRef] [PubMed]
114. Chaiwatanasirikul, K.A.; Sala, A. The tumour-suppressive function of CLU is explained by its localisation and interaction with HSP60. *Cell Death Dis.* **2011**, *2*, e219. [CrossRef]
115. Campanella, C.; Bucchieri, F.; Ardizzone, N.M.; Marino Gammazza, A.; Montalbano, A.; Ribbene, A.; Di Felice, V.; Bellafiore, M.; David, S.; Rappa, F.; et al. Upon oxidative stress, the antiapoptotic Hsp60/procaspase-3 complex persists in mucoepidermoid carcinoma cells. *Eur. J. Histochem.* **2008**, *52*, 221–228. [CrossRef]
116. Caruso Bavisotto, C.; Nikolic, D.; Marino Gammazza, A.; Barone, R.; Lo Cascio, F.; Mocciaro, E.; Zummo, G.; Conway de Macario, E.; Macario, A.J.L.; Cappello, F.; et al. The dissociation of the Hsp60/pro-Caspase-3 complex by bis(pyridyl)oxadiazole copper complex (CubipyOXA) leads to cell death in NCI-H292 cancer cells. *J. Inorg. Biochem.* **2017**, *170*, 8–16. [CrossRef]
117. Tang, H.; Chen, Y.; Liu, X.; Wang, S.; Lv, Y.; Wu, D.; Wang, Q.; Luo, M.; Deng, H. Downregulation of HSP60 disrupts mitochondrial proteostasis to promote tumorigenesis and progression in clear cell renal cell carcinoma. *Oncotarget* **2016**, *7*, 38822–38834. [CrossRef] [PubMed]
118. Zhang, J.; Zhou, X.; Chang, H.; Huang, X.; Guo, X.; Du, X.; Tian, S.; Wang, L.; Lyv, Y.; Yuan, P.; et al. Hsp60 exerts a tumor suppressor function by inducing cell differentiation and inhibiting invasion in hepatocellular carcinoma. *Oncotarget* **2016**, *7*, 68976–68989. [CrossRef] [PubMed]
119. Cappello, F.; Di Stefano, A.; David, S.; Rappa, F.; Anzalone, R.; La Rocca, G.; D'Anna, S.E.; Magno, F.; Donner, C.F.; Balbi, B.; et al. Hsp60 and Hsp10 down-regulation predicts bronchial epithelial carcinogenesis in smokers with chronic obstructive pulmonary disease. *Cancer* **2006**, *107*, 2417–2424. [CrossRef]
120. Hamelin, C.; Cornut, E.; Poirier, F.; Pons, S.; Beaulieu, C.; Charrier, J.P.; Haïdous, H.; Cotte, E.; Lambert, C.; Piard, F.; et al. Identification and verification of heat shock protein 60 as a potential serum marker for colorectal cancer. *FEBS J.* **2011**, *278*, 4845–4859. [CrossRef] [PubMed]
121. Lv, L.H.; Wan, Y.L.; Lin, Y.; Zhang, W.; Yang, M.; Li, G.N.; Lin, H.M.; Shang, C.Z.; Chen, Y.J.; Min, J. Anticancer drugs cause release of exosomes with heat shock proteins from human hepatocellular carcinoma cells that elicit effective natural killer cell antitumor responses in vitro. *J. Biol. Chem.* **2012**, *287*, 15874–15885. [CrossRef]
122. Lebret, T.; Watson, R.W.G.; Molinié, V.; O'Neill, A.; Gabriel, C.; Fitzpatrick, J.M.; Botto, H. Heat shock proteins HSP27, HSP60, HSP70, and HSP90: Expression in bladder carcinoma. *Cancer* **2003**, *98*, 970–977. [CrossRef]
123. Campanella, C.; Rappa, F.; Sciumè, C.; Marino Gammazza, A.; Barone, R.; Bucchieri, F.; David, S.; Curcurù, G.; Caruso Bavisotto, C.; Pitruzzella, A.; et al. Heat shock protein 60 levels in tissue and circulating exosomes in human large bowel cancer before and after ablative surgery. *Cancer* **2015**, *121*, 3230–3239. [CrossRef]
124. Théry, C.; Witwer, K.W.; Aikawa, E.; Alcaraz, M.J.; Anderson, J.D.; Andriantsitohaina, R.; Antoniou, A.; Arab, T.; Archer, F.; Atkin-Smith, G.K.; et al. Minimal information for studies of extracellular vesicles 2018 (MISEV2018): A position statement of the International Society for Extracellular Vesicles and update of the MISEV2014 guidelines. *J. Extracell. Vesicles* **2018**, *7*, 1535750. [CrossRef]
125. Stoorvogel, W.; Kleijmeer, M.J.; Geuze, H.J.; Raposo, G. The biogenesis and functions of exosomes. *Traffic* **2002**, *3*, 321–330. [CrossRef]
126. Camussi, G.; Deregibus, M.C.; Cantaluppi, V. Role of stem-cell-derived microvesicles in the paracrine action of stem cells. *Biochem. Soc. Trans.* **2013**, *41*, 283–287. [CrossRef]
127. Liu, Q.; Rojas-Canales, D.M.; Divito, S.J.; Shufesky, W.J.; Stolz, D.B.; Erdos, G.; Sullivan, M.L.G.; Gibson, G.A.; Watkins, S.C.; Larregina, A.T.; et al. Donor dendritic cell-derived exosomes promote allograft-targeting immune response. *J. Clin. Investig.* **2016**, *126*, 2805–2820. [CrossRef] [PubMed]
128. Long, Q.; Upadhya, D.; Hattiangady, B.; Kim, D.K.; An, S.Y.; Shuai, B.; Prockop, D.J.; Shetty, A.K. Intranasal MSC-derived A1-exosomes ease inflammation, and prevent abnormal neurogenesis and memory dysfunction after status epilepticus. *Proc. Natl. Acad. Sci. USA* **2017**, *114*, E3536–E3545. [CrossRef] [PubMed]
129. Whiteside, T.L. Exosome and mesenchymal stem cell cross-talk in the tumor microenvironment. *Semin. Immunol.* **2018**, *35*, 69–79. [CrossRef] [PubMed]
130. Tai, Y.L.; Chen, K.C.; Hsieh, J.T.; Shen, T.L. Exosomes in cancer development and clinical applications. *Cancer Sci.* **2018**, *109*, 2364–2374. [CrossRef]
131. Caruso Bavisotto, C.; Marino Gammazza, A.; Rappa, F.; Fucarino, A.; Pitruzzella, A.; David, S.; Campanella, C. Exosomes: Can doctors still ignore their existence? *EuroMediterr. Biomed. J.* **2013**, *8*, 136–139.
132. Vesiclepedia: Browse Results. Available online: http://microvesicles.org/browse_results?org_name=Homosapiens&cont_type=&tissue=&gene_symbol=&ves_type= (accessed on 10 January 2021).

133. ExoCarta: Browse Results. Available online: http://exocarta.org/browse_results?org_name=Homosapiens&cont_type=&tissue=&gene_symbol= (accessed on 10 January 2021).
134. Maacha, S.; Bhat, A.A.; Jimenez, L.; Raza, A.; Haris, M.; Uddin, S.; Grivel, J.C. Extracellular vesicles-mediated intercellular communication: Roles in the tumor microenvironment and anti-cancer drug resistance. *Mol. Cancer* **2019**, *18*, 55. [CrossRef]
135. Dickman, C.T.; Lawson, J.; Jabalee, J.; MacLellan, S.A.; LePard, N.E.; Bennewith, K.L.; Garnis, C. Selective extracellular vesicle exclusion of miR-142-3p by oral cancer cells promotes both internal and extracellular malignant phenotypes. *Oncotarget* **2017**, *8*, 15252–15266. [CrossRef]
136. Lawson, J.; Dickman, C.; MacLellan, S.; Towle, R.; Jabalee, J.; Lam, S.; Garnis, C. Selective secretion of microRNAs from lung cancer cells via extracellular vesicles promotes CAMK1D-mediated tube formation in endothelial cells. *Oncotarget* **2017**, *8*, 83913–83924. [CrossRef]
137. Fais, S.; O'Driscoll, L.; Borras, F.E.; Buzas, E.; Camussi, G.; Cappello, F.; Carvalho, J.; Da Silva, A.C.; Del Portillo, H.; El Andaloussi, S.; et al. Evidence-Based Clinical Use of Nanoscale Extracellular Vesicles in Nanomedicine. *ACS Nano* **2016**, *10*, 3886–3899. [CrossRef]
138. Campanella, C.; Caruso Bavisotto, C.; Logozzi, M.; Marino Gammazza, A.; Mizzoni, D.; Cappello, F.; Fais, S. On the Choice of the Extracellular Vesicles for Therapeutic Purposes. *Int. J. Mol. Sci.* **2019**, *20*, 236. [CrossRef]
139. Caruso Bavisotto, C.; Cappello, F.; Macario, A.J.L.; Conway de Macario, E.; Logozzi, M.; Fais, S.; Campanella, C. Exosomal HSP60: A potentially useful biomarker for diagnosis, assessing prognosis, and monitoring response to treatment. *Expert Rev. Mol. Diagn.* **2017**, *17*, 815–822. [CrossRef] [PubMed]
140. Bartel, D.P. MicroRNAs: Genomics, Biogenesis, Mechanism, and Function. *Cell* **2004**, *116*, 281–297. [CrossRef]
141. Lewis, B.P.; Burge, C.B.; Bartel, D.P. Conserved seed pairing, often flanked by adenosines, indicates that thousands of human genes are microRNA targets. *Cell* **2005**, *120*, 15–20. [CrossRef] [PubMed]
142. Friedman, R.C.; Farh, K.K.H.; Burge, C.B.; Bartel, D.P. Most mammalian mRNAs are conserved targets of microRNAs. *Genome Res.* **2009**, *19*, 92–105. [CrossRef]
143. Kumar, M.S.; Lu, J.; Mercer, K.L.; Golub, T.R.; Jacks, T. Impaired microRNA processing enhances cellular transformation and tumorigenesis. *Nat. Genet.* **2007**, *39*, 673–677. [CrossRef]
144. Liu, M.; Zhu, K.; Qian, X.; Li, W. Identification of miRNA/mRNA-negative regulation pairs in nasopharyngeal carcinoma. *Med. Sci. Monit.* **2016**, *22*, 2215–2234. [CrossRef]
145. Peng, Y.; Croce, C.M. The role of microRNAs in human cancer. *Signal Transduct. Target. Ther.* **2016**, *1*, 15004. [CrossRef]
146. Svoronos, A.A.; Engelman, D.M.; Slack, F.J. OncomiR or tumor suppressor? The duplicity of MicroRNAs in cancer. *Cancer Res.* **2016**, *76*, 3666–3670. [CrossRef]
147. Lan, H.; Lu, H.; Wang, X.; Jin, H. MicroRNAs as potential biomarkers in cancer: Opportunities and challenges. *BioMed Res. Int.* **2015**, *2015*, 125094. [CrossRef]
148. Callegari, E.; Elamin, B.K.; Giannone, F.; Milazzo, M.; Altavilla, G.; Fornari, F.; Giacomelli, L.; D'Abundo, L.; Ferracin, M.; Bassi, C.; et al. Liver tumorigenicity promoted by microRNA-221 in a mouse transgenic model. *Hepatology* **2012**, *56*, 1025–1033. [CrossRef]
149. Gwak, J.M.; Kim, H.J.; Kim, E.J.; Chung, Y.R.; Yun, S.; Seo, A.N.; Lee, H.J.; Park, S.Y. MicroRNA-9 is associated with epithelial-mesenchymal transition, breast cancer stem cell phenotype, and tumor progression in breast cancer. *Breast Cancer Res. Treat.* **2014**, *147*, 39–49. [CrossRef]
150. Tanaka, R.; Tomosugi, M.; Horinaka, M.; Sowa, Y.; Sakai, T. Metformin causes G1-phase arrest via down-regulation of MIR-221 and enhances TRAIL sensitivity through DR5 up-regulation in pancreatic cancer cells. *PLoS ONE* **2015**, *10*, e0125779. [CrossRef]
151. Munoz, J.L.; Rodriguez-Cruz, V.; Rameshwar, P. High expression of miR-9 in CD133 + glioblastoma cells in chemoresistance to temozolomide. *J. Cancer Stem Cell Res.* **2015**, *3*, 1. [CrossRef] [PubMed]
152. Cheng, C.-W.; Yu, J.-C.; Hsieh, Y.-H.; Liao, W.-L.; Shieh, J.-C.; Yao, C.-C.; Lee, H.-J.; Chen, P.-M.; Wu, P.-E.; Shen, C.-Y. Increased Cellular Levels of MicroRNA-9 and MicroRNA-221 Correlate with Cancer Stemness and Predict Poor Outcome in Human Breast Cancer. *Cell. Physiol. Biochem.* **2018**, *48*, 2205–2218. [CrossRef] [PubMed]
153. Kao, C.J.; Martiniez, A.; Shi, X.B.; Yang, J.; Evans, C.P.; Dobi, A.; Devere White, R.W.; Kung, H.J. MiR-30 as a tumor suppressor connects EGF/Src signal to ERG and EMT. *Oncogene* **2014**, *33*, 2495–2503. [CrossRef] [PubMed]
154. Ozgur, A.; Tutar, L.; Tutar, Y. Regulation of Heat Shock Proteins by miRNAs in Human Breast Cancer. *MicroRNA* **2014**, *3*, 118–135. [CrossRef]
155. Choghaei, E.; Khamisipour, G.; Falahati, M.; Naeimi, B.; Mossahebi-Mohammadi, M.; Tahmasebi, R.; Hasanpour, M.; Shamsian, S.; Hashemi, Z.S. Knockdown of microRNA-29a changes the expression of heat shock proteins in breast carcinoma MCF-7 cells. *Oncol. Res.* **2016**, *23*, 69–78. [CrossRef] [PubMed]
156. Liang, W.; Liao, Y.; Li, Z.; Wang, Y.; Zheng, S.; Xu, X.; Ran, F.; Tang, B.; Wang, Z. MicroRNA-644a promotes apoptosis of hepatocellular carcinoma cells by downregulating the expression of heat shock factor 1. *Cell Commun. Signal.* **2018**, *16*, 30. [CrossRef] [PubMed]
157. Ban, H.S.; Han, T.S.; Hur, K.; Cho, H.S. Epigenetic alterations of Heat Shock Proteins (HSPs) in cancer. *Int. J. Mol. Sci.* **2019**, *20*, 4758. [CrossRef] [PubMed]
158. Shan, Z.X.; Lin, Q.X.; Deng, C.Y.; Zhu, J.N.; Mai, L.P.; Liu, J.L.; Fu, Y.H.; Liu, X.Y.; Li, Y.X.; Zhang, Y.Y.; et al. MiR-1/miR-206 regulate Hsp60 expression contributing to glucose-mediated apoptosis in cardiomyocytes. *FEBS Lett.* **2010**, *584*, 3592–3600. [CrossRef] [PubMed]

159. Hu, Y.; Chen, X.; Li, X.; Li, Z.; Diao, H.; Liu, L.; Zhang, J.; Ju, J.; Wen, L.; Liu, X.; et al. MicroRNA-1 downregulation induced by carvedilol protects cardiomyocytes against apoptosis by targeting heat shock protein 60. *Mol. Med. Rep.* **2019**, *49*, 3527–3536. [CrossRef]
160. Wang, L.L.; Dong, J.J.; An, B.Z.; Liang, J.; Cai, K.R.; Jin, Z.S.; Jin, H.S.; Hu, J.P. Has-miR-17 increases the malignancy of gastric lymphoma by HSP60/TNFR2 pathway. *J. Biol. Regul. Homeost. Agents* **2020**, *34*, 1317–1324. [PubMed]
161. Ciocca, D.R.; Calderwood, S.K. Heat shock proteins in cancer: Diagnostic, prognostic, predictive, and treatment implications. *Cell Stress Chaperones* **2005**, *10*, 86–103. [CrossRef]
162. Lianos, G.D.; Alexiou, G.A.; Mangano, A.; Mangano, A.; Rausei, S.; Boni, L.; Dionigi, G.; Roukos, D.H. The role of heat shock proteins in cancer. *Cancer Lett.* **2015**, *360*, 114–118. [CrossRef]
163. Qian-Lin, Z.; Ting-Feng, W.; Qi-Feng, C.; Min-Hua, Z.; Ai-Guo, L. Inhibition of cytosolic chaperonin CCTζ-1 expression depletes proliferation of colorectal carcinoma in vitro. *J. Surg. Oncol.* **2010**, *102*, 419–423. [CrossRef]
164. Dong, D.; Stapleton, C.; Luo, B.; Xiong, S.; Ye, W.; Zhang, Y.; Jhaveri, N.; Zhu, G.; Ye, R.; Liu, Z.; et al. A critical role for GRP78/BiP in the tumor microenvironment for neovascularization during tumor growth and metastasis. *Cancer Res.* **2011**, *71*, 2848–2857. [CrossRef] [PubMed]
165. Wang, J.; Cui, S.; Zhang, X.; Wu, Y.; Tang, H. High Expression of Heat Shock Protein 90 Is Associated with Tumor Aggressiveness and Poor Prognosis in Patients with Advanced Gastric Cancer. *PLoS ONE* **2013**, *8*, e62876. [CrossRef] [PubMed]
166. Jagadish, N.; Agarwal, S.; Gupta, N.; Fatima, R.; Devi, S.; Kumar, V.; Suri, V.; Kumar, R.; Suri, V.; Sadasukhi, T.C.; et al. Heat shock protein 70-2 (HSP70-2) overexpression in breast cancer. *J. Exp. Clin. Cancer Res.* **2016**, *35*, 1–14. [CrossRef]
167. Huang, C.-Y.; Wei, P.-L.; Chen, W.-Y.; Chang, W.-C.; Chang, Y.-J. Silencing Heat Shock Protein 27 Inhibits the Progression and Metastasis of Colorectal Cancer (CRC) by Maintaining the Stability of Stromal Interaction Molecule 1 (STIM1) Proteins. *Cells* **2018**, *7*, 262. [CrossRef] [PubMed]
168. Rappa, F.; Farina, F.; Zummo, G.; David, S.; Campanella, C.; Carini, F.; Tomasello, G.; Damiani, P.; Cappello, F.; Conway de Macario, E.; et al. HSP-molecular chaperones in cancer biogenesis and tumor therapy: An overview. *AntiCancer Res.* **2012**, *32*, 5139–5150.
169. Chatterjee, S.; Burns, T.F. Targeting heat shock proteins in cancer: A promising therapeutic approach. *Int. J. Mol. Sci.* **2017**, *18*, 1978. [CrossRef] [PubMed]
170. Desmetz, C.; Bibeau, F.; Boissière, F.; Bellet, V.; Rouanet, P.; Maudelonde, T.; Mangé, A.; Solassol, J. Proteomics-based identification of HSP60 as a tumor-associated antigen in early-stage breast cancer and ductal carcinoma in situ. *J. Proteome Res.* **2008**, *7*, 3830–3837. [CrossRef] [PubMed]
171. Abdalla, M.A.K.; Haj-Ahmad, Y. Promising urinary protein biomarkers for the early detection of hepato-cellular carcinoma among high-risk hepatitis C virus egyptian patients. *J. Cancer* **2012**, *3*, 390–403. [CrossRef] [PubMed]
172. Chang, C.-L.; Hsu, Y.-T.; Wu, C.-C.; Yang, Y.-C.; Wang, C.; Wu, T.-C.; Hung, C.-F. Immune Mechanism of the Antitumor Effects Generated by Bortezomib. *J. Immunol.* **2012**, *189*, 3209–3220. [CrossRef] [PubMed]
173. Murshid, A.; Gong, J.; Stevenson, M.A.; Calderwood, S.K. Heat shock proteins and cancer vaccines: Developments in the past decade and chaperoning in the decade to come. *Expert Rev. Vaccines* **2011**, *10*, 1553–1568. [CrossRef]
174. Wiechmann, K.; Müller, H.; König, S.; Wielsch, N.; Svatoš, A.; Jauch, J.; Werz, O. Mitochondrial Chaperonin HSP60 Is the Apoptosis-Related Target for Myrtucommulone. *Cell Chem. Biol.* **2017**, *24*, 614–623.e6. [CrossRef]

applied sciences

MDPI

Review

The Neurochaperonopathies: Anomalies of the Chaperone System with Pathogenic Effects in Neurodegenerative and Neuromuscular Disorders

Federica Scalia [1,2], Alessandra Maria Vitale [1,2], Radha Santonocito [1], Everly Conway de Macario [3], Alberto J. L. Macario [2,3] and Francesco Cappello [1,2,*]

1 Department of Biomedicine, Neuroscience and Advanced Diagnostics (BIND), University of Palermo, 90127 Palermo, Italy; federica.scalia02@unipa.it (F.S.); alessandramaria.vitale@unipa.it (A.M.V.); radha.santonocito@unipa.it (R.S.)
2 Euro-Mediterranean Institute of Science and Technology (IEMEST), 90139 Palermo, Italy; AJLMacario@som.umaryland.edu
3 Department of Microbiology and Immunology, School of Medicine, University of Maryland at Baltimore-Institute of Marine and Environmental Technology (IMET), Baltimore, MD 21202, USA; econwaydemacario@som.umaryland.edu
* Correspondence: francesco.cappello@unipa.it or francapp@hotmail.com

Abstract: The chaperone (or chaperoning) system (CS) constitutes molecular chaperones, co-chaperones, and chaperone co-factors, interactors and receptors, and its canonical role is protein quality control. A malfunction of the CS may cause diseases, known as the chaperonopathies. These are caused by qualitatively and/or quantitatively abnormal molecular chaperones. Since the CS is ubiquitous, chaperonopathies are systemic, affecting various tissues and organs, playing an etiologic-pathogenic role in diverse conditions. In this review, we focus on chaperonopathies involved in the pathogenic mechanisms of diseases of the central and peripheral nervous systems: the neurochaperonopathies (NCPs). Genetic NCPs are linked to pathogenic variants of chaperone genes encoding, for example, the small Hsp, Hsp10, Hsp40, Hsp60, and CCT-BBS (chaperonin-containing TCP-1- Bardet–Biedl syndrome) chaperones. Instead, the acquired NCPs are associated with malfunctional chaperones, such as Hsp70, Hsp90, and VCP/p97 with aberrant post-translational modifications. Awareness of the chaperonopathies as the underlying primary or secondary causes of disease will improve diagnosis and patient management and open the possibility of investigating and developing chaperonotherapy, namely treatment with the abnormal chaperone as the main target. Positive chaperonotherapy would apply in chaperonopathies by defect, i.e., chaperone insufficiency, and consist of chaperone replacement or boosting, whereas negative chaperonotherapy would be pertinent when a chaperone actively participates in the initiation and progression of the disease and must be blocked and eliminated.

Keywords: chaperone system; molecular chaperones; chaperonopathies; nervous system; neurochaperonopathies; Hsps; neurodegeneration; neuromuscular disorders; chaperonotherapy

Citation: Scalia, F.; Vitale, A.M.; Santonocito, R.; Conway de Macario, E.; Macario, A.J.L.; Cappello, F. The Neurochaperonopathies: Anomalies of the Chaperone System with Pathogenic Effects in Neurodegenerative and Neuromuscular Disorders. *Appl. Sci.* **2021**, *11*, 898. https://doi.org/10.3390/app11030898

Received: 28 December 2020
Accepted: 12 January 2021
Published: 20 January 2021

1. Introduction

A better understanding of neurodegenerative diseases may be achieved by examining them as part of the broad area of protein quality control, since they show signs of protein pathology. Indeed, a common feature in neuropathology is protein misfolding with subsequent formation of protein aggregates and precipitates, and various neurological disorders are proteinopathies or have a proteinopathy component [1–4]. In these, one or more proteins are pathogenic and prone to aggregate because of a genetic or acquired abnormality. In genetic proteinopathies, the gene encoding the pathologic protein bears a variant which makes its product pathogenic, whereas in the acquired proteinopathies the gene is normal, but the encoded protein may undergo a post-translation modification (PTM). In either case, aggregates formed by some of the abnormal proteins are cytotoxic and cause disease. Through evolution, humans, as well as other organisms, have developed mechanisms to deal with proteinopathies. These mechanisms include protein-degradation and autophagic machineries, for instance the ubiquitin-proteasome system (UPS) [5–7] and chaperone-mediated autophagy (CMA) [8–10]. In addition, organisms have developed means to ensure that all proteins fold correctly, translocate to the place in which they function, resist the damaging effect of stressors, and regain activity after reversible damage and aggregation, namely all mechanisms that maintain protein homeostasis. Major players in protein-quality control are a variety of molecular chaperones, some of which are called Hsp (heat shock protein) that form groups of phylogenetically related proteins, such as the small-heat shock protein (sHsp), Hsp40(DnaJ), Hsp70(DnaK), Hsp90, and heavy Hsp families. In addition, there is a class of protein chaperones of about 60 kDa, also called chaperonins, of which two groups are distinguished, I (Hsp60, or Cpn60 for chaperonin 60) and II (CCT for chaperonin-containing TCP-1; also called TriC, for TCP-1 ring complex). All these chaperones typically function as multimolecular complexes of various sizes and degrees of complexity [11–17] and are part of the chaperone system (CS), which also includes co-chaperones, chaperone co-factors, and chaperone interactors and receptors [18–20]. The main functional partners of the CS in protein-quality control are the UPS and the CMA machineries. When the CS malfunctions, for example because of an abnormal chaperone, disease may ensue: a chaperonopathy [18,21] (free access updates at http://www.chaperones-pathology.org/). In many instances chaperonopathies have a pathogenic impact on the nervous system and contribute to the mechanisms underpinning neuropathies [20,22–24]. In this article, we will briefly discuss examples of neuropathies in which chaperonopathies play an etiologic-pathogenic role. These conditions we will call neurochaperonopathies (NCPs).

2. Types of Chaperonopathies in NCPs

NCPs are a heterogeneous group of diseases with complicated etiological and pathogenic features, among which failure of the CS can contribute to various degrees, depending on the disease and the patient considered. Here, we will briefly discuss some salient aspects of NCPs we considered useful to pathologists and physicians for understanding these conditions and for managing these patients.

The canonical functions of the CS pertain to maintenance of protein homoeostasis together with the CMA and the UPS, while its non-canonical functions are unrelated with the former and mostly co-involve the immune system (IS) [20,21]. A deficiency in a component of the CS, for example a chaperone, may cause disease, a chaperonopathy, which can be genetic or acquired [18,21]. In the former, a genetic variant, for example a mutation, results in the production of a defective, malfunctional chaperone, which may lead to tissue lesions and disease (illustrative examples and pertinent bibliography are provided in Table 1). Instead, in acquired chaperonopathies the gene encoding the affected chaperone is normal, but the gene product, namely the protein chaperone, is not. Chaperone abnormalities, genetic and acquired can also be classified considering the level of the chaperone's concentration and/or physiological activity by defect, excess, and mistake. In the chaperonopathies by defect, a chaperone deficit is the distinctive feature, which can be qualitative, quantitative or both. In the former, the affected chaperone may be at the right concentration, but its function is impaired because of a structural defect, for example; or the chaperone may be at a concentration below the normal range and it cannot reach the physiological level of functionality; or both deficits may coincide, below-normal concentration and functional ability, a situation often occurring in genetic chaperonopathies. In this regard, it should be emphasized that in neuropathies with abundant protein aggregates and precipitates, the level of chaperones available to deal with the excessive demand from misfolding clients may not be sufficient. This generates a chaperonopathy by defect caused by a disproportionate demand even when chaperone production is normal. The pool of useful chaperone molecules may be even further depleted by their being sequestered in the aggregates and precipitates [21] (free access updates at http://www.chaperones-pathology.org/). The opposite parameters occur in the chaperonopathies by excess, namely the affected chaperone is either quantitatively above the normal range of concentrations or is qualitatively abnormal, displaying abnormal functions (gain of function).

The chaperonopathies by mistake include disorders in which a chaperone, apparently normal according to the results of tests currently available, plays a pathogenic role: it participates in the mechanism of disease as it occurs in various types of cancer and inflammatory and autoimmune conditions. In these instances, the pathogenic chaperone may have been structurally modified post-transcriptionally, for example by a post-translation modification (PTM), and thereby its intrinsic properties and functions changed. This is a field that calls for intensive research to develop means to detect PTMs and measure their impact on the chaperone molecule's properties and functions in single cells or using minimal amounts of chaperone. A case in point is the chaperone Hsp60, which plays a variety of roles, canonical and non-canonical, physiological, and pathogenic, even though there is only a single *Hsp60* gene in the human genome. How this multifunctionality is generated when there is only a single Hsp60 molecular type? One mechanism could be through post-translational modifications, a possibility that deserves further investigation [25].

Table 1. Examples of genetic neurochaperonopathies.

Mol. Chap.	Mutation	Disease	M. I.	Ref.
HSPA9	Homozygous c.376C-T transition in exon 4 resulting in p.Arg126Trp substitution	EVEN-plus syndrome	AR	[26]
	Compound heterozygous for a c.383A-G transition in exon 4 resulting in p.Tyr128Cys substitution, and a 2-bp deletion in exon 8 (c.882_883delAG) causing a frameshift and resulting in a premature termination codon at amino acid 296	EVEN-plus syndrome	AR	[26]
HSPB8	Heterozygous c.421A-G transition in exon 1 resulting in p.Lys141Glu substitution	dHMN2A	ADo	[27,28]
	Heterozygous c.423G-C transversion in exon 1 resulting in p.Lys141Asn substitution	dHMN2A	ADo	[28,29]
	c.423G-T transversion in exon 1 resulting in p.Lys141Asn substitution	CMT2L	ADo	[30,31]
HSPB1	c.100G-A transition in exon 1 resulting in p.Gly34Arg substitution	dHMN2B	Sporadic	[32]
	c.116C-T transition in exon 1 resulting in p.Pro39Leu substitution	dHMN2B/CMT2F	ADo	[32,33]
	c.121G-A transition in exon 1 resulting in p.Glu41Lys substitution	dHMN2B	ADo	[32]
	c.250G-C transversion in exon 1 resulting in p.Gly84Arg substitution	dHMN2B/CMT2F	ADo	[33–35]
	Homozygous c.295C-A transversion in exon 1 resulting in p.Leu99Met substitution	dHMN2B/CMT2F	AR	[33]
	c.379C-T transition in exon 1 resulting in p.Arg127Trp substitution	dHMN2B/CMT2F	ADo	[36,37]
	c.404C-G transversion in exon 1 resulting in p.Ser135Cys substitution	dHMN2B/CMT2F	ADo	[38,39]
	c.404C-T transition in exon 1 resulting in p.Ser135Phe substitution	dHMN2B/CMT2F	ADo	[33,36,40]
	c.406C-T transition in exon 2 resulting in p.Arg136Trp substitution	CMT2F	ADo	[36]
	c.407G-T transversion in exon 2 resulting in p.Arg136Leu substitution	dHMN2B/CMT2F	Sporadic	[32]

Table 1. *Cont.*

Mol. Chap.	Mutation	Disease	M. I.	Ref.
	c.418C-G transversion in exon 2 resulting in p.Arg140Gly substitution	dHMN2B/CMT2F	ADo or sporadic	[33]
	c.421A-C transversion in exon 2 resulting in p.Lys141Gln substitution	dHMN2B	ADo	[41]
	c.452C-T transition in exon 2 resulting in p.Thr151Ile substitution	dHMN2B	ADo	[36]
	c.490A-G transition in exon resulting in exon 3 resulting in p.Thr164Ala substitution	CMT2F	ADo	[42]
	c.539C-T transition in exon 3 resulting in p.Thr180Ile substitution	dHMN2B/CMT2F	ADo	[32,43]
	c.544C-T transition in exon 3 resulting in p.Pro182Ser substitution	dHMN2B	ADo	[44]
	c.545C-T transition in exon 3 resulting in p.Pro182Leu substitution	dHMN2B	ADo	[36]
	c.562C-T transition in exon 3 resulting in p.Arg188Trp substitution	CMT2F	Sporadic	[32]
HSPE1	Heterozygous c.217C-T transition in exon 2 resulting in p.Leu73Phe substitution	Undefined neurologic disorder	Sporadic	[45]
DNAJB2	Homozygous c.14A-G transition in exon 1 resulting in p.Tyr5Cys substitution	CMT2F/DSMA5	AR	[46]
	Homozygous G-A transition in intron 4 (c.229+1G-A)	dHMN/DSMA5	AR recessive	[46]
	Homozygous G-A transition in the donor splice site of exon 5 (c.352+1G-A)	DSMA5	AR	[47]
DNAJC5	Heterozygous c.344T-G transversion in exon 3 resulting in p.Leu115Arg substitution	CLN4B	ADo	[48–51]
	Heterozygous 3-bp deletion in exon 3 (c.346_348del) resulting in p.Leu116del	CLN4B	ADo	[48–51]
HSPD1	Homozygous c.86A-G transition in exon 2 resulting in p.Asp29Gly substitution	HLD4 or MitCHAP-60 disease	AR	[52,53]
	Heterozygous c.292G-A transition in exon 3 resulting in p.Val98Ile substitution	SPG13	ADo	[54]
	Heterozygous c.1381C-G transversion in exon 10 resulting in p.Gln461Glu substitution	SPG13	ADo	[55]
CCT5	Homozygous c.440A-G transition in exon 4 resulting in a His147Arg substitution	Hereditary sensory neuropathy with spastic paraplegia	AR	[56,57]
	Homozigous c.670C>G transversion in exon 5 resulting in Leu224Val substitution	Demyelinating neuropathy with severe motor disability.	AR	[58]
BBS6	c.110A-G transition in exon 1 resulting in p.Tyr37Cys substitution	BBS	AR	[59,60]
	c.155G-A in exon 1 transition resulting in p.Gly52Asp substitution	BBS	AR	[61]
	c.169A-G transition in exon 1 resulting in Thr57Ala substitution	BBS	AR	[59]
	Homozygous 1-bp deletion (c.281del) in exon 2 resulting in a frameshift after amino acid Phe94 (p.Phe94fs), terminating the protein at amino acid 103	BBS	AR	[59,61]

Table 1. *Cont.*

Mol. Chap.	Mutation	Disease	M. I.	Ref.
	Homozygosity for a complex 2-bp deletion (c.429_430del and c.433_434del) in exon 2 resulting in a frameshift and a premature termination of the protein at amino acid 157	BBS	AR	[59,61]
	Nonsense mutation leading to premature termination (c.442C-T transition in exon 2 resulting in p.Gln148Ter)	BBS	AR	[60]
	c.792T-A transversion in exon 3 resulting in a premature termination (p.Tyr264Ter)	BBS	AR	[61]
	c.830C-T transition in exon 3 resulting in p.Leu277Pro substitution	BBS	AR	[59]
	c.1496G-C transversion in exon 3 resulting in p.Cys499Ser substitution	BBS	AR	[60]
BBS10	c.32T-G transversion in exon 1 resulting in p.Val11Gly substitution	BBS	AR	[62]
	c.101G-C transversion in exon 1 resulting in p.Arg34Pro substitution	BBS	AR	[63]
	1-bp insertion in exon 2 (c.271dupT) leading to premature termination (p.Cys91fsTer95)	BBS	AR	[63,64]
	c.273C-G transversion in exon 2 resulting in p.Cys91Trp substitution	BBS	AR	[64]
	4-bp deletion in exon 2 (c.909_912del) resulting in premature termination (p.S303fsTer305)	BBS	AR	[63]
	c.931T-G transversion in exon 2 resulting in p.Ser311Ala substitution	BBS	AR	[62,63]
	2-bp deletion in exon 2 (c.1044_1045del) resulting in a frameshift and premature termination (p.Pro350fs)	BBS	AR	[64]
BBS12	Homozygous 3-bp deletion in exon 2 (c.337_339del) resulting in p.Val113del	BBS	AR	[65]
	Homozygous c.865G-C transversion in exon 3 resulting in p.Ala289Pro substitution	BBS	AR	[65]
	c.1063C-T transition in exon 3 resulting in a nonsense mutation (p.Arg355Ter)	BBS	AR	[65]
	2-bp deletion in exon 3 (c.1115_1116del) resulting in frameshift and premature termination of the protein (p.Phe372fsTer373)	BBS	AR	[65]
	2-bp deletion in exon 3 (c.1483_1484del) resulting in frameshift and premature termination (p.Glu495fsTer498)	BBS	AR	[65]

Abbreviations: Mol. Chap., molecular chaperones; M.I., mode of inheritance; Ref., reference; ADo, autosomal dominant; AR, autosomal recessive; dHMN2A/2B, distal Hereditary Motor Neuropathy type 2A/2B; CMT2L/2F, Charcot-Marie-Tooth type 2L/2F; DSMA5, Distal Spinal Muscular Atrophy-5, CLN4B, Ceroid Lipofuscinosis Neuronal 4B; SPG13, Spastic Paraplegia 13; HLD4, Hypomyelinating leukodystrophy-4; BBS, Bardet–Biedl syndrome.

All the above-mentioned types of chaperonopathies, involving the chaperones listed in Table 1 have been implicated in neurological disorders and some examples of these NCPs are discussed below.

3. Genetic NCPs

3.1. Small Heat Shock Proteins

Diverse neurological and muscular disorders have been associated with mutations in sHsps. The sHsp α-crystallin family includes 10 members and one related protein (HSP16.2/HSPB11) [66] that, in addition to the chaperoning function, participate in cytoskeleton stabilization and possess anti-aggregation and anti-apoptotic activities [67–71].

Most of the mutations found in the sHsps are located in the highly conserved α-crystallin domain, which is an 80–100 amino-acid-long domain responsible for the association/dissociation of sHsp dimers and for the formation and stabilization of large multisubunit homo- and hetero-oligomers [72]. These mutations lead to protein aggregation and cell death. Two hypotheses have been proposed to explain the pathologic findings: the mutated sHsp molecules may acquire an intrinsic toxicity (gain of toxic function), or they may suffer a loss of function with abolition of their protein quality control activity [73]. Two representative examples are the mutations occurring in the *HSPB8* and *HSPB1* genes that have been associated with the neuromuscular disorders Charcot-Marie-Tooth (CMT) disease and distal hereditary motor neuropathies (dHMN) (Table 1). The term CMT is used to indicate a clinically and genetically heterogeneous group of hereditary motor and sensory neuropathies, characterized by degeneration of peripheral nerves and subdivided in two subgroups: demyelinating (CMT1) and axonal (CMT2) [74]. dHMNs are genetically heterogenous characterized by degeneration of distal lower motor neurons, resulting in muscle weakness and atrophy [75]. However, in many forms of dHMN there may also be minor sensory abnormalities and/or a significant upper-motor-neuron involvement. For this reason, there is often an overlap not only with axonal forms of CMT, i.e., CMT2, but also with juvenile forms of amyotrophic lateral sclerosis, Kennedy's disease, spinal muscular atrophy, and hereditary spastic paraplegia [76].

Various types of chaperonopathies can occur, as follows: CMT type 2L is caused by mutations in *HSPB8* and CMT type 2F is caused by mutations in *HSPB1*, while dHMN type IIA is caused by mutations in *HSPB8* and dHMN type IIB is caused by mutations in *HSPB1* [27–44]. Often, a single mutation in *HSPB1* or *HSPB8* gives rise to one of these conditions. However, in many cases the same mutation was found associated with both diseases, such as the p.(Lys141Asn) substitution in HSPB8 [28,31], and the p.(Ser135Phe) [38,54] and p.(Arg127Trp) substitutions in HSPB1 [36,37,77]. Moreover, there are also examples of mutations found in patients with an unclear phenotype, as shown in Table 1.

For both *HSPB8* and *HSPB1*, the identified mutations in most cases occur in the highly conserved α-crystallin domain, likely destabilizing the proteins structurally and functionally [78]. To elucidate the molecular mechanisms responsible for the association between mutated forms of HSPB8 and HSPB1 and neuromuscular disorders, in vitro studies have been performed. For instance, it was reported that the expression of HSPB8 mutants p.(Lys141Asn) and p.(Lys141Glu) in motor neurons resulted in a reduction in average length and number of neurites per cell, without inducing cell death. In contrast, these abnormalities were very moderate in sensory neurons, and absent in cortical neurons and glial cells [79]. Thus, motor neurons appear to be more sensitive to HSPB8 dysfunction compared to sensory neurons, as indicated by the predominant motor neuron phenotype in dHMN and CMT2L. Moreover, both HSPB8 mutant forms tend to associate with HSPB1 mutant proteins, forming intracellular aggregates [28].

Studies in vitro with HSPB1 mutants demonstrated that some of the substitutions affecting the α-crystallin domain reduced motor neuron viability, impaired neurofilament assembly, destabilized microtubules, and disrupted the axonal transport of specific cellular cargoes and of mitochondria [36,80–84], events that can be responsible for the premature axonal degeneration, typical of both CMT and dHMN [85]. In addition, in vivo studies have shown that overexpression of HSPB1 mutants in neurons is sufficient to cause pathological and electrophysiological changes typically observed in patients with motor neuropathy [86].

3.2. *DnaJ(Hsp40)*

Mutations in members of the Hsp40(DnaJ) family have been associated with neurologic and muscular disorders. Two examples are variants of the *DNJAB2* and *DNAJC5* genes, encoding members of the subfamilies B and C, respectively. *DNAJB2* is mainly expressed in neurons and plays anti-protein aggregation and neuroprotective roles, as shown in models of neurodegenerative disorders [87–89]. The mutation c.352+1G-A in

DNAJB2 was found associated with autosomal recessive distal spinal muscular atrophy-5 (DSMA5), a type of dHMN with young adult onset and characterized by progressive distal muscle weakness and atrophy with gait impairment and loss of reflexes [47] (Table 1). This genetic variant reduced or abolished chaperone expression, with a consequent accumulation and aggregation of misfolded proteins, which would cause lower motor neurons degeneration [47]. The same *DNAJB2* variant was found in patients diagnosed with CMT2, and careful examination of phenotype and clinical evolution showed that pure motor impairment occurred early in the disease course, with no initial sensory symptoms, but sensory disturbances in the lower limbs appeared as the disease progressed [90]. Two other *DNAJB2* gene variants were found in two families (Table 1). One (c.229+1G-A) showed a dHMN phenotype, with signs of distal muscular atrophy and paresis; the other, causing the missense mutation p.(Tyr5Cys), showed a CMT2 phenotype, with a noticeable sensory involvement, extending the group of *DNAJB2*-related diseases to include sensory neuropathy [46]. However, another evaluation of the phenotypes associated with these two latter genetic variations has classified them as DSMA5 (for details see the corresponding page on ClinVar database).

DNAJC5, also known as cysteine string protein (CSP), is abundant in neuronal cells and is a key element of the synaptic molecular machinery [48]. Two different genetic variations, an amino acid deletion and an amino acid substitution, were found associated with autosomal dominant adult-onset neuronal ceroid lipofuscinosis-4B (CLN4B), also known as Kufs' disease [48–51]. These mutations affect two evolutionarily conserved leucine residues located in a conserved region of the cysteine-string domain that is involved in palmitoylation and membrane targeting of the protein [91]. CLN4B belongs to the group of neuronal ceroid lipofuscinosis (NCLs), a genetically heterogeneous group of at least nine neurodegenerative disorders, clinically characterized by progressive cognitive and motor impairment, visual impairment, epileptic seizures, and premature death. Despite the different ages of onset, all forms presented a common feature, i.e., the lysosomal accumulation of auto-fluorescent lipo-pigment in neuronal cells and peripheral tissues, which caused progressive and selective neurodegeneration and gliosis with secondary white matter lesions [92,93]. To clarify the molecular mechanisms responsible for the negative effects of the identified CSP mutations in neuronal cells, functional in vitro studies were performed. It was shown that both mutants had an abnormal intracellular localization and were less efficiently palmitoylated compared to the wild type protein. Moreover, they formed aggregates, causing CSP depletion that, in turn, could be responsible for the inhibition of synapse formation and synaptic transmission [49,94].

3.3. Hsp70(DnaK)

Pathogenic mutations in one of the members of the Hsp70(DnaK) family have been reported [26]. This chaperone family is composed of at least 17 members [95] but thus far only one of them, the mitochondrial mtHsp70, also named mortalin or HSPA9, has been found mutated and causing disease. Two mutations of this chaperone were associated with a syndrome very similar to CODAS (Cerebral, Ocular, Dental, Auricular, Skeletal Syndrome), which is caused by a mutation in the LONP1 gene, however it differs in that it also shows severe microtia, nasal hypoplasia, and other malformations. These characteristics are indicated by the name given to this new condition: EVEN-PLUS (from Epiphyseal, Vertebral, Ear, Nose, Plus associated findings) syndrome.

3.4. The Chaperonins

The best-known examples of genetic neurochaperonopathies are those involving members belonging to the family of chaperonins, including the group I chaperonin Hsp60 (HSPD1, of which a few pathogenic variants have been identified in humans) and the group II chaperonin CCT.

Chaperonins of Group I. The chaperonin HSPD1 is typically expressed within the mitochondrion where it assists the folding of proteins destined to the matrix together with the co-chaperonin Hsp10 (HSPE1) [45,96,97]. HSPD1 plays a key role not only in the maintenance of protein homeostasis in mitochondria, but also in sustaining cellular viability, since complete loss of the protein impaired mammalian development and postnatal survival, as demonstrated by an in vivo model of mice homozygous for an inactivation of *HSPD1* gene [98]. In humans, missense mutations of the *HSPD1* gene have been found associated with severe and chronic nervous system diseases. These include: 1) a dominantly inherited form of Spastic Paraplegia (SPG13) affecting motor neurons with the longest axons in the spinal cord and caused by either one of two missense mutations occurring at different positions, i.e., p.(Val98Ile) and p.(Gln461Glu), [54,55,99] (Table 1); and 2) a recessively inherited hypo-myelinating leukodystrophy (HLD4), also known as MitCHAP60 disease, which is a fatal early-onset neurodegenerative disorder characterized by pronounced cerebral hypomyelination and is caused by a single specific missense mutation, i.e., p.(Asp29Gly) [52,53] (Table 1). In vitro and in vivo studies have shown that these mutations destabilized the chaperonin oligomeric complex and reduced its ATPase and folding activities compared to the wild type form [100,101]. In addition to these two well-known neurological conditions, many others of various degrees of severity may be caused by other *HSPD1* mutations that are known to occur in human genomes [97,102].

A point missense mutation, p.(Leu73Phe), in the gene encoding the co-chaperonin Hsp10 or HSPE1 (*hsp10* or *HSPE1* gene), has been found in a patient with a history of infantile spasms, hypotonia, developmental delay, a slightly enlarged liver, macrocephaly, and mild non-specific dysmorphic features [45] (Table 1).

The association between neurological disorders and *HSPD1/E1* genes mutations is related to the great sensitivity of neuronal cells to mitochondrial dysfunction, since they have a metabolism mainly based on oxidative phosphorylation [103]. In vitro studies have shown that the expression of HSPD1 variants alters mitochondrial morphology, dynamics, and functionality [104,105].

Chaperonins of Group II. This group includes the CCT family, which in humans is composed of nine subunits: 1 through 8, with two versions of subunit 6; plus, three chaperones named BBS6, BBS10, and BBS12, which are involved in cilia biogenesis; and two other evolutionarily related molecules, CCT8L1 and CCT8L2, of which little is known [106] (Figure 1).

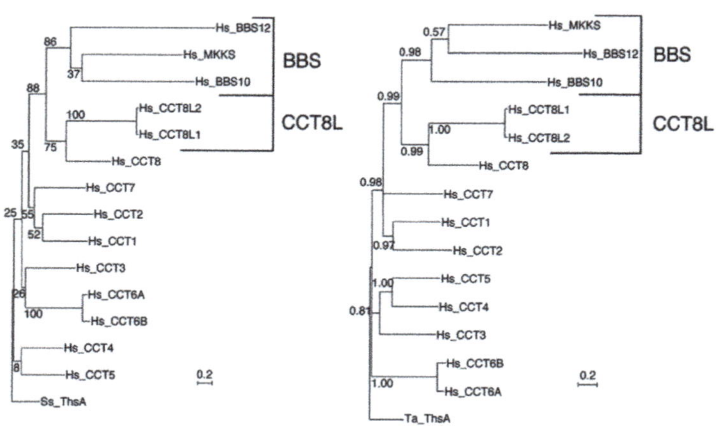

Figure 1. Evolutionary trees of chaperonin-containing TCP-1 (CCT) chaperonins. (**Left**): Maximum-likelihood tree of human chaperonins, including CCT monomers, MKKS, BBS10, and BBS12 as well as the members of the CCT8L class, CCT8L1 and CCT8L2. Numbers associated with each branch indicate bootstrap support from 100 replicates. Tree rooted by the archaeal thermosome α subunit of *Sulfolobus solfataricus* (Ss_ThsA). (**Right**): Bayesian tree of the same sequences. The numbers assigned to each branch indicate posterior probabilities. Tree rooted by the thermosome α subunit of *Thermoplasma acidophilum* (Ta_ThsA). Scale bars, number of substitutions per position for a unit branch length. Reproduced from Reference [106]

Here, we will consider chaperonopathies of the subunit CCT5 and those of the BBS cluster. BBS stands for Bardet–Biedl syndrome, which is the pathological condition caused by abnormalities in any one of these three chaperones. BBS6 is also called MKKS, because the disease associated with mutations in this gene is also called McKusick–Kaufmann syndrome.

The CCT subunits form functional hexadecamers with two octameric rings joined by their openings, which results in a barrel-like structure with a central cavity inside which polypeptide folding occurs [15]. This complex is also named TRiC, for TCP-1 ring complex, where TCP-1 stands for Tailless Complex Polypeptide 1. The integrity and functionality of this complicated multimolecular machine greatly depends on the integrity and functionality of all its individual components, the CCT subunits. For instance, a mutation in one of them may disrupt ring formation and hexadecamer assemblage or make the assemblage unstable and prone to dissociation, especially under the action of stressors.

A mutation occurring in subunit 5 of CCT was found to be associated with mutilating, distal sensory neuropathy [56,57] (Table 1). The functional CCT complex is composed of two hetero-octameric rings associated in a hexadecamer. The reported mutation, p.(His147Arg), occurred in subunit 5 (CCT5), and the disease was characterized by severe atrophy of the posterior tract of the spinal cord [56,57]. In vitro studies with human and archaeal purified molecules showed that the mutation, which occurs in the equatorial domain of the chaperonin subunit, impairs some of its properties and functions, resulting in poor oligomeric assembly [107–109].

More recently, a new disease was described associated with a different missense mutation, p.(Leu224Val), in CCT5 [58]. While the previously described mutation His147Arg caused a distal sensory neuropathy more pronounced in adulthood, the newly found mutation is associated with a motor neuropathy of early onset. Noteworthy is the different location of the mutations: His147Arg is in the equatorial domain, whereas Leu224Val is in the intermediate domain and appears to have an impact on the conformation of the apical domain, which is involved in substrate recognition and binding, namely functions quite distinct from those of the equatorial domain.

The disorders caused by mutations in the BBS genes affect ciliogenesis. Cilia are built with the participation of the BBSsome, an octameric complex composed of seven BBS proteins—BBS1, BBS2, BBS4, BBS5, BBS7, BBS8, and BBS9—and the protein BBIP10 [110]. BBS6, BBS10, and BBS12 serve as chaperones for BBSome assembly. Defects in the process of ciliogenesis and, consequently, in cilia structure and/or function, lead to a group of disorders called ciliopathies [24,111]. The Bardet–Biedl syndrome is a genetic disorder with autosomal recessive inheritance, highly prevalent in inbred/consanguineous populations, and with numerous primary and secondary clinical manifestations, including obesity, retinal degeneration, olfactory deficits, nephropathy, polydactyly, development delay, and cognitive or other neurological impairments [24,112]. About 70–80% of all cases of Bardet–Biedl syndrome result from mutations in *BBS1* to *BBS18* genes, with frequent mutations in *BBS1* and *BBS10* genes, especially in populations of European and Caucasian descent [113].

The BBS proteins excluded from the BBSome, i.e., BBS6, BBS10, and BBS12, have been classified as chaperonins, since they have high sequence similarity with members of the CCT family [63,65,106,114] (Figure 1). Even if these proteins are not components of the BBSome, mutations in their encoding genes also lead to the same phenotype resulting from the loss or abnormalities of BBSome subunits [24,65,115]. Moreover, BBS patients bearing pathogenic variants in the *BBS6*, *BBS10*, and *BBS12* genes develop a more severe phenotype with earlier disease onset and greater prevalence of all BBS primary diagnostic features typical of ciliopathies than patients with variants only in the non-chaperonin BBS genes [24,116]. The reason for this increased pathogenic severity associated with variants of the BBS chaperonin genes could be the malfunctioning of these three proteins, which are critical for BBSome assembly [24,117].

The Bardet–Biedl syndrome caused by mutations in the genes encoding for the chaperonin-like BBS proteins can be included in the group of genetic chaperonopathies [24], and in particular within the neurochaperonopathies if we consider the associated neurological abnormalities. In a mouse model of BBS, the morphological evaluation of brain neuroanatomy revealed ventriculomegaly of the lateral and third ventricles, thinning of the cerebral cortex, and reduced volume of the corpus striatum and hippocampus [118]. These abnormalities could be related to defects in the cilia of the ependymal cells, affecting cilia assembly, structure, and/or function [118,119].

Numerous genetic mutations in the genes encoding BBS6, BBS10, and BBS12 have been found [24,59–65,120]. Illustrative examples are listed in Table 1. Mutations in the *BBS6* gene have been associated with typical BBS and with another similar disorder, the McKusick–Kaufman syndrome, and it was suggested that both syndromes are different allelic forms of the same clinical entity [59,121,122]. Moreover, in different families affected by BBS, many of the genetic variants in the *BBS6* gene were found in the same individual [61] or associated with mutations in genes encoding BBS2 and other BBS proteins, suggesting a triallelic inheritance model for penetrance of the BBS phenotype [60,120]. *BBS10* is, together with *BBS1*, the major contributor to BBS [113]. The most common mutation in the *BBS10* is the 271dupT, occurring in 46% of mutant alleles [63].

Many of the *BBS10* genetic variants were found in the same individual [62,64] or associated with mutations in the *BBS1* gene [63]. Variants of *BBS12* account for 8%–11% of the total cases of BBS in most of the cohorts reported [115].

4. Acquired Neurochaperonopathies

4.1. Hsp70(DnaK)

The Hsp70 family is composed of at least 17 members [95], and they are targets of a variety of PTMs such as phosphorylation, acetylation, ubiquitination, AMPylation, and ADP-ribosylation [123]. The only one human Hsp70 AMPylator enzyme containing Fic (Filamentation induced by cAMP) domain is the FICD enzyme, also known as Huntingtin yeast partner E (HYPE), able to catalyze the transfer of AMP onto a serine, threonine, or tyrosine residue of the substrate protein [124]. The consequences of Hsp70 AMPylation are poorly understood and, although there is no information on its impact on neurodegeneration to date, some interest in this regard has developed because of observations made using *Saccharomyces cerevisiae*, a eukaryote free of endogenous AMPylation machinery, and *Caenorhabditis elegans* models manipulated for the FIC-1 enzyme orthologous of human FICD (or HYPE) [125]. Induction of *C. elegans* FIC-1 (E274G) and *Homo sapiens* HYPE (E234G) in *S. cerevisiae*, in absence of stress, inhibited Hsp70 activity leading to decreased cell growth, toxic protein aggregation, and misfolding. These results suggest that AMPylation somehow modulates the Hsp70 protein folding machinery. Furthermore, other experiments confirmed AMPylation of a human Hsp70 family member, HSPA5, in a FICD-dependent manner, indicating that a high number of neuronal proteins involved in neuronal differentiation are AMPylated by FICD enzyme [126]. In addition, the fly *Drosophila*'s dFic AMPylator was shown to be essential for maintaining the levels of AMPylated Hsp70 BiP (the Hsp70 family member that resides in the endoplasmic reticulum (ER)), which are crucial for the photoreceptor's resistance against stress; flies deprived of dFic are blind and unable of postsynaptic activation [123].

On the other hand, *C. elegans* models for Alzheimer's disease (AD) and Parkinson 's disease (PD) showed the formation of large protein aggregates when AMPylation of three Hsp70 family members was FIC-1 induced [127] (Table 2). Strikingly, the larger aggregates were more strongly cytoprotective than the smaller ones, extending the nematode's life.

Table 2. Effects of PTM on molecular chaperones produced by in vivo and in vitro wild type and disease models.

Model	Status	Molecular Chaperone	Human Ortholog	PTM	Effect	Ref.
Drosophila	Wt	Bip	Grp78 (HSPA5)	AMPylation	Blindness	[123]
Saccha-romyces cerevisiae	Wt	Hsp70	HSPA	AMPylation	Decreased cell growth; protein misfolding; toxic protein aggregation	[125]
Caeno-rhabditis elegans	AD; PD	HSP-1; HSP-3; HSP-4	HSPA	AMPylation	Large cytoprotective protein aggregates	[127]
PC12 cells	Wt	Hsp90	HSPC	Nitration	ALS	[128,129]
PC12 cells	Wt	VCP	p97	Phosphorylation and acetylation	Neurite retraction and shrinkage	[130]

Abbreviations: PTM, post-translation modification; Ref., reference; Wt, wild type; Hsp or HSP, heat shock protein; AD, Alzheimer's disease; PD, Parkinson's disease; ALS, amyotrophic lateral sclerosis; VCP, valosin-containing protein.

4.2. *Hsp90 (HSPC)*

The human Hsp90 (HSPC) family is composed of various members: two are cytosolic, one resides in the ER, and the other in the mitochondrion (references in [21]). Hsp90 chaperones have multiple sites for phosphorylation, acetylation, SUMOylation, methylation, O-GlcNAcylation, ubiquitination, and other PTMs [128]. Nitration of tyrosine has been associated with apoptosis of motor neurons in amyotrophic lateral sclerosis (ALS) patients and superoxide dismutase (SOD)-transgenic animal models [129] (Table 2). Nitration of five of the 24 tyrosine residues of HSPC2 and HSPC3 impaired ATPase chaperone activity. Nitrated tyrosine at position 56 of HSPC3 was identified in spinal cord sections of an ALS patient and of a mutated mouse model. The nitrated HSPC induced cell apoptosis by activation of P2 × 7 receptor-mediated calcium influx which induced Fas pathway in PC12 cells [129]. Several chronic and acute conditions causing oxidative stress induced the nitration of Hsp90 family members. However, the impact of Hsp90 nitration depends on the position of the nitrated tyrosine. In PC12 cells and motor neurons from rat embryo, it was observed that nitration of tyrosine at position 56 causes a gain of function, which is cytotoxic, whereas nitration of tyrosine at position 33 impairs mitochondrial activity [131].

Hsp90 collaborates with the Cell Division Cycle 37 (Cdc37) co-chaperone, forming a HSPC/Cdc37 complex, which binds almost 60% of the kinome preventing aggregation of activated regulators [132]. The co-chaperone Cdc37 is also a target of PTMs [133]. Phosphorylation at serine 13 of Cdc37 by CK2 (casein kinase II) affects the Hsp90/Cdc37 complex in performing its kinases homeostasis activity. In addition, the casein kinases 1 and 2 which involved AD, PD, and ALS pathogenesis have been reported to impair the interaction between Hsp90 and Cdc37 [134,135].

4.3. *VCP/p97*

The valosin-containing protein (p97 or VCP) is a homohexameric AAA+ ATPase protein of about 97 kDA. It is an abundant and ubiquitous molecule involved in cell cycle regulation [136], autophagy [137], DNA repair [138], and ER-associated and ubiquitin-mediated protein degradation [139]. VCP has been found within protein aggregates in PD, ALS, AD, and in polyglutamine aggregates [130]. Polyglutamine disease is characterized by deacetylation of histones H3 and H4 and a correlation of this deacetylation with VCP activity has been suggested [130]. VCP was found in the PC12 cell nucleus in protein aggregates with polyglutamine expansions, showing PTM, such as phosphorylation at Ser-612 and Thr-613, and acetylation at Lys-614 [130]. Nuclear, post-translationally modified VCP is associated with neurite retraction and shrinkage, features which have been found also in PC12 cells expressing protein aggregates with polyglutamine expansions [130]. Therefore, it has been suggested that VCP could be a mediator of histones H3 and H4 deacetylation, even if the underlined mechanism remains to be elucidated. VCP appears to be essential for the clearance of "occasionally" aggregated proteins, promoting the deacetylation of H3 and H4 and, in turn, reducing the transcription of de novo proteins; however, during chronic and prolonged protein aggregation, nuclear translocation and post-translational modification of VCP could induce neurodegeneration and cause an atrophic phenotype as observed in the *Drosophila* experimental model [130].

5. Discussion

In the last few years, an increasing number of scientific reports have highlighted the involvement of members of the CS in the pathogenesis of diverse neurological disorders. It was found that genetic and/or acquired malfunction (or lack of function) of members of the CS are implicated in the pathogenic mechanisms underlying NCPs (Figure 2).

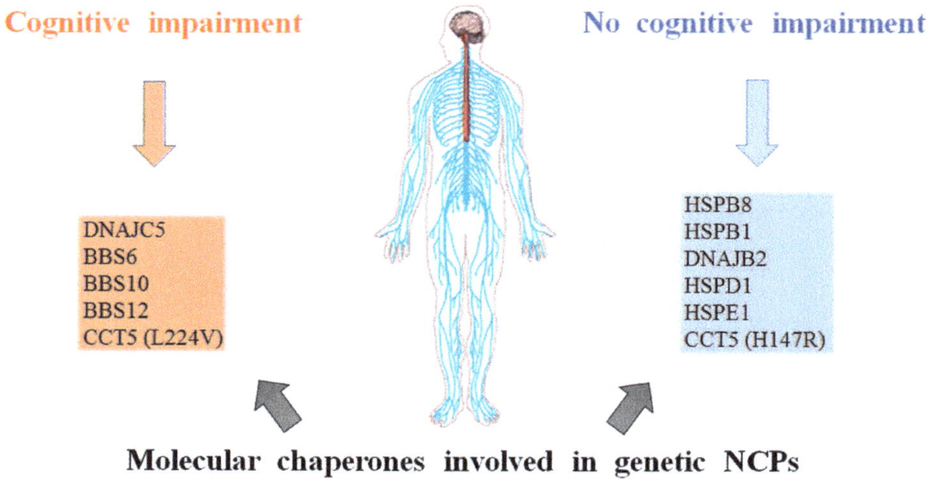

Figure 2. Molecular chaperones involved in genetic neurochaperonopathies (NCPs). Various molecular chaperones are involved in genetic NCPs, but only some (orange) are associated with cognitive impairment, while others (blue) are not.

The heterogeneity of these disorders and the difficulties implied in the study of the nervous system in humans are the major challenges for physicians and other investigators willing to elucidate the relevant molecular mechanisms and to develop efficacious treatment strategies. The data reported thus far, some of which has been briefly discussed in the previous sections, indicate that pathogenic variants of members of the sHsp, DnaJ(Hsp40), and CCT families, and of the Hsp60 gene, tend to cause predominantly motor disorders, although sensory and mental functions may also be affected [24,28,32,33,36,46,49,52,54]. For example, a missense mutation of the subunit number 5 of CCT, CCT5, was reported to be accompanied by a phenotype characterized mostly by distal sensory mutilating neuropathy [56,57]. In contrast, another recently discovered mutant located in a different structural domain of CCT5 was associated with a motor distal neuropathy without mutilation [58]. Thus, the impact of the mutation on the properties and functions of the chaperone molecule, and the accompanying tissue abnormalities and lesions observed in patients may greatly differ, depending on the location of the mutant amino acid, but other elements, e.g., environmental and nutritional factors that may also contribute to the generation of diverse phenotypes cannot be excluded.

Other genetic NCPs discussed are caused by variants of a subfamily of CCT genes composed of the BBS6 (MKKS), BBS10, and BBS12 genes [24,59–61,63–65]. Cognitive deficit was recorded only in patients with pathogenic variants of the CCT5 (the newly reported mutation p.(Leu224Val)), *DNAJC*, and *BBS* genes (Figure 2).

Only one of the members of the Hsp70(DnaK) family, the mitochondrial Hsp70 (mtHsp70 also named mortalin or HSPA9) has been reported with pathogenic mutations causing a syndrome that can be classified as NCP. Other members of this family, such as the HSPA1A and HSPA1B, whose genes are organized in tandem within the so called *HSPA1*-cluster [140] are stress-inducible and have been implicated in the pathogenesis of several diseases, such as cancer and neurodegenerative conditions, but their genes were normal [141,142]. This pathogenicity of the Hsp70 molecules fits the description of acquired chaperonopathies by mistake or collaborationism, in which apparently normal chaperones help the disease rather than protect from it. The lack of genetic NCPs associated with mutations in *HSPA1A* and *HSPA1B* genes could be related to their vital roles suggested also by their evolutionary conservation [95]. This is also suggested by the fact that the majority of polymorphisms identified in the human *HSPA1*-cluster are in the UTR regions, and the most common SNPs occurring in the coding regions are synonymous and thus, do not affect the proteins' amino acid sequences [143,144]. A few studies experimentally investigated the impact of human *HSPA1A-1B* mutations on the function/expression of the encoded proteins [145–148]. The functional analysis of six naturally occurring variants of the *HSPA1A* gene revealed that these variants result in significant functional alterations, affecting ATP-hydrolysis rate and substrate-binding ability, likely by altering the allosteric communication between its two major domains. However, they are rare mutations, with very low penetrance in the human population. All these results support the notion that the *HSPA1A* gene is under strong selection and that its functions must be conserved [149], which would explain why no patients with pathogenic mutations of this gene have been found.

The Hsp90 (HSPC) family, including five members and various isoforms expressed in specific cellular organelles, is highly conserved from early eukaryotes to humans and plays important roles in protein homeostasis (references in [21]). Hsp90 chaperones have a wide array of client proteins and, together with their co-chaperones and co-factors, participate in cell survival, cell-cycle control mechanisms, hormone signaling and other signaling pathways, and apoptosis [150]. In addition, Hsp90 has been implicated in the pathogenesis of Alzheimer's disease and Parkinson's disease [151,152]. However, it is noteworthy that no Hsp90 mutations have been described yet as an etiological-pathogenic factor in NCPs. The same message is given by analysis of Hsp90 genetic polymorphisms. Polymorphic variants in introns and exons of the human Hsp90beta (*HSPC3)* gene have been detected with no apparent effects on phenotype [153].

In recent years, along with genetic NCPs, attention has also been directed to acquired NCPs associated with molecular chaperones with PTMs. For instance, in vivo studies demonstrated that changes in the AMPylation balance of molecular chaperones could influence cell fate in physiological [125] and pathological conditions, such as PD and AD [127]. Human FICD AMPylates various amino acids in the Hsp70, DNAJB1, and Hsp90 chaperones. AMPylation levels regulate the correct neurogenesis and neuronal conduction [123,126]. So far there are no data suggesting that chaperone AMPylation may be disruptive and contribute to the pathogenic mechanism of neurological disorders.

Nitration of Hsp90 (HSPC) family members in relation to ALS disease is an example of the differential outcomes one can expect depending on the site and level of the PTM. Furthermore, considering the importance of mitochondrial activity in neuronal cells, one may assume that nitrated Hsp90 may be associated with acquired NCPs in addition to those for ALS. Acquired NCPs could be the result of PTMs in molecular chaperones and of PTMs in co-chaperones, chaperone co-factors or chaperone interactors, as illustrated by the consequences of phosphorylation of Hsp90/Cdc37 complex [133,135].

Two coexisting PTMs, phosphorylation and acetylation, in the VCP in the nucleus may promote histone deacetylation [138]. VCP could also undergo methylation and S-glutathionylation, two modifications that occur during oxidative stress with impairment of ATPase activity [139]. Since oxidative stress is a prevalent feature of neurodegenerative diseases, it is possible that modified VCP also plays a pathogenic role in NCPs.

Pathogenic chaperones with aberrant PTMs associated with NCPs, other than those described above, have not been reported. Most of the naturally occurring or induced PTMs in Hsp60 (HSPD1) with a pathogenic role have been implicated in carcinogenesis [154–157], apoptosis signaling regulation [158,159], and immune system regulation [160], but none has been associated with NCPs [25,161]. However, an involvement of Hsp60 (HSPD1) in AD and PD is under scrutiny. It is possible that Hsp60 with aberrant PTMs and/or with abnormal levels and location, i.e., intra- or extracellular, and when intracellular intra- or extramitochondrial, may be pathogenic in neurological disorders [25].

The number of multiple PTMs that can occur on a molecular chaperone is strikingly high, e.g., 60% and 56% of the T, Y, and S residues of Hsp70 (HSPA8) and Hsp90, respectively, could be phosphorylated [123] (Figure 3).

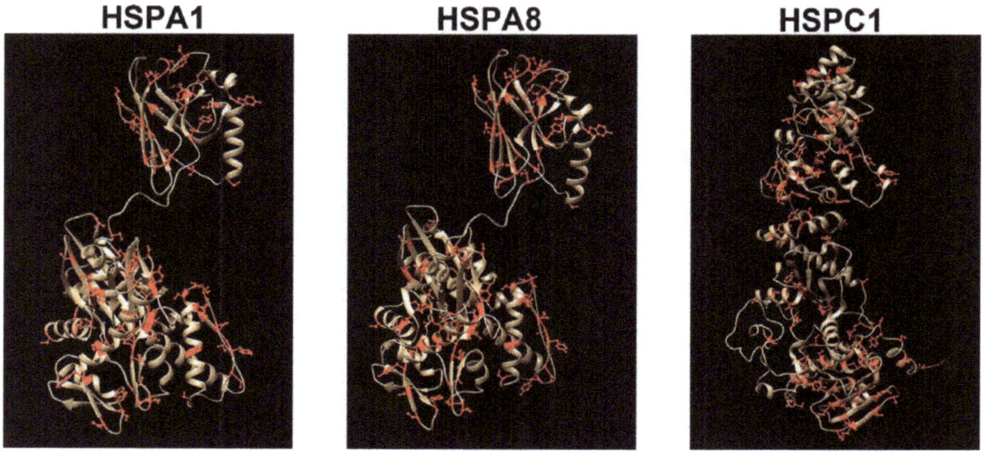

Figure 3. Examples of Hsp70(DnaK) and Hsp90 (HSPC) molecular chaperones rendered as light-brown ribbon models, showing the amino acids serine, threonine, and tyrosine in red (ball-and-stick). The models were built using SWISS-MODEL, developed by the Computational Structural Biology Group at the SIB Swiss Institute of Bioinformatics at the Biozentrum, University of Basel (https://swissmodel.expasy.org/, [162]), and were visualized with UCSF Chimera, developed by the Resource for Biocomputing, Visualization, and Informatics at the University of California, San Francisco, with support from NIH P41-GM103311 (https://www.cgl.ucsf.edu/chimera/, [163]). The HSPA1 and HSPA8 models were built using the NP_005336.3 and NP_006588.1, respectively, amino acid sequences reported in the NCBI Protein database, and as template the B chain of the crystal structure PDB ID 3C7N in the RCSB Protein Databank. The structures shown cover the HSPA1 and the HSPA8 sequences between amino acids 3 and 547. The HSPC1 model was built using the amino acid sequence NP_001017963.2 reported in the NCBI Protein database, and as template the B chain of the crystal structure PDB ID 5FWM in the RCSB Protein Databank. The structure shown covers the HSPC1 sequence between amino acids 137 and 821.

In addition, the same residue could be subjected to different PTMs, and several PTMs can also occur on the relevant co-chaperones and collaborators. All these modifications create the "Chaperone Code" that regulates chaperone functions, physiologically and pathologically (e.g., in cancer, viral infections, chronic inflammation, and neurodegenerative disorders). This flexibility and diversity of potential functions in health and disease of a single chaperone type, together with the variety of chaperone types, opens exciting novel avenues for investigating the mechanisms involved in acquired NCPs, for example those characteristics of ageing and AD and PD [135].

6. Conclusions and Future Perspective

Why is it worthwhile to determine if any given neurological patient bears a chaperonopathy? Firstly, the CS is present in all cells, tissues, and organs, therefore, a deficiency in any of its components may have a pathogenic impact in various anatomic areas and physiological systems, including the nervous system, and the probability of causing signs and symptoms is high. Secondly, the CS participates in a variety of physiological mechanisms, both as guardian of protein homeostasis and protector against stressors, and as the effector of many other cellular processes unrelated to protein quality control, i.e., non-canonical functions, including critical interactions with the immune system. This also increases the probability of deficiencies in the CS having clinical manifestations. Thirdly, detection of a chaperonopathy opens the door to investigate the possibility of applying chaperonotherapy in any of its modes, which is a significant prospect considering that most chaperonopathies are serious diseases still awaiting efficacious treatment [20,164–167]. It would be beneficial for many patients if physicians would bear in mind the concept of hidden chaperonopathies, particularly in those cases in which the clinical signs and symptoms do not quite fit within the expected clinical picture for any given neurological disorder [168]. To unveil this hidden pathogenic factor, physicians should add a search for a chaperonopathy to the differential diagnosis algorithm.

Author Contributions: Project planning, F.C. and F.S.; literature searches, A.M.V., R.S., F.S., E.C.d.M. and A.J.L.M.; molecular modeling, F.S. and A.M.V.; writing, F.S., A.M.V., R.S., E.C.d.M. and A.J.L.M. All authors have read and agreed to the published version of the manuscript.

Funding: A.J.L.M and E.C.d.M. were partially supported by IMET and IEMEST. This is IMET contribution number IMET 21-002.

Institutional Review Board Statement: Not applicable.

Informed Consent Statement: Not applicable.

Data Availability Statement: Not applicable.

Conflicts of Interest: The authors declare no conflict of interest.

References

1. Briston, T.; Hicks, A.R. Mitochondrial dysfunction and neurodegenerative proteinopathies: Mechanisms and prospects for therapeutic intervention. *Biochem. Soc. Trans.* **2018**, *46*, 829–842. [CrossRef] [PubMed]
2. Dugger, B.N.; Dickson, D.W. Pathology of neurodegenerative diseases. *Cold Spring Harb. Perspect. Biol.* **2017**, *9*, a028035. [CrossRef] [PubMed]
3. Gao, J.; Wang, L.; Yan, T.; Perry, G.; Wang, X. TDP-43 proteinopathy and mitochondrial abnormalities in neurodegeneration. *Mol. Cell. Neurosci.* **2019**, *100*, 103396. [CrossRef] [PubMed]
4. Robinson, J.L.; Lee, E.B.; Xie, S.X.; Rennert, L.; Suh, E.; Bredenberg, C.; Caswell, C.; Van Deerlin, V.M.; Yan, N.; Yousef, A.; et al. Neurodegenerative disease concomitant proteinopathies are prevalent, age-related and APOE4-associated. *Brain* **2018**, *141*, 2181–2193. [CrossRef] [PubMed]
5. Carlisle, C.; Prill, K.; Pilgrim, D. Chaperones and the proteasome system: Regulating the construction and demolition of striated muscle. *Int. J. Mol. Sci.* **2018**, *19*, 32. [CrossRef] [PubMed]
6. Dikic, I. Proteasomal and autophagic degradation systems. *Annu. Rev. Biochem.* **2017**, *86*, 193–224. [CrossRef]
7. Nam, T.; Han, J.H.; Devkota, S.; Lee, H.W. Emerging paradigm of crosstalk between autophagy and the ubiquitin-proteasome system. *Mol. Cells* **2017**, *40*, 897–905.
8. Catarino, S.; Pereira, P.; Girao, H. Molecular control of chaperone-mediated autophagy. *Essays Biochem.* **2017**, *61*, 663–674.
9. Tekirdag, K.; Cuervo, A.M. Chaperone-mediated autophagy and endosomal microautophagy: Joint by a chaperone. *J. Biol. Chem.* **2018**, *293*, 5414–5424. [CrossRef]
10. Kocaturk, N.M.; Gozuacik, D. Crosstalk between mammalian autophagy and the ubiquitin-proteasome system. *Front. Cell Dev. Biol.* **2018**, *6*, 128. [CrossRef]
11. Horwich, A.L.; Fenton, W.A.; Chapman, E.; Farr, G.W. Two families of chaperonin: Physiology and mechanism. *Annu. Rev. Cell Dev. Biol.* **2007**, *23*, 115–145. [CrossRef] [PubMed]
12. Kim, Y.E.; Hipp, M.S.; Bracher, A.; Hayer-Hartl, M.; Ulrich Hartl, F. Molecular Chaperone Functions in Protein Folding and Proteostasis. *Annu. Rev. Biochem.* **2013**, *82*, 323–355. [CrossRef] [PubMed]
13. Finka, A.; Sharma, S.K.; Goloubinoff, P. Multi-layered molecular mechanisms of polypeptide holding, unfolding and disaggregation by HSP70/HSP110 chaperones. *Front. Mol. Biosci.* **2015**, *2*, 29. [CrossRef] [PubMed]
14. Mogk, A.; Bukau, B.; Kampinga, H.H. Cellular Handling of Protein Aggregates by Disaggregation Machines. *Mol. Cell* **2018**, *69*, 214–226. [CrossRef] [PubMed]
15. Willison, K.R. The structure and evolution of eukaryotic chaperonin-containing TCP-1 and its mechanism that folds actin into a protein spring. *Biochem. J.* **2018**, *475*, 3009–3034. [CrossRef]
16. Adams, B.M.; Oster, M.E.; Hebert, D.N. Protein Quality Control in the Endoplasmic Reticulum. *Protein J.* **2019**, *38*, 317–329. [CrossRef]

17. Dahiya, V.; Buchner, J. Functional principles and regulation of molecular chaperones. In *Advances in Protein Chemistry and Structural Biology*; Academic Press Inc.: Cambridge, MA, USA, 2019; Volume 114, pp. 1–60, ISBN 9780128155578.
18. Macario, A.J.L.; Conway de Macario, E. Sick chaperones, cellular stress, and disease. *N. Engl. J. Med.* **2005**, *353*, 1489–1501. [CrossRef]
19. Macario, A.J.L.; Conway de Macario, E. The chaperoning system: Physiology and pathology. In *Experimental Medicine Reviews*; Plumelia: Bagheria, Italy, 2008; Volumes 2–3, pp. 9–21.
20. Macario, A.J.L.; Conway de Macario, E. Chaperone proteins and chaperonopathies. In *Handbook of Stress*; Elsevier; Academic Press: Cambridge, MS, USA, 2019.
21. Macario, A.J.L.; Conway de Macario, E.; Cappello, F. *The Chaperonopathies. Diseases with Defective Molecular Chaperones*; Springer: Berlin, Germany, 2013; ISBN 978-94-007-4666-4.
22. Lanfranco, M.; Vassallo, N.; Cauchi, R.J. Spinal muscular atrophy: From defective chaperoning of snRNP assembly to neuromuscular dysfunction. *Front. Mol. Biosci.* **2017**, *4*, 41. [CrossRef]
23. Lupo, V.; Aguado, C.; Knecht, E.; Espinós, C. Chaperonopathies: Spotlight on hereditary motor neuropathies. *Front. Mol. Biosci.* **2016**, *3*, 81. [CrossRef]
24. Álvarez-Satta, M.; Castro-Sánchez, S.; Valverde, D. Bardet-Biedl syndrome as a chaperonopathy: Dissecting the major role of chaperonin-like BBS proteins (BBS6-BBS10-BBS12). *Front. Mol. Biosci.* **2017**, *4*, 55. [CrossRef]
25. Caruso Bavisotto, C.; Alberti, G.; Vitale, A.M.; Paladino, L.; Campanella, C.; Rappa, F.; Gorska, M.; Conway de Macario, E.; Cappello, F.; Macario, A.J.L.; et al. Hsp60 Post-translational Modifications: Functional and Pathological Consequences. *Front. Mol. Biosci.* **2020**, *7*, 95. [CrossRef] [PubMed]
26. Royer-Bertrand, B.; Castillo-Taucher, S.; Moreno-Salinas, R.; Cho, T.J.; Chae, J.H.; Choi, M.; Kim, O.H.; Dikoglu, E.; Campos-Xavier, B.; Girardi, E.; et al. Mutations in the heat-shock protein A9 (HSPA9) gene cause the EVEN-PLUS syndrome of congenital malformations and skeletal dysplasia. *Sci. Rep.* **2015**, *5*, 17154. [CrossRef] [PubMed]
27. Ghaoui, R.; Palmio, J.; Brewer, J.; Lek, M.; Needham, M.; Evilä, A.; Hackman, P.; Jonson, P.H.; Penttilä, S.; Vihola, A.; et al. Mutations in HSPB8 causing a new phenotype of distal myopathy and motor neuropathy. *Neurology* **2016**, *86*, 391–398. [CrossRef] [PubMed]
28. Irobi, J.; Van Impe, K.; Seeman, P.; Jordanova, A.; Dierick, I.; Verpoorten, N.; Michalik, A.; De Vriendt, E.; Jacobs, A.; Van Gerwen, V.; et al. Hot-spot residue in small heat-shock protein 22 causes distal motor neuropathy. *Nat. Genet.* **2004**, *36*, 597–601. [CrossRef]
29. Timmerman, V.; Raeymaekers, P.; Nelis, E.; De Jonghe, P.; Muylle, L.; Ceuterick, C.; Martin, J.J.; Van Broeckhoven, C. Linkage analysis of distal hereditary motor neuropathy type II (distal HMN II) in a single pedigree. *J. Neurol. Sci.* **1992**, *109*, 41–48. [CrossRef]
30. Tang, B.S.; Luo, W.; Xia, K.; Xiao, J.F.; Jiang, H.; Shen, L.; Tang, J.G.; Zhao, G.H.; Cai, F.; Pan, Q.; et al. A new locus for autosomal dominant Charcot-Marie-Tooth disease type 2 (CMT2L) maps to chromosome 12q. *Hum. Genet.* **2004**, *114*, 527–533.
31. Tang, B.S.; Zhao, G.; Luo, W.; Xia, K.; Cai, F.; Pan, Q.; Zhang, R.; Zhang, F.; Liu, X.; Chen, B.; et al. Small heat-shock protein 22 mutated in autosomal dominant Charcot-Marie-Tooth disease type 2L. *Hum. Genet.* **2005**, *116*, 222–224. [CrossRef]
32. Capponi, S.; Geroldi, A.; Fossa, P.; Grandis, M.; Ciotti, P.; Gulli, R.; Schenone, A.; Mandich, P.; Bellone, E. HSPB1 and HSPB8 in inherited neuropathies: Study of an Italian cohort of dHMN and CMT2 patients. *J. Peripher. Nerv. Syst.* **2011**, *16*, 287–294. [CrossRef]
33. Houlden, H.; Laura, M.; Wavrant-De Vrièze, F.; Blake, J.; Wood, N.; Reilly, M.M. Mutations in the HSP27 (HSPB1) gene cause dominant, recessive, and sporadic distal HMN/CMT type. *Neurology* **2008**, *71*, 1660–1668. [CrossRef]
34. James, P.A.; Rankin, J.; Talbot, K. Asymmetrical late onset motor neuropathy associated with a novel mutation in the small heat shock protein HSPB1 (HSP27). *J. Neurol. Neurosurg. Psychiatry* **2008**, *79*, 461–463. [CrossRef]
35. Ho, C.C.; Tai, S.M.; Lee, E.C.N.; Mak, T.S.H.; Liu, T.K.T.; Tang, V.W.L.; Poon, W.T. Rapid identification of pathogenic variants in two cases of Charcot-Marie-tooth disease by gene-panel sequencing. *Int. J. Mol. Sci.* **2017**, *18*, 770. [CrossRef] [PubMed]
36. Evgrafov, O.V.; Mersiyanova, I.; Irobi, J.; Van Den Bosch, L.V.; Dierick, I.; Leung, C.L.; Schagina, O.; Verpoorten, N.; Van Impe, K.; Fedotov, V.; et al. Mutant small heat-shock protein 27 causes axonal Charcot-Marie-Tooth disease and distal hereditary motor neuropathy. *Nat. Genet.* **2004**, *36*, 602–606. [CrossRef] [PubMed]
37. Tang, B.; Liu, X.; Zhao, G.; Luo, W.; Xia, K.; Pan, Q.; Cai, F.; Hu, Z.; Zhang, C.; Chen, B.; et al. Mutation analysis of the small heat shock protein 27 gene in Chinese patients with Charcot-Marie-Tooth disease. *Arch. Neurol.* **2005**, *62*, 1201–1207. [CrossRef] [PubMed]
38. Benedetti, S.; Previtali, S.C.; Coviello, S.; Scarlato, M.; Cerri, F.; Di Pierri, E.; Piantoni, L.; Spiga, I.; Fazio, R.; Riva, N.; et al. Analyzing histopathological features of rare Charcot-Marie-Tooth neuropathies to unravel their pathogenesis. *Arch. Neurol.* **2010**, *67*, 1498–1505. [CrossRef] [PubMed]
39. Oberstadt, M.; Mitter, D.; Classen, J.; Baum, P. Late onset dHMN II caused by c.404C>G mutation in HSPB1 gene. *J. Peripher. Nerv. Syst.* **2016**, *21*, 111–113. [CrossRef]
40. Ismailov, S.M.; Fedotov, V.P.; Dadali, E.L.; Polyakov, A.V.; Van Broeckhoven, C.; Ivanov, V.I.; De Jonghe, P.; Timmerman, V.; Evgrafov, O.V. A new locus for autosomal dominant Charcot-Marie-Tooth disease type 2 (CMT2F) maps to chromosome 7q11-q. *Eur. J. Hum. Genet.* **2001**, *9*, 646–650. [CrossRef]
41. Ikeda, Y.; Abe, A.; Ishida, C.; Takahashi, K.; Hayasaka, K.; Yamada, M. A clinical phenotype of distal hereditary motor neuronopathy type II with a novel HSPB1 mutation. *J. Neurol. Sci.* **2009**, *277*, 9–12. [CrossRef]

42. Lin, K.-P.; Soong, B.-W.; Yang, C.-C.; Huang, L.-W.; Chang, M.-H.; Lee, I.-H.; Antonellis, A.; Lee, Y.-C. The Mutational Spectrum in a Cohort of Charcot-Marie-Tooth Disease Type 2 among the Han Chinese in Taiwan. *PLoS ONE* **2011**, *6*, e29393. [CrossRef]
43. Luigetti, M.; Fabrizi, G.M.; Madia, F.; Ferrarini, M.; Conte, A.; Del Grande, A.; Tasca, G.; Tonali, P.A.; Sabatelli, M. A novel HSPB1 mutation in an Italian patient with CMT2/dHMN phenotype. *J. Neurol. Sci.* **2010**, *298*, 114–117. [CrossRef]
44. Kijima, K.; Numakura, C.; Goto, T.; Takahashi, T.; Otagiri, T.; Umetsu, K.; Hayasaka, K. Small heat shock protein 27 mutation in a Japanese patient with distal hereditary motor neuropathy. *J. Hum. Genet.* **2005**, *50*, 473–476. [CrossRef]
45. Bie, A.S.; Fernandez-Guerra, P.; Birkler, R.I.D.; Nisemblat, S.; Pelnena, D.; Lu, X.; Deignan, J.L.; Lee, H.; Dorrani, N.; Corydon, T.J.; et al. Effects of a mutation in the HSPE1 gene encoding the mitochondrial co-chaperonin HSP10 and its potential association with a neurological and developmental disorder. *Front. Mol. Biosci.* **2016**, *3*, 65. [CrossRef] [PubMed]
46. Gess, B.; Auer-Grumbach, M.; Schirmacher, A.; Strom, T.; Zitzelsberger, M.; Rudnik-Schöneborn, S.; Röhr, D.; Halfter, H.; Young, P.; Senderek, J. HSJ1-related hereditary neuropathies Novel mutations and extended clinical spectrum. *Neurology* **2014**, *83*, 1726–1732. [CrossRef] [PubMed]
47. Blumen, S.C.; Astord, S.; Robin, V.; Vignaud, L.; Toumi, N.; Cieslik, A.; Achiron, A.; Carasso, R.L.; Gurevich, M.; Braverman, I.; et al. A rare recessive distal hereditary motor neuropathy with HSJ1 chaperone mutation. *Ann. Neurol.* **2012**, *71*, 509–519.
48. Benitez, B.A.; Alvarado, D.; Cai, Y.; Mayo, K.; Chakraverty, S.; Norton, J.; Morris, J.C.; Sands, M.S.; Goate, A.; Cruchaga, C. Exome-sequencing confirms DNAJC5 mutations as cause of adult neuronal ceroid-lipofuscinosis. *PLoS ONE* **2011**, *6*, 26741. [CrossRef] [PubMed]
49. Nosková, L.; Stránecký, V.; Hartmannová, H.; Přistoupilová, A.; Barešová, V.; Ivánek, R.; Hlková, H.; Jahnová, H.; Van Der Zee, J.; Staropoli, J.F.; et al. Mutations in DNAJC5, encoding cysteine-string protein alpha, cause autosomal-dominant adult-onset neuronal ceroid lipofuscinosis. *Am. J. Hum. Genet.* **2011**, *89*, 241–252. [CrossRef]
50. Velinov, M.; Dolzhanskaya, N.; Gonzalez, M.; Powell, E.; Konidari, I.; Hulme, W.; Staropoli, J.F.; Xin, W.; Wen, G.Y.; Barone, R.; et al. Mutations in the Gene DNAJC5 Cause Autosomal Dominant Kufs Disease in a Proportion of Cases: Study of the Parry Family and 8 Other Families. *PLoS ONE* **2012**, *7*, e29729. [CrossRef]
51. Cadieux-Dion, M.; Andermann, E.; Lachance-Touchette, P.; Ansorge, O.; Meloche, C.; Barnabé, A.; Kuzniecky, R.I.; Andermann, F.; Faught, E.; Leonberg, S.; et al. Recurrent mutations in DNAJC5 cause autosomal dominant Kufs disease. *Clin. Genet.* **2013**, *83*, 571–575. [CrossRef]
52. Magen, D.; Georgopoulos, C.; Bross, P.; Ang, D.; Segev, Y.; Goldsher, D.; Nemirovski, A.; Shahar, E.; Ravid, S.; Luder, A.; et al. Mitochondrial Hsp60 Chaperonopathy Causes an Autosomal-Recessive Neurodegenerative Disorder Linked to Brain Hypomyelination and Leukodystrophy. *Am. J. Hum. Genet.* **2008**, *83*, 30–42. [CrossRef]
53. Kusk, M.S.; Damgaard, B.; Risom, L.; Hansen, B.; Østergaard, E. Hypomyelinating Leukodystrophy due to HSPD1 Mutations: A New Patient. *Neuropediatrics* **2016**, *47*, 332–335. [CrossRef]
54. Hansen, J.J.; Dürr, A.; Cournu-Rebeix, I.; Georgopoulos, C.; Ang, D.; Nielsen, M.N.; Davoine, C.S.; Brice, A.; Fontaine, B.; Gregersen, N.; et al. Hereditary spastic paraplegia SPG13 is associated with a mutation in the gene encoding the mitochondrial chaperonin Hsp60. *Am. J. Hum. Genet.* **2002**, *70*, 1328–1332. [CrossRef]
55. Hansen, J.; Svenstrup, K.; Ang, D.; Nielsen, M.N.; Christensen, J.H.; Gregersen, N.; Nielsen, J.E.; Georgopoulos, C.; Bross, P. A novel mutation in the HSPD1 gene in a patient with hereditary spastic paraplegia. *J. Neurol.* **2007**, *254*, 897–900. [CrossRef] [PubMed]
56. Bouhouche, A.; Benomar, A.; Bouslam, N.; Chkili, T.; Yahyaoui, M. Mutation in the epsilon subunit of the cytosolic chaperonin-containing t-complex peptide-1 (Cct5) gene causes autosomal recessive mutilating sensory neuropathy with spastic paraplegia. *J. Med. Genet.* **2006**, *43*, 441–443. [CrossRef] [PubMed]
57. Bouhouche, A.; Benomar, A.; Bouslam, N.; Ouazzani, R.; Chkili, T.; Yahyaoui, M. Autosomal recessive mutilating sensory neuropathy with spastic paraplegia maps to chromosome 5p15.31-14. *Eur. J. Hum. Genet.* **2006**, *14*, 249–252. [CrossRef] [PubMed]
58. Antona, V.; Scalia, F.; Giorgio, E.; Radio, F.C.; Brusco, A.; Oliveri, M.; Corsello, G.; Lo Celso, F.; Vadalà, M.; Conway de Macario, E.; et al. A novel cct5 missense variant associated with early onset motor neuropathy. *Int. J. Mol. Sci.* **2020**, *21*, 7631. [CrossRef] [PubMed]
59. Katsanis, N.; Beales, P.L.; Woods, M.O.; Lewis, R.A.; Green, J.S.; Parfrey, P.S.; Ansley, S.J.; Davidson, W.S.; Lupski, J.R. Mutations in MKKS cause obesity, retinal dystrophy and renal malformations associated with Bardet-Biedl syndrome. *Nat. Genet.* **2000**, *26*, 67–70. [CrossRef] [PubMed]
60. Katsanis, N.; Lupski, J.R.; Beales, P.L. Exploring the molecular basis of Bardet-Biedl syndrome. *Hum. Mol. Genet.* **2001**, *10*, 2293–2299. [CrossRef] [PubMed]
61. Slavotinek, A.M.; Stone, E.M.; Mykytyn, K.; Heckenlively, J.R.; Green, J.S.; Heon, E.; Musarella, M.A.; Parfrey, P.S.; Sheffield, V.C.; Biesecker, L.G. Mutations in MKKS cause Bardet-Biedl syndrome. *Nat. Genet.* **2000**, *26*, 15–16. [CrossRef]
62. Laurier, V.; Stoetzel, C.; Muller, J.; Thibault, C.; Corbani, S.; Jalkh, N.; Salem, N.; Chouery, E.; Poch, O.; Licaire, S.; et al. Pitfalls of homozygosity mapping: An extended consanguineous Bardet-Biedl syndrome family with two mutant genes (BBS2, BBS10), three mutations, but no triallelism. *Eur. J. Hum. Genet.* **2006**, *14*, 1195–1203. [CrossRef]
63. Stoetzel, C.; Laurier, V.; Davis, E.E.; Muller, J.; Rix, S.; Badano, J.L.; Leitch, C.C.; Salem, N.; Chouery, E.; Corbani, S.; et al. BBS10 encodes a vertebrate-specific chaperonin-like protein and is a major BBS locus. *Nat. Genet.* **2006**, *38*, 521–524. [CrossRef]
64. Putoux, A.; Mougou-Zerelli, S.; Thomas, S.; Elkhartoufi, N.; Audollent, S.; Le Merrer, M.; Lachmeijer, A.; Sigaudy, S.; Buenerd, A.; Fernandez, C.; et al. BBS10 mutations are common in 'Meckel'-type cystic kidneys. *J. Med. Genet.* **2010**, *47*, 848–852. [CrossRef]

65. Stoetzel, C.; Muller, J.; Laurier, V.; Davis, E.E.; Zaghloul, N.A.; Vicaire, S.; Jacquelin, C.; Plewniak, F.; Leitch, C.C.; Sarda, P.; et al. Identification of a novel BBS gene (BBS12) highlights the major role of a vertebrate-specific branch of chaperonin-related proteins in Bardet-Biedl syndrome. *Am. J. Hum. Genet.* **2007**, *80*, 1–11. [CrossRef] [PubMed]
66. Garrido, C.; Paul, C.; Seigneuric, R.; Kampinga, H.H. The small heat shock proteins family: The long forgotten chaperones. *Int. J. Biochem. Cell Biol.* **2012**, *44*, 1588–1592. [CrossRef] [PubMed]
67. Bruey, J.M.; Ducasse, C.; Bonniaud, P.; Ravagnan, L.; Susin, S.A.; Diaz-Latoud, C.; Gurbuxani, S.; Arrigo, A.P.; Kroemer, G.; Solary, E.; et al. Hsp27 negatively regulates cell death by interacting with cytochrome c. *Nat. Cell Biol.* **2000**, *2*, 645–652. [CrossRef] [PubMed]
68. Pandey, P.; Farber, R.; Nakazawa, A.; Kumar, S.; Bharti, A.; Nalin, C.; Weichselbaum, R.; Kufe, D.; Kharbanda, S. Hsp27 functions as a negative regulator of cytochrome c-dependent activation of procaspase. *Oncogene* **2000**, *19*, 1975–1981. [CrossRef] [PubMed]
69. Chowdary, T.K.; Raman, B.; Ramakrishna, T.; Rao, C.M. Mammalian Hsp22 is a heat-inducible small heat-shock protein with chaperone-like activity. *Biochem. J.* **2004**, *381*, 379–387. [CrossRef]
70. Vos, M.J.; Zijlstra, M.P.; Kanon, B.; van Waarde-Verhagen, M.A.W.H.; Brunt, E.R.P.; Oosterveld-Hut, H.M.J.; Carra, S.; Sibon, O.C.M.; Kampinga, H.H. HSPB7 is the most potent polyQ aggregation suppressor within the HSPB family of molecular chaperones. *Hum. Mol. Genet.* **2010**, *19*, 4677–4693. [CrossRef]
71. Kawano, F.; Fujita, R.; Nakai, N.; Terada, M.; Ohira, T.; Ohira, Y. HSP25 can modulate myofibrillar desmin cytoskeleton following the phosphorylation at Ser15 in rat soleus muscle. *J. Appl. Physiol.* **2012**, *112*, 176–186. [CrossRef]
72. Van Montfort, R.; Slingsby, C.; Vierling, E. Structure and function of the small heat shock protein/α-crystallin family of molecular chaperones. *Adv. Protein Chem.* **2001**, *59*, 105–156.
73. Boncoraglio, A.; Minoia, M.; Carra, S. The family of mammalian small heat shock proteins (HSPBs): Implications in protein deposit diseases and motor neuropathies. *Int. J. Biochem. Cell Biol.* **2012**, *44*, 1657–1669. [CrossRef]
74. Berger, P.; Young, P.; Suter, U. Molecular cell biology of Charcot-Marie-Tooth disease. *Neurogenetics* **2002**, *4*, 1–15. [CrossRef]
75. Harding, A.E. Inherited Neuronal Atrophy and Degeneration Predominantly of Lower Motor Neurons. In *Peripheral Neuropathy*; Elsevier: Amsterdam, The Netherlands, 2005; Volume 2, pp. 1603–1621. ISBN 9780721694917.
76. Rossor, A.M.; Kalmar, B.; Greensmith, L.; Reilly, M.M. The distal hereditary motor neuropathies. *J. Neurol. Neurosurg. Psychiatry* **2012**, *83*, 6–14. [CrossRef] [PubMed]
77. Solla, P.; Vannelli, A.; Bolino, A.; Marrosu, G.; Coviello, S.; Murru, M.R.; Tranquilli, S.; Corongiu, D.; Benedetti, S.; Marrosu, M.G. Heat shock protein 27 R127W mutation: Evidence of a continuum between axonal Charcot-Marie-Tooth and distal hereditary motor neuropathy. *J. Neurol. Neurosurg. Psychiatry* **2010**, *81*, 958–962. [CrossRef] [PubMed]
78. Weeks, S.D.; Muranova, L.K.; Heirbaut, M.; Beelen, S.; Strelkov, S.V.; Gusev, N.B. Characterization of human small heat shock protein HSPB1 α-crystallin domain localized mutants associated with hereditary motor neuron diseases. *Sci. Rep.* **2018**, *8*, 688. [CrossRef] [PubMed]
79. Irobi, J.; Almeida-Souza, L.; Asselbergh, B.; de Winter, V.; Goethals, S.; Dierick, I.; Krishnan, J.; Timmermans, J.P.; Robberecht, W.; de Jonghe, P.; et al. Mutant HSPB8 causes motor neuron-specific neurite degeneration. *Hum. Mol. Genet.* **2010**, *19*, 3254–3265. [CrossRef] [PubMed]
80. Ackerley, S.; James, P.A.; Kalli, A.; French, S.; Davies, K.E.; Talbot, K. A mutation in the small heat-shock protein HSPB1 leading to distal hereditary motor neuronopathy disrupts neurofilament assembly and the axonal transport of specific cellular cargoes. *Hum. Mol. Genet.* **2006**, *15*, 347–354. [CrossRef]
81. Zhai, J.; Lin, H.; Julien, J.P.; Schlaepfer, W.W. Disruption of neurofilament network with aggregation of light neurofilament protein: A common pathway leading to motor neuron degeneration due to Charcot-Marie-Tooth disease-linked mutations in NFL and HSPB. *Hum. Mol. Genet.* **2007**, *16*, 3103–3116. [CrossRef]
82. Almeida-Souza, L.; Asselbergh, B.; d'Ydewalle, C.; Moonens, K.; Goethals, S.; de Winter, V.; Azmi, A.; Irobi, J.; Timmermans, J.P.; Gevaert, K.; et al. Small heat-shock protein HSPB1 mutants stabilize microtubules in Charcot-Marie-Tooth neuropathy. *J. Neurosci.* **2011**, *31*, 15320–15328. [CrossRef]
83. Kalmar, B.; Innes, A.; Wanisch, K.; Kolaszynska, A.K.; Pandraud, A.; Kelly, G.; Abramov, A.Y.; Reilly, M.M.; Schiavo, G.; Greensmith, L. Mitochondrial deficits and abnormal mitochondrial retrograde axonal transport play a role in the pathogenesis of mutant Hsp27-induced charcot marie tooth disease. *Hum. Mol. Genet.* **2017**, *26*, 3313–3326. [CrossRef]
84. Tanabe, H.; Higuchi, Y.; Yuan, J.H.; Hashiguchi, A.; Yoshimura, A.; Ishihara, S.; Nozuma, S.; Okamoto, Y.; Matsuura, E.; Ishiura, H.; et al. Clinical and genetic features of Charcot-Marie-Tooth disease 2F and hereditary motor neuropathy 2B in Japan. *J. Peripher. Nerv. Syst.* **2018**, *23*, 40–48. [CrossRef]
85. Lee, J.; Jung, S.-C.; Joo, J.; Choi, Y.-R.; Moon, H.W.; Kwak, G.; Yeo, H.K.; Lee, J.-S.; Ahn, H.-J.; Jung, N.; et al. Overexpression of mutant HSP27 causes axonal neuropathy in mice. *J. Biomed. Sci.* **2015**, *22*, 43. [CrossRef]
86. Srivastava, A.K.; Renusch, S.R.; Naiman, N.E.; Gu, S.; Sneh, A.; Arnold, W.D.; Sahenk, Z.; Kolb, S.J. Mutant HSPB1 overexpression in neurons is sufficient to cause age-related motor neuronopathy in mice. *Neurobiol. Dis.* **2012**, *47*, 163–173. [CrossRef] [PubMed]
87. Muchowski, P.J.; Schaffar, G.; Sittler, A.; Wanker, E.E.; Hayer-Hartl, M.K.; Hartl, F.U. Hsp70 and Hsp40 chaperones can inhibit self-assembly of polyglutamine proteins into amyloid-like fibrils. *Proc. Natl. Acad. Sci. USA* **2000**, *97*, 7841–7846. [CrossRef] [PubMed]

88. Jana, N.R.; Tanaka, M.; Wang, G.H.; Nukina, N. Polyglutamine length-dependent interaction of Hsp40 and Hsp70 family chaperones with truncated N-terminal huntingtin: Their role in suppression of aggregation and cellular toxicity. *Hum. Mol. Genet.* **2000**, *9*, 2009–2018. [CrossRef] [PubMed]
89. Howarth, J.L.; Kelly, S.; Keasey, M.P.; Glover, C.P.J.; Lee, Y.B.; Mitrophanous, K.; Chapple, J.P.; Gallo, J.M.; Cheetham, M.E.; Uney, J.B. Hsp40 molecules that target to the ubiquitin-proteasome system decrease inclusion formation in models of polyglutamine disease. *Mol. Ther.* **2007**, *15*, 1100–1105. [CrossRef] [PubMed]
90. Frasquet, M.; Chumillas, M.J.; Vílchez, J.J.; Márquez-Infante, C.; Palau, F.; Vázquez-Costa, J.F.; Lupo, V.; Espinós, C.; Sevilla, T. Phenotype and natural history of inherited neuropathies caused by HSJ1 c.352+1G>A mutation. *J. Neurol. Neurosurg. Psychiatry* **2016**, *87*, 1265–1268. [CrossRef]
91. Greaves, J.; Chamberlain, L.H. Dual role of the cysteine-string domain in membrane binding and palmitoylation-dependent sorting of the molecular chaperone cysteine-string protein. *Mol. Biol. Cell* **2006**, *17*, 4748–4759. [CrossRef]
92. Martin, J.J. Adult type of neuronal ceroid-lipofuscinosis. *J. Inherit. Metab. Dis.* **1993**, *16*, 237–240. [CrossRef]
93. Haltia, M. The neuronal ceroid-lipofuscinoses. *J. Neuropathol. Exp. Neurol.* **2003**, *62*, 1–13. [CrossRef]
94. Xu, F.; Proft, J.; Gibbs, S.; Winkfein, B.; Johnson, J.N.; Syed, N.; Braun, J.E.A. Quercetin targets cysteine string protein (CSPα) and impairs synaptic transmission. *PLoS ONE* **2010**, *5*, e11045. [CrossRef]
95. Brocchieri, L.; Conway de Macario, E.; Macario, A.J.L. hsp70 genes in the human genome: Conservation and differentiation patterns predict a wide array of overlapping and specialized functions. *BMC EVolume Biol.* **2008**, *8*, 19. [CrossRef]
96. Cheng, M.Y.; Hartl, F.U.; Martin, J.; Pollock, R.A.; Kalousek, F.; Neuper, W.; Hallberg, E.M.; Hallberg, R.L.; Horwich, A.L. Mitochondrial heat-shock protein hsp60 is essential for assembly of proteins imported into yeast mitochondria. *Nature* **1989**, *337*, 620–625. [CrossRef] [PubMed]
97. Bross, P.; Fernandez-Guerra, P. Disease-associated mutations in the HSPD1 gene encoding the large subunit of the mitochondrial HSP60/HSP10 chaperonin complex. *Front. Mol. Biosci.* **2016**, *3*, 49. [CrossRef] [PubMed]
98. Christensen, J.H.; Nielsen, M.N.; Hansen, J.; Füchtbauer, A.; Füchtbauer, E.M.; West, M.; Corydon, T.J.; Gregersen, N.; Bross, P. Inactivation of the hereditary spastic paraplegia-associated Hspd1 gene encoding the Hsp60 chaperone results in early embryonic lethality in mice. *Cell Stress Chaperones* **2010**, *15*, 851–863. [CrossRef] [PubMed]
99. Fontaine, B.; Davoine, C.S.; Dürr, A.; Paternotte, C.; Feki, I.; Weissenbach, J.; Hazan, J.; Brice, A. A new locus for autosomal dominant pure spastic paraplegia, on chromosome 2q24-q. *Am. J. Hum. Genet.* **2000**, *66*, 702–707. [CrossRef] [PubMed]
100. Bross, P.; Naundrup, S.; Hansen, J.; Nielsen, M.N.; Christensen, J.H.; Kruhøffer, M.; Palmfeldt, J.; Corydon, T.J.; Gregersen, N.; Ang, D.; et al. The Hsp60-(p.V98I) mutation associated with hereditary spastic paraplegia SPG13 compromises chaperonin function both in vitro and in vivo. *J. Biol. Chem.* **2008**, *283*, 15694–15700. [CrossRef] [PubMed]
101. Parnas, A.; Nadler, M.; Nisemblat, S.; Horovitz, A.; Mandel, H.; Azem, A. The MitCHAP-60 disease is due to entropic destabilization of the human mitochondrial Hsp60 oligomer. *J. Biol. Chem.* **2009**, *284*, 28198–28203. [CrossRef]
102. Vitale, A.M.; Conway de Macario, E.; Alessandro, R.; Cappello, F.; Macario, A.J.L.; Marino Gammazza, A. Missense Mutations of Human Hsp60: A Computational Analysis to Unveil Their Pathological Significance. *Front. Genet.* **2020**, *11*, 969. [CrossRef]
103. Bross, P.; Magnoni, R.; Sigaard Bie, A. Molecular Chaperone Disorders: Defective Hsp60 in Neurodegeneration. *Curr. Top. Med. Chem.* **2013**, *12*, 2491–2503. [CrossRef]
104. Miyamoto, Y.; Eguchi, T.; Kawahara, K.; Hasegawa, N.; Nakamura, K.; Funakoshi-Tago, M.; Tanoue, A.; Tamura, H.; Yamauchi, J. Hypomyelinating leukodystrophy-associated missense mutation in HSPD1 blunts mitochondrial dynamics. *Biochem. Biophys. Res. Commun.* **2015**, *462*, 275–281. [CrossRef]
105. Miyamoto, Y.; Megumi, F.T.; Hasegawa, N.; Eguchi, T.; Tanoue, A.; Tamura, H.; Yamauchi, J. Data supporting mitochondrial morphological changes by SPG13-associated HSPD1 mutants. *Data Br.* **2016**, *6*, 482–488. [CrossRef]
106. Mukherjee, K.; Conway de Macario, E.; Macario, A.J.L.; Brocchieri, L. Chaperonin genes on the rise: New divergent classes and intense duplication in human and other vertebrate genomes. *BMC EVolume Biol.* **2010**, *10*, 64. [CrossRef] [PubMed]
107. Min, W.; Angileri, F.; Luo, H.; Lauria, A.; Shanmugasundaram, M.; Almerico, A.M.; Cappello, F.; Conway de Macario, E.; Lednev, I.K.; Macario, A.J.L.; et al. A human CCT5 gene mutation causing distal neuropathy impairs hexadecamer assembly in an archaeal model. *Sci. Rep.* **2014**, *4*, 1–9. [CrossRef]
108. Sergeeva, O.A.; Tran, M.T.; Haase-Pettingell, C.; King, J.A. Biochemical characterization of mutants in chaperonin proteins CCT4 and CCT5 associated with hereditary sensory neuropathy. *J. Biol. Chem.* **2014**, *289*, 27470–27480. [CrossRef] [PubMed]
109. Spigolon, D.; Gallagher, D.T.; Velazquez-Campoy, A.; Bulone, D.; Narang, J.; San Biagio, P.L.; Cappello, F.; Macario, A.J.L.; Conway de Macario, E.; Robb, F.T. Quantitative analysis of the impact of a human pathogenic mutation on the CCT5 chaperonin subunit using a proxy archaeal ortholog. *Biochem. Biophys. Rep.* **2017**, *12*, 66–71. [CrossRef] [PubMed]
110. Nachury, M.V.; Loktev, A.V.; Zhang, Q.; Westlake, C.J.; Peränen, J.; Merdes, A.; Slusarski, D.C.; Scheller, R.H.; Bazan, J.F.; Sheffield, V.C.; et al. A Core Complex of BBS Proteins Cooperates with the GTPase Rab8 to Promote Ciliary Membrane Biogenesis. *Cell* **2007**, *129*, 1201–1213. [CrossRef] [PubMed]
111. Hildebrandt, F.; Benzing, T.; Katsanis, N. Ciliopathies. *N. Engl. J. Med.* **2011**, *364*, 1533–1543. [CrossRef] [PubMed]
112. Priya, S.; Nampoothiri, S.; Sen, P.; Sripriya, S. Bardet-Biedl syndrome: Genetics, molecular pathophysiology, and disease management. *Indian J. Ophthalmol.* **2016**, *64*, 620–627. [CrossRef] [PubMed]
113. M'Hamdi, O.; Ouertani, I.; Chaabouni-Bouhamed, H. Update on the genetics of bardet-biedl syndrome. *Mol. Syndromol.* **2014**, *5*, 51–56. [CrossRef]

114. Kim, J.C.; Ou, Y.Y.; Badano, J.L.; Esmail, M.A.; Leitch, C.C.; Fiedrich, E.; Beales, P.L.; Archibald, J.M.; Katsanis, N.; Rattner, J.B.; et al. MKKS/BBS6, a divergent chaperonin-like protein linked to the obesity disorder Bardet-Biedl syndrome, is a novel centrosomal component required for cytokinesis. *J. Cell Sci.* **2005**, *118*, 1007–1020. [CrossRef]
115. Muller, J.; Stoetzel, C.; Vincent, M.C.; Leitch, C.C.; Laurier, V.; Danse, J.M.; Hellé, S.; Marion, V.; Bennouna-Greene, V.; Vicaire, S.; et al. Identification of 28 novel mutations in the Bardet-Biedl syndrome genes: The burden of private mutations in an extensively heterogeneous disease. *Hum. Genet.* **2010**, *127*, 583–593. [CrossRef]
116. Billingsley, G.; Bin, J.; Fieggen, K.J.; Duncan, J.L.; Gerth, C.; Ogata, K.; Wodak, S.S.; Traboulsi, E.I.; Fishman, G.A.; Paterson, A.; et al. Mutations in chaperonin-like BBS genes are a major contributor to disease development in a multiethnic Bardet-Biedl syndrome patient population. *J. Med. Genet.* **2010**, *47*, 453–463. [CrossRef] [PubMed]
117. Seo, S.; Baye, L.M.; Schulz, N.P.; Beck, J.S.; Zhang, Q.; Slusarski, D.C.; Sheffield, V.C. BBS6, BBS10, and BBS12 form a complex with CCT/TRiC family chaperonins and mediate BBSome assembly. *Proc. Natl. Acad. Sci. USA* **2010**, *107*, 1488–1493. [CrossRef] [PubMed]
118. Davis, R.E.; Swiderski, R.E.; Rahmouni, K.; Nishimura, D.Y.; Mullins, R.F.; Agassandian, K.; Philp, A.R.; Searby, C.C.; Andrews, M.P.; Thompson, S.; et al. A knockin mouse model of the Bardet-Biedl syndrome 1 M390R mutation has cilia defects, ventriculomegaly, retinopathy, and obesity. *Proc. Natl. Acad. Sci. USA* **2007**, *104*, 19422–19427. [CrossRef]
119. Swiderski, R.E.; Agassandian, K.; Ross, J.L.; Bugge, K.; Cassell, M.D.; Yeaman, C. Structural defects in cilia of the choroid plexus, subfornical organ and ventricular ependyma are associated with ventriculomegaly. *Fluids Barriers CNS* **2012**, *9*, 22. [CrossRef] [PubMed]
120. Badano, J.L.; Kim, J.C.; Hoskins, B.E.; Lewis, R.A.; Ansley, S.J.; Cutler, D.J.; Castellan, C.; Beales, P.L.; Leroux, M.R.; Katsanis, N. Heterozygous mutations in BBS1, BBS2 and BBS6 have a potential epistatic effect on Bardet-Beidl patients with two mutations at a second BBS locus. *Hum. Mol. Genet.* **2003**, *12*, 1651–1659. [CrossRef]
121. Stone, D.L.; Slavotinek, A.; Bouffard, G.G.; Banerjee-Basu, S.; Baxevanis, A.D.; Barr, M.; Biesecker, L.G. Mutation of a gene encoding a putative chaperonin causes McKusick- Kaufman syndrome. *Nat. Genet.* **2000**, *25*, 79–82. [CrossRef] [PubMed]
122. Schaefer, E.; Durand, M.; Stoetzel, C.; Doray, B.; Viville, B.; Hellé, S.; Danse, J.M.; Hamel, C.; Bitoun, P.; Goldenberg, A.; et al. Molecular diagnosis reveals genetic heterogeneity for the overlapping MKKS and BBS phenotypes. *Eur. J. Med. Genet.* **2011**, *54*, 157–160. [CrossRef]
123. Nitika; Porter, C.M.; Truman, A.W.; Truttmann, M.C. Post-translational modifications of Hsp70 family proteins: Expanding the chaperone code. *J. Biol. Chem.* **2020**, *295*, 10689–10708. [CrossRef]
124. Engel, P.; Goepfert, A.; Stanger, F.V.; Harms, A.; Schmidt, A.; Schirmer, T.; Dehio, C. Adenylylation control by intra-or intermolecular active-site obstruction in Fic proteins. *Nature* **2012**, *482*, 107–110. [CrossRef]
125. Truttmann, M.C.; Zheng, X.; Hanke, L.; Damon, J.R.; Grootveld, M.; Krakowiak, J.; Pincus, D.; Ploegh, H.L. Unrestrained AMPylation targets cytosolic chaperones and activates the heat shock response. *Proc. Natl. Acad. Sci. USA* **2017**, *114*, E152–E160. [CrossRef]
126. Kielkowski, P.; Buchsbaum, I.Y.; Kirsch, V.C.; Bach, N.C.; Drukker, M.; Cappello, S.; Sieber, S.A. FICD activity and AMPylation remodelling modulate human neurogenesis. *Nat. Commun.* **2020**, *11*, 517. [CrossRef] [PubMed]
127. Truttmann, M.C.; Pincus, D.; Ploegh, H.L. Chaperone AMPylation modulates aggregation and toxicity of neurodegenerative disease-associated polypeptides. *Proc. Natl. Acad. Sci. USA* **2018**, *115*, E5008–E5017. [CrossRef] [PubMed]
128. Backe, S.J.; Sager, R.A.; Woodford, M.R.; Makedon, A.M.; Mollapour, M. Post-translational modifications of Hsp90 and translating the chaperone code. *J. Biol. Chem.* **2020**, *295*, 11099–11117. [CrossRef] [PubMed]
129. Franco, M.C.; Ye, Y.; Refakis, C.A.; Feldman, J.L.; Stokes, A.L.; Basso, M.; De Mera, R.M.M.F.; Sparrow, N.A.; Calingasan, N.Y.; Kiaei, M.; et al. Nitration of Hsp90 induces cell death. *Proc. Natl. Acad. Sci. USA* **2013**, *110*, E1102–E1111. [CrossRef] [PubMed]
130. Koike, M.; Fukushi, J.; Ichinohe, Y.; Higashimae, N.; Fujishiro, M.; Sasaki, C.; Yamaguchi, M.; Uchihara, T.; Yagishita, S.; Ohizumi, H.; et al. Valosin-containing protein (VCP) in novel feedback machinery between abnormal protein accumulation and transcriptional suppression. *J. Biol. Chem.* **2010**, *285*, 21736–21749. [CrossRef]
131. Franco, M.C.; Ricart, K.C.; Gonzalez, A.S.; Dennys, C.N.; Nelson, P.A.; Janes, M.S.; Mehl, R.A.; Landar, A.; Estévez, A.G. Nitration of Hsp90 on Tyrosine 33 regulates mitochondrial metabolism. *J. Biol. Chem.* **2015**, *290*, 19055–19066. [CrossRef]
132. Verba, K.A.; Agard, D.A. How Hsp90 and Cdc37 Lubricate Kinase Molecular Switches. *Trends Biochem. Sci.* **2017**, *42*, 799–811. [CrossRef]
133. Gracia, L.; Lora, G.; Blair, L.J.; Jinwal, U.K. Therapeutic Potential of the Hsp90/Cdc37 Interaction in Neurodegenerative Diseases. *Front. Neurosci.* **2019**, *13*, 1263. [CrossRef]
134. Miyata, Y.; Nishida, E. CK2 Controls Multiple Protein Kinases by Phosphorylating a Kinase-Targeting Molecular Chaperone, Cdc37. *Mol. Cell. Biol.* **2004**, *24*, 4065–4074. [CrossRef]
135. Perez, D.I.; Gil, C.; Martinez, A. Protein kinases CK1 and CK2 as new targets for neurodegenerative diseases. *Med. Res. Rev.* **2011**, *31*, 924–954. [CrossRef]
136. Zhang, S.H.; Liu, J.; Kobayashi, R.; Tonks, N.K. Identification of the cell cycle regulator VCP (p97/CDC48) as a substrate of the band 4.1-related protein-tyrosine phosphatase PTPH. *J. Biol. Chem.* **1999**, *274*, 17806–17812. [CrossRef] [PubMed]
137. Ju, J.S.; Fuentealba, R.A.; Miller, S.E.; Jackson, E.; Piwnica-Worms, D.; Baloh, R.H.; Weihl, C.C. Valosin-containing protein (VCP) is required for autophagy and is disrupted in VCP disease. *J. Cell Biol.* **2009**, *187*, 875–888. [CrossRef] [PubMed]

138. Torrecilla, I.; Oehler, J.; Ramadan, K. The role of ubiquitin-dependent segregase p97 (VCP or Cdc48) in chromatin dynamics after DNA double strand breaks. *Philos. Trans. R. Soc. B Biol. Sci.* **2017**, *372*. [CrossRef] [PubMed]
139. Cloutier, P.; Coulombe, B. Regulation of molecular chaperones through post-translational modifications: Decrypting the chaperone code. *Biochim. Biophys. Acta Gene Regul. Mech.* **2013**, *1829*, 443–454. [CrossRef] [PubMed]
140. Astakhova, L.N.; Zatsepina, O.G.; Funikov, S.Y.; Zelentsova, E.S.; Schostak, N.G.; Orishchenko, K.E.; Evgen'Ev, M.B.; Garbuz, D.G. Activity of heat shock genes' promoters in thermally contrasting animal species. *PLoS ONE* **2015**, *10*, 115536. [CrossRef]
141. Murphy, M.E. The HSP70 family and cancer. *Carcinogenesis* **2013**, *34*, 1181–1188. [CrossRef]
142. Sun, Y.; Zheng, J.; Xu, Y.; Zhang, X. Paraquat-induced inflammatory response of microglia through HSP60/TLR4 signaling. *Hum. Exp. Toxicol.* **2018**, *37*, 1161–1168. [CrossRef]
143. Nguyen, P.; Kdeiss, B.; Ord, S.; Hess, K.; Oliverio, R.; Nikolaidis, N. Evolution and natural variation of HSPA1A, the major stress inducible gene, in humans. *FASEB J.* **2018**, *31*, 604–618.
144. Hess, K.; Oliverio, R.; Nguyen, P.; Le, D.; Ellis, J.; Kdeiss, B.; Ord, S.; Chalkia, D.; Nikolaidis, N. Concurrent action of purifying selection and gene conversion results in extreme conservation of the major stress-inducible Hsp70 genes in mammals. *Sci. Rep.* **2018**, *8*, 5082. [CrossRef]
145. Singh, R.; Kølvraa, S.; Bross, P.; Jensen, U.B.; Gregersen, N.; Tan, Q.; Knudsen, C.; Rattan, S.I.S. Reduced heat shock response in human mononuclear cells during aging and its association with polymorphisms in HSP70 genes. *Cell Stress Chaperones* **2006**, *11*, 208–215. [CrossRef]
146. He, M.; Guo, H.; Yang, X.; Zhang, X.; Zhou, L.; Cheng, L.; Zeng, H.; Hu, F.B.; Tanguay, R.M.; Wu, T. Functional SNPs in HSPA1A Gene Predict Risk of Coronary Heart Disease. *PLoS ONE* **2009**, *4*, e4851. [CrossRef]
147. Konings, A.; Van Laer, L.; Michel, S.; Pawelczyk, M.; Carlsson, P.I.; Bondeson, M.L.; Rajkowska, E.; Dudarewicz, A.; Vandevelde, A.; Fransen, E.; et al. Variations in HSP70 genes associated with noise-induced hearing loss in two independent populations. *Eur. J. Hum. Genet.* **2009**, *17*, 329–335. [CrossRef]
148. Maugeri, N.; Radhakrishnan, J.; Knight, J.C. Genetic determinants of HSP70 gene expression following heat shock. *Hum. Mol. Genet.* **2010**, *19*, 4939–4947. [CrossRef] [PubMed]
149. Oliverio, R.; Nguyen, P.; Kdeiss, B.; Ord, S.; Daniels, A.J.; Nikolaidis, N. Functional characterization of natural variants found on the major stress inducible 70-kDa heat shock gene, HSPA1A, in humans. *Biochem. Biophys. Res. Commun.* **2018**, *506*, 799–804. [CrossRef] [PubMed]
150. Schopf, F.H.; Biebl, M.M.; Buchner, J. The HSP90 chaperone machinery. *Nat. Rev. Mol. Cell Biol.* **2017**, *18*, 345–360. [CrossRef] [PubMed]
151. Evans, C.G.; Wisén, S.; Gestwicki, J.E. Heat shock proteins 70 and 90 inhibit early stages of amyloid β-(1-42) aggregation in vitro. *J. Biol. Chem.* **2006**, *281*, 33182–33191. [CrossRef] [PubMed]
152. Luo, W.; Sun, W.; Taldone, T.; Rodina, A.; Chiosis, G. Heat shock protein 90 in neurodegenerative diseases. *Mol. Neurodegener.* **2010**, *5*, 24. [CrossRef]
153. Chen, B.; Piel, W.H.; Gui, L.; Bruford, E.; Monteiro, A. The HSP90 family of genes in the human genome: Insights into their divergence and evolution. *Genomics* **2005**, *86*, 627–637. [CrossRef] [PubMed]
154. Barazi, H.O.; Zhou, L.; Templeton, N.S.; Krutzsch, H.C.; Roberts, D.D. Identification of heat shock protein 60 as a molecular mediator of alpha 3 beta 1 integrin activation. *Cancer Res.* **2002**, *62*, 1541–1548.
155. Gorska, M.; Marino Gammazza, A.; Zmijewski, M.A.; Campanella, C.; Cappello, F.; Wasiewicz, T.; Kuban-Jankowska, A.; Daca, A.; Sielicka, A.; Popowska, U.; et al. Geldanamycin-Induced Osteosarcoma Cell Death Is Associated with Hyperacetylation and Loss of Mitochondrial Pool of Heat Shock Protein 60 (Hsp60). *PLoS ONE* **2013**, *8*, e71135. [CrossRef]
156. Lin, C.Y.; Hu, C.T.; Cheng, C.C.; Lee, M.C.; Pan, S.M.; Lin, T.Y.; Wu, W.S. Oxidation of heat shock protein 60 and protein disulfide isomerase activates ERK and migration of human hepatocellular carcinoma HepG. *Oncotarget* **2016**, *7*, 11067–11082. [CrossRef] [PubMed]
157. Marino Gammazza, A.; Campanella, C.; Barone, R.; Caruso Bavisotto, C.; Gorska, M.; Wozniak, M.; Carini, F.; Cappello, F.; D'Anneo, A.; Lauricella, M.; et al. Doxorubicin anti-tumor mechanisms include Hsp60 post-translational modifications leading to the Hsp60/p53 complex dissociation and instauration of replicative senescence. *Cancer Lett.* **2017**, *385*, 75–86. [CrossRef] [PubMed]
158. Kim, H.S.; Kim, E.M.; Lee, J.; Yang, W.H.; Park, T.Y.; Kim, Y.M.; Cho, J.W. Heat shock protein 60 modified with O-linked N-acetylglucosamine is involved in pancreatic β-cell death under hyperglycemic conditions. *FEBS Lett.* **2006**, *580*, 2311–2316. [CrossRef] [PubMed]
159. Chattopadhyay, S.; Mukherjee, A.; Patra, U.; Bhowmick, R.; Basak, T.; Sengupta, S.; Chawla-Sarkar, M. Tyrosine phosphorylation modulates mitochondrial chaperonin Hsp60 and delays rotavirus NSP4-mediated apoptotic signaling in host cells. *Cell. Microbiol.* **2017**, *19*. [CrossRef] [PubMed]
160. Hayoun, D.; Kapp, T.; Edri-Brami, M.; Ventura, T.; Cohen, M.; Avidan, A.; Lichtenstein, R.G. HSP60 is transported through the secretory pathway of 3-MCA-induced fibrosarcoma tumour cells and undergoes N-glycosylation. *FEBS J.* **2012**, *279*, 2083–2095. [CrossRef]
161. Baron, B. *Role of the Post-Translational Modifications of HSP60 in Disease*; Springer: Cham, Switzerland, 2019; Volume 18, pp. 69–94.

162. Waterhouse, A.; Bertoni, M.; Bienert, S.; Studer, G.; Tauriello, G.; Gumienny, R.; Heer, F.T.; De Beer, T.A.P.; Rempfer, C.; Bordoli, L.; et al. SWISS-MODEL: Homology modelling of protein structures and complexes. *Nucleic Acids Res.* **2018**, *46*, W296–W303. [CrossRef]
163. Pettersen, E.F.; Goddard, T.D.; Huang, C.C.; Couch, G.S.; Greenblatt, D.M.; Meng, E.C.; Ferrin, T.E. UCSF Chimera—A visualization system for exploratory research and analysis. *J. Comput. Chem.* **2004**, *25*, 1605–1612. [CrossRef]
164. Macario, A.J.L.; Conway de Macario, E. Chaperonopathies and chaperonotherapy. *FEBS Lett.* **2007**, *581*, 3681–3688. [CrossRef]
165. Macario, A.J.L.; Conway de Macario, E. Molecular mechanisms in chaperonopathies: Clues to understanding the histopathological abnormalities and developing novel therapies. *J. Pathol.* **2020**, *250*, 9–18. [CrossRef]
166. Cappello, F.; Marino Gammazza, A.; Palumbo Piccionello, A.; Campanella, C.; Pace, A.; Conway de Macario, E.; Macario, A.J.L. Hsp60 chaperonopathies and chaperonotherapy: Targets and agents. *Expert Opin. Ther. Targets* **2014**, *18*, 185–208. [CrossRef]
167. Meng, Q.; Li, B.X.; Xiao, X. Toward developing chemical modulators of Hsp60 as potential therapeutics. *Front. Mol. Biosci.* **2018**, *5*, 35. [CrossRef] [PubMed]
168. Macario, A.J.L.; Conway de Macario, E. Hidden chaperonopathies: Alerting physicians and pathologists on the possibility that uncharacteristic, baffling clinical features in otherwise known diseases may be due to failure of the chaperoning system. *Life Saf. Secur.* **2020**. [CrossRef]

applied sciences

MDPI

Review

Functions and Therapeutic Potential of Extracellular Hsp60, Hsp70, and Hsp90 in Neuroinflammatory Disorders

Giusi Alberti [1], Letizia Paladino [1], Alessandra Maria Vitale [1], Celeste Caruso Bavisotto [1], Everly Conway de Macario [2], Claudia Campanella [1], Alberto J. L. Macario [2,3] and Antonella Marino Gammazza [1,*]

1 Department of Biomedicine, Neurosciences and Advanced Diagnostics (BiND), University of Palermo, 90127 Palermo, Italy; giusi.alberti@unipa.it (G.A.); letizia.paladino@unipa.it (L.P.); alessandramaria.vitale@unipa.it (A.M.V.); celeste.carusobavisotto@unipa.it (C.C.B.); claudia.campanella@unipa.it (C.C.)

2 Department of Microbiology and Immunology, School of Medicine, University of Maryland at Baltimore-Institute of Marine and Environmental Technology (IMET), Baltimore, MD 21202, USA; econwaydemacario@som.umaryland.edu (E.C.d.M.); AJLMacario@som.umaryland.edu or albertomacario@iemest.eu (A.J.L.M.)

3 Euro-Mediterranean Institute of Science and Technology (IEMEST), 90139 Palermo, Italy

* Correspondence: antonella.marinogammazza@unipa.it

Abstract: Neuroinflammation is implicated in central nervous system (CNS) diseases, but the molecular mechanisms involved are poorly understood. Progress may be accelerated by developing a comprehensive view of the pathogenesis of CNS disorders, including the immune and the chaperone systems (IS and CS). The latter consists of the molecular chaperones; cochaperones; and chaperone cofactors, interactors, and receptors of an organism and its main collaborators in maintaining protein homeostasis (canonical function) are the ubiquitin–proteasome system and chaperone-mediated autophagy. The CS has also noncanonical functions, for instance, modulation of the IS with induction of proinflammatory cytokines. This deserves investigation because it may be at the core of neuroinflammation, and elucidation of its mechanism will open roads toward developing efficacious treatments centered on molecular chaperones (i.e., chaperonotherapy). Here, we discuss information available on the role of three members of the CS—heat shock protein (Hsp)60, Hsp70, and Hsp90—in IS modulation and neuroinflammation. These three chaperones occur intra- and extracellularly, with the latter being the most likely involved in neuroinflammation because they can interact with the IS. We discuss some of the interactions, their consequences, and the molecules involved but many aspects are still incompletely elucidated, and we hope that this review will encourage research based on the data presented to pave the way for the development of chaperonotherapy. This may consist of blocking a chaperone that promotes destructive neuroinflammation or replacing or boosting a defective chaperone with cytoprotective activity against neurodegeneration.

Keywords: chaperone system; molecular chaperones; Alzheimer's disease; Parkinson's disease; Huntington's disease; amyotrophic lateral sclerosis; multiple sclerosis; chaperonotherapy; chaperonopathies

Citation: Alberti, G.; Paladino, L.; Vitale, A.M.; Caruso Bavisotto, C.; Conway de Macario, E.; Campanella, C.; Macario, A.J.L.; Marino Gammazza, A. Functions and Therapeutic Potential of Extracellular Hsp60, Hsp70, and Hsp90 in Neuroinflammatory Disorders. *Appl. Sci.* **2021**, *11*, 736. https://doi.org/10.3390/app11020736

Received: 19 December 2020
Accepted: 10 January 2021
Published: 14 January 2021

1. Introduction

Neuroinflammation occurs in brain injury and chronic neurodegenerative diseases affecting the central nervous system (CNS) [1–5]. The CNS is characterized by two main types of cells: neurons and neuroglia. The former's function is impulse transmission and signaling, while the latter play other roles [6]. For instance, microglia and astrocytes, resident antigen-presenting cells (APCs), rapidly respond to tissue damage that compromises the homeostasis of the local brain parenchyma [7]. Microglia activation is a highly regulated process involved in the generation of different and complex phenotypes, the reorganization of cell surface markers, and the release of soluble pro-and anti-inflammatory factors. Neuroinflammation is a complex cellular and biochemical response that increases inflammatory mediators (such as cytokines and chemokines) and activates glial cells and leukocyte invasion of brain tissue. These events have been correlated with an increased permeability of the blood–brain barrier (BBB). Microglial cells can remain activated for long periods, which causes the release of large amounts of cytokines and neurotoxic molecules that contribute to neurodegeneration [8]. It is important to bear in mind that inflammation is not necessarily deleterious because moderate inflammatory reactions are involved in diverse phenomena that protect cells and tissues from a variety of noxae [9]. Whether inflammation is good or bad for the organism depends mostly on the intensity and duration of the inflammatory reaction: the more intense and long lasting the reaction, the higher the probability of disease development or aggravation. In inflammatory and immune reactions, molecular chaperones interact with the immune system, especially when they are activated under stress conditions in different organs, including the brain. Molecular chaperones, many of which are heat shock proteins (Hsps), are the main components of the chaperone system [10,11]. They are ubiquitously expressed, and their canonical role is to assist in the folding of nascent polypeptides avoiding protein misfolding and aggregation, and to deliver damaged proteins to protein degradation machineries [10,12]. The levels of some chaperones change in response to stressors, for example, oxidative stress and DNA damage [13]. Typically, chaperones are cytoprotective, but they can also be pathogenic when they are structurally and/or functionally abnormal and can contribute to the mechanism of diseases termed chaperonopathies [10]. Chaperonopathies are involved in the development of some neurodegenerative diseases in which neuroinflammation is implicated. The role played by chaperones in neuroinflammation is under scrutiny and constitutes a promising area of research because it may lead to the discovery of novel treatment strategies centered on chaperonotherapy, namely, the use of chaperones as therapeutic targets or agents [14,15]. Here, we discuss molecular chaperones within the context of neurodegenerative diseases/neuroinflammation and the interactions between the immune system and the chaperoning system, focusing on extracellular Hsp60, Hsp70, and Hsp90 in Alzheimer's disease (AD), amyotrophic lateral sclerosis (ALS), Parkinson's disease (PD), Huntington's disease (HD), and multiple sclerosis (MS).

2. Immunomodulatory Function of Extracellular Hsp60, Hsp70, and Hsp90

Hsp60, Hsp70, and Hsp90 interact with the immune system in many ways and thereby have an impact on neurodegenerative diseases. Extracellular Hsp60, Hsp70, and Hsp90 influence both the innate and the adaptive immune responses. Generally, extracellular Hsp–receptor interaction involves specific receptors expressed on macrophages and dendritic and microglia cells, including toll-like receptors (TLRs), scavenger receptors (SR), and other molecules [16]. For example, Hsp70 and Hsp90 can interact with the SR LOX-1 [17], and Hsp70 interacts also with multiple members of the SR family [18]. The SR are expressed on different types of cells and they are involved in the binding and internalization of stress proteins [18]. Extracellular Hsp60, Hsp70, and Hsp90 can modulate the innate immune response, causing the secretion of proinflammatory cytokines by APCs [19]. This interaction elicits a proinflammatory response that involves mainly nuclear factor-kappa B (NF-kB). These chaperones are endogenous ligands for TLRs, and by interacting also with CD14 molecules, they can induce the production of cytokines (e.g., interleukin 1 beta (IL-1β), IL-6, inducible isoform of nitric oxide synthase (iNOS)) [20,21]. TLR4 is a receptor expressed on the microglia plasma cell membrane with a key role in the generation of immune responses in the nervous system, responses that are implicated in the development of neurodegenerative disorders [22]. For instance, Hsp60 can mediate neuroinflammation through a MyD88-dependent pathway by interacting with TLR4 on the microglia surface [21] and by inducing the production of proinflammatory factors via microglial LOX-1 [23]. Intrathecal injection of Hsp60 lead to neurodegeneration and demyelination by the activation of TLR4-MyD88 signaling in microglial cells [24]. Hsp70 can interact with microglia, dendritic cells, and macrophages through TLR2 and TLR4, leading to proinflammatory NF-kB activation and its associated pathways [25]. Hsp90 interacts with an extensive list of key mediators involved in pathways regulating inflammatory and immune responses. For example, among the protein clients of Hsp90, there is the receptor-interacting protein (RIP) kinase, which is involved in the innate immune response and in the cell-death signaling pathway. [26] RIP, following TLR4 activation, induces the expression of proinflammatory cytokines by NF-kB signaling [27] (Figure 1).

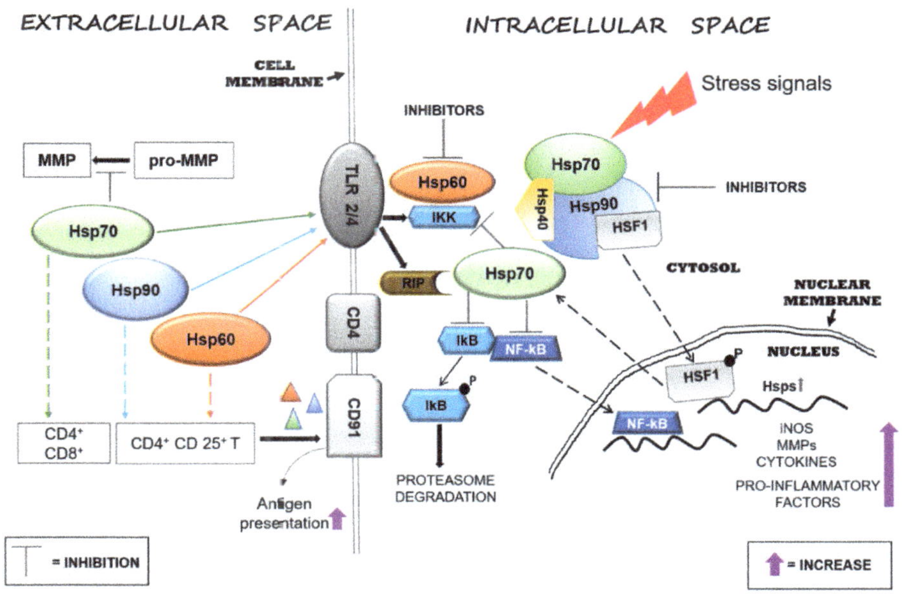

Figure 1. Heat shock protein (Hsp)60, Hsp70, and Hsp90 modulate inflammatory reactions by interacting with factors involved in the regulation of innate and adaptive immune responses. Stressors can activate the immune system and, in turn, promote neurodegeneration by inducing Hsps in brain tissue as a mechanism of protection. Extracellular Hsp60, Hsp70, and Hsp90 interact with receptors present on the surface of cells of the neural tissue's immune compartment (e.g., microglia) and elicit pro- or anti-inflammatory responses, depending on the local cellular status. The interaction of extracellular Hsp60, Hsp70, and Hsp90 with toll-like receptor (TLR)2/4 induces the activation of the nuclear factor-kappa B (NF-kB) inhibitor protein, which in turn triggers the activation of the NF-kB pathway, promoting an inflammatory response. Hsp60, Hsp70, and Hsp90 form complexes with antigens (represented by triangles) mediating their presentation via the CD91 cell surface receptor on antigen-presenting cells (APCs) [28]. Hsp90 plays a proinflammatory role through the interaction with its client proteins, such as members of the receptor-interacting protein (RIP) kinases and, thereby, activates the NF-kB pathway. Under physiological conditions, intracellular Hsp90, by blocking heat shock factor (HSF)1, prevents the transcription of Hsp genes, such as Hsp70, or other genes that code for anti-inflammatory molecules. The pharmacological inhibition of Hsp90 can lead to upregulation of the transcription of intracellular Hsp70 and of anti-inflammatory molecules by its release. Abbreviations: MMP, matrix metalloproteinase; TLR, toll-like receptor; IKK, inhibitor of κB kinase; RIP, receptor interaction protein; CD, cluster of differentiation; lkB, inhibitory subunit I kappa B-alpha; NF-kB, nuclear factor kappa-light-chain-enhancer of activated B cells; HSF1, heat shock factor 1; P, phosphate; Hsp, heat shock protein; iNOS, inducible isoform of nitric oxide synthase.

Extracellular Hsp60, Hsp70, and Hsp90 can also help antigen presentation in the adaptive immune responses by upregulating the expression of major histocompatibility complex (MHC) molecules and their load [29,30]. Extracellular Hsp70 and Hsp90 complexed with antigens elicit the responses of cluster of differentiation (CD)8+ or CD4+ T cells by adaptive receptors [16], while Hsp60 by itself can stimulate regulatory CD4+ and

CD 25+ T cells (Tregs), leading to an immunosuppressive adaptive response without APC participation [16,31]. In addition, the chaperone–peptide complexes can also recognize the CD91 receptor of macrophages/dendritic cells and facilitate antigen presentation [28] (Figure 1). The activation of the adaptive response via Hsp70 might represent a negative reaction for the cell, but it could be considered advantageous for the development of immunological memory in preparation for rapid reaction against subsequent insults [32]. In contrast to the proinflammatory function of extracellular Hsp70, intracellular Hsp70 has an anti-inflammatory effect in the brain, especially when overexpressed following brain damage. Thus, Hsp70 can be anti-inflammatory because it can block the expression of proinflammatory molecules, such as matrix metalloproteinases [33], and it can also promote the reduction or the inhibition of NF-kB activity [34,35] (Figure 1). In addition, intracellular Hsp70 interferes also with genes involved in various neuronal pathways such as transmission of nerve impulses [36]. Therefore, extracellular Hsp70 could in principle have anti-inflammatory and neuroprotective effects similar to those of the intracellular counterpart [37]. Consequently, it is likely that an increase of intracellular Hsp70 will lead to an increase of functional extracellular Hsp70, contributing to the reduction of the inflammation associated with neurodegeneration. Pharmacological increase of the Hsp70 level in neurons and microglia by 17-N-allylamino-17-demethoxygeldanamycin (17-AAG) reduced the hemorrhagic volume in a mouse model of traumatic brain injury [38]. Likewise, 17-AAG inhibition of Hsp90 induced the expression of Hsp70 and Hsp60 [39]. Thus, it may be said that Hsp70, Hsp60, and Hsp90 promote inflammatory responses and, consequently, neuronal damage and are implicated in neuroinflammation and neurotoxicity. Cytosolic Hsp60 has been shown to directly interact with the inhibitor of κB kinase (IKK), promoting activation of NF-kB-dependent gene transcription by tumor necrosis factor-α (TNFα) [40] (Figure 1). Hsp90 can induce a proinflammatory response in different ways, for example, by sequestering the regulator transcriptional factor heat shock factor (HSF)1 and thereby inhibiting the expression of Hsps (e.g., Hsp70) or activating the NF-kB pathway through the activation of its protein clients RIP [41] (Figure 1). In view of these results, Hsp60 and Hsp90 modulators appear as potentially useful agents for controlling inflammation in the nervous system [42,43]. Currently, numerous compounds have been designed to inhibit Hsp90 activity, but few have been developed for Hsp60 [15]. Hsp90 inhibitors have been developed to directly act on the chaperone or on its client proteins. Some inhibitors block the Hsp90 folding activity linked to adenosine triphosphate (ATP)-dependent conformation changes [44], while others inactivate its client proteins via proteasomal degradation [45]. For example, geldanamycin induces the degradation of Hsp90 client proteins of the RIP family with the consequent inhibition of TNF-mediated IkB kinase and NF-kB activation [41]. Furthermore, Hsp90 forms a complex with HSF1, blocking its translocation to the nucleus and, thereby, impedes the upregulation of Hsp70 and other anti-inflammatory molecules [46] (Figure 1). Most of the compounds that inhibit Hsp70 function by targeting its ATP hydrolysis activity or specific cysteine residues [47].

3. Extracellular Hsp60, Hsp70, and Hsp90 in Acute Nervous System Injury and Chronic Neurodegenerative Diseases

Neurodegenerative diseases are accompanied by inflammatory responses aimed at eliminating dead and damaged neuronal cells to restore the compromised area to its normal status [7]. It should be borne in mind that while short-lived inflammatory responses generally have a beneficial effect, excessive and persistent release of inflammatory mediators can be harmful to brain tissue [9]. Moreover, prolonged activation of microglia and astrocytes could also lead to the alteration of their beneficial functions, which they display under normal conditions [48]. Therefore, it is not surprising that neuroinflammation contributes to CNS diseases [49]. Although different in their origins, many neurodegenerative conditions are characterized by shared cellular responses that promote the upregulation of molecular chaperones as the first line of defense against misfolded, dysfunctional, and aggregation-prone proteins [50]. There is increasing evidence for the release of Hsp60, Hsp70, and Hsp90 into the extracellular environment, with functions that are complementary or independent of those of their intracellular counterparts. Since these chaperones lack a secretion signal in their sequences, the mechanisms by which they are released are poorly understood. In vitro and in vivo studies with Hsp60 have unveiled secretion pathways, involving lipid rafts and exosomes, which would explain the presence of Hsp60 in extramitochondrial sites such as interstitial space, cellular membrane, and biological fluids [51]. Similarly, nontraditional secretion mechanisms participate in the membrane delivery and release of Hsp70, involving lipid rafts [52] and lysosomes [53], in line with its role as a lysosomal stabilizer [54]. Secretion of Hsp90 via exosomes depends on its ATPase function and on the open or closed conformational state of the Hsp90 dimer: the open state promotes Hsp90 release via exosomes, whereas the closed state blocks this process [55]. Different types of CNS cells, including neurons and glial cells, can release exosomes with their cargo of specific molecules that could affect the function of acceptor cells [56]. At the extracellular level, Hsp60 is known to contribute to neuroinflammation with possible negative implications: this chaperone is highly expressed in activated microglia, and when released extracellularly, it induces neuroinflammation with neuronal cell death [57]. For this reason, inhibition of Hsp60 expression and its release represents a possible therapeutic mechanism applicable to neurodegenerative diseases. The pro- and anti-inflammatory effects of extracellular Hsp60, Hsp70, and Hsp90 in AD, PD, ALS, HD, and MS are summarized in Table 1 and discussed in the following paragraphs.

Table 1. Anti- and proinflammatory effects of extracellular Hsp60, Hsp70, and Hsp90 in neurodegenerative diseases.

Disease	Hsp60	Hsp70	Hsp90
Alzheimer's disease (AD)	Anti-	Anti-	Pro-
Parkinson's disease (PD)	Pro-	Anti-	Pro-
Huntington's disease (HD)	Anti-	Anti-	Pro-
Amyotrophic lateral sclerosis (ALS)	Not reported	Anti-	Not reported
Multiple sclerosis (MS)	Pro-	Pro-	Pro-

3.1. *Alzheimer's Disease*

AD is a neurodegenerative disorder in which the amyloid-β peptide (Aβ) accumulates in extracellular deposits named plaque, whereas neurofibrillary tangles (NFTs) occur intracellularly with hyperphosphorylated tau [12,58]. Under pro-aggregating conditions (37 °C and stirring), extracellular Hsp60 inhibits the onset of Aβ cross-β-structure formation that typically accompanies the peptide assembly toward higher ordered structures [59]. The hypotheses formulated on the possible role of Hsp60 in the formation of protein deposits are mainly based on its holding activity. For instance, Hsp60 could act as a noncatalytic inhibitor of polypeptide aggregation by sequestering unfolded monomers via hydrophobic interactions. However, the stoichiometric ratio of the Aβpeptide/Hsp60 and the limitation of the methods applied for these measurements put a question mark on the validity of the results. In fact, the inhibition of amyloid formation appears discontinuous when passing from a 75:1 to 50:1 molar ratio. Furthermore, the method used, size-exclusion chromatography, cannot distinguish between Aβ monomers or peptide oligomers of very low molecular weight, such as dimers or trimers, nor can it discriminate between on-pathway and off-pathway species. These data suggest that Hsp60 exerts its inhibitory action only under stress conditions and, in particular, in the presence of other factors such as high temperature and stirring, which favor the formation of on-pathway seeding species. [59]. Higher levels of Hsp60 were found in lymphocytes isolated from AD subjects [60,61]. αβ immunization with peptides derived from Hsp60 induced a decrease of cerebral amyloid burden in a mouse model [62]. Like Hsp60, extracellular Hsp70 also interacts with Aβ oligomers, blocking their oligomerization into fibers and reducing their toxicity [63]. The engineered form of secreted Hsp70 (secHsp70) in *Drosophila* protects against the toxicity induced by Aβ42 deposits in the extracellular milieu [64]. Exogenous Hsp90 was found to induce microglial activation and to facilitate phagocytosis and clearance of Aβ directly via the TLR4 pathway, but when bound to the Aβ oligomers, it induced the production of IL-6 and TNF-α [65]. In another work, it was revealed that Hsp90 modulates the formation of the STIP1 (or Hsp70/Hsp90 organizing protein (HOP))/PrPC complex, which inhibits the neuroprotective role of STIP1 against amyloid-beta peptide [66]. However, it is still unclear whether extracellular Hsp70/Hsp90/STIP1 in AD brain exists separately or as a complex with the Aβ aggregate [66]. All these observations indicate that the understanding of Hsp90′s role in neurodegeneration deserves further investigation.

3.2. Parkinson's Disease

PD is characterized by movement disorders and loss of dopaminergic neurons in the brain's substantia nigra pars compacta [67,68]. The disease is also characterized by aggregated α-synuclein that forms nuclear inclusions called Lewy bodies [69]. A study in yeast cells has shown that null mutations in the Hsp60 gene are linked with defects in the folding of mitochondrial proteins, with accumulations of misfolded peptides analogous to the α-synuclein aggregates of PD [70]. Hsp60, Hsp70, and Hsp90 interact with α-synuclein in the Lewy bodies in PD patients. These inclusions consist not only of α-synuclein aggregates but also contain molecular chaperones which have been sequestered in the aggregates while attempting to impede or correct protein misfolding and aggregation [71,72]. This sequestration leads to a deficit of chaperones available for maintaining protein homeostasis, namely, a chaperonopathy by defect occurs, which contributes to the aggravation of the pathologic process leading to neurodegeneration. The interaction between Hsp70 and α-synuclein involves the central hydrophobic region of the pathological protein and the substrate-binding domain Hsp70 and is crucial for inhibiting assembly before the elongation stage [73]. The neuroprotective function of overexpressed Hsp70 has been confirmed in experimental models in vivo [74]. There is less information regarding the protective role of Hsp90 in the regulation of α-synuclein aggregation. Like in AD, Hsp60, Hsp70, and Hsp90 contribute to neuronal toxicity in PD. Hsp90 abolishes the binding of α-synuclein to vesicles and promotes the formation of fibrils [75]. In in vivo and in vitro models of PD, it was found that Hsp60 expression gradually decreased after 6-hydroxydopamine (6-OHDA) injection into dopaminergic neurons (DA). This result may be explained by the release of Hsp60 by the damaged neurons, as suggested by its presence in the cell culture medium [76]. In PD models and patients, activation of microglia plays a key role in the release of proinflammatory factors that aggravate the loss of DA neurons [77]. Astrocytes, which are the predominant glial cell type in the CNS, are also critically affected by stressors. The expression of Hsp60 on the surface of activated microglia suggests that Hsp60 is involved in the progression of PD. Extracellular release of Hsp60 from CNS cells undergoing necrotic or apoptotic death activates microglia in a TLR4- and MyD88-dependent manner [21]. Hsp60 was released from degenerated neurons to activate microglia in a rat PD model, providing a novel idea for developing a therapeutic strategy to slow or stop PD progression by preventing the release of Hsp60 or interfering with the interaction between Hsp60 and microglia [76].

3.3. Amyotrophic Lateral Sclerosis

ALS is a chronic inflammatory demyelinating disease that affects motor neurons and is characterized by atrophy and paralysis of muscles, with progressive aggravation over the years [78]. This disease occurs sporadically, but a small percentage is familial with mutations in specific genes, such as the gene encoding the free-radical-scavenging enzyme superoxide dismutase-1 (SOD1) [79]. An important aspect of SOD1-associated ALS is the deposition of SOD1 in large insoluble aggregates in motor neurons. The SOD1 mutated protein mediates the induction of the disease through the dysregulation of the heat shock response (HSR)–apoptosis axis [80]. The development of ALS is linked to the formation of intracellular aggregates of misfolded proteins [78]. Few data are available regarding the involvement of molecular chaperones in ALS onset. Motor neurons of ALS patients have an intrinsic deficit in the ability to activate the HSR and, consequently, do not readily regulate Hsp expression, as shown, for example, for Hsp70 [81]. It has been observed that the Hsp70/Hsp40 pair is complexed with the mutant form of the SOD1 protein in cultured neuronal cells [82]. However, data indicate that the increase of Hsp70 level alone is not sufficient to ameliorate mutant SOD1-protein-mediated toxicity in mouse models [83]. Histamine is neuroprotective through the HSR in motor neurons and microglia cell cultures, and in vivo in spinal cord and cortex from symptomatic SOD1-G93A mice [84]. These results emphasize the relevance of histidine-induced Hsp70 stimulation for preserving motor function [84]. Further, the intraperitoneal administration of human recombinant exogenous Hsp70 increased lifespan, delayed the onset of symptoms, preserved locomotor function, and prolonged motoneuron survival in a mouse model of ALS [85]. Extracellular Hsp70 stimulates the survival of neurons following injury [86] and overexpressed Hsp70 induces the survival of astrocytes [87]. Under stress, astrocytes increase the release of exosomes enriched in Hsp70, with positive implications on the survival of nearby neurons [88]. Interestingly, exosomes derived from cancer cells express Hsp70 on their surface, which allows their interaction with target cells carrying surface Hsp receptors [15]. Hsp70 (DnaJC5/Hsc70 complex) is also believed to be involved in the extracellular release of proteins associated with neurodegenerative disease as part of its chaperoning functions [89]. Exogenous Hsp70 protects from oxidative damage death in motor neurons through binding and sequestration of toxic proteins [90].

3.4. Huntington's Disease

HD is a progressive neurodegenerative disease caused by excess repeats of glutamine residues, called polyQ repeats, in the huntingtin (Htt) protein, causing protein misfolding [91]. The accumulation of misfolded Htt is associated with cognitive decline and motor defects [92]. Few studies have investigated the involvement of Hsp60 in HD. For instance, Hsp60 plays a protective role in HD by inhibiting polyglutamine aggregate formation and toxicity in vitro [93]. In HD, using confocal microscopy, it was observed that exogenous Hsp70 helps to reduce the number and size of polyQ inclusions [94]. Under normal conditions, the Htt proteins are under the quality control of the chaperone system, particularly Hsp90 and Hsp70. Hsp90 co-immunoprecipitates with both mutant and wild-type forms of Htt, and its inhibition blocks the interaction [95]. Hsp90 preferentially binds the mutant huntingtin (mHtt) rather than normal Htt proteins and also binds other proteins, such as the transcriptional repressor RE1-silencing transcription factor (REST) [96]. REST is normally quiescent in differentiated neurons, but its levels and activity increase as a consequence of neuronal damage [97]. Under physiological conditions, Htt indirectly regulates REST nuclear traffic through the formation of a complex that causes REST retention in the cytoplasm, whereas under pathological conditions, the binding of mHtt to this complex induces a conformational change that leads to the release of REST and its subsequent translocation to the nucleus [96]. A direct effect of this pathological transport of REST is the repression of neuronal genes containing RE1 sequences, including the brain-derived neurotrophic factor (BDNF), a survival factor for striatal neurons. Hsp90-specific inhibitors dramatically reduce Htt stability and REST levels, providing neuroprotective activity [96]. Given the complexity of the mechanism regulating REST expression and the way by which REST modulates the expression of its target genes, further studies are needed to understand the Hsp90–mHtt and Hsp90–REST interactions. Other studies report the involvement of Hsp70 in the pathogenesis of HD. Hsp70 has not only a neuroprotective role as an intracellular chaperone but it has also important extracellular functions. Extracellular Hsp70 can reach, for example, the hippocampus, leading to the initiation and propagation of generalized tonic–clonic seizures [98,99].

3.5. Multiple Sclerosis

MS is a disease of the CNS with autoimmune components that provoke the damage of myelin around nerves and axons, impairing the transmission of information between brain and the rest of the body [100]. There is little information on the role that chaperones might play in MS. For example, it is not yet established if Hsp60 plays a role in the immunopathogenesis of MS. Serum and cerebrospinal fluid (CSF) samples with untreated, relapsing–remitting MS showed antibody signatures targeting epitopes of various proteins, including Hsp60 [101]. Although extracellular Hsp70 is associated with neuroprotective functions in AD and PD as it helps in lowering the levels of misfolded proteins, in MS, it may intensify the immune response. Hsp70 has been found in MS lesions, often in association with the two major myelin proteins of the myelin sheath, namely, myelin basic protein (MBP) and proteolipid protein (PLP) [102]. In the experimental autoimmune encephalomyelitis (EAE) model, Hsp70 promotes an immunological response mediated by its myelin peptide adjuvant capacity [103]. Moreover, Hsp70 overexpression in vitro leads to enhanced presentation of MBP in an MHC class-II-dependent manner [104]. It may be hypothesized that the association of Hsp70 with myelin proteins would be required also for remyelination during the repair process, and that its deficiency could compromise this process. This view is supported by findings with autopsy tissue of MS lesions which show a quantitative reduction of Hsp70 compared with normal brain tissues, a reduction that parallels the impairment of the remyelination process [105]. Oligodendrocyte precursor cells (OPCs) are the targets of autoimmune attack in MS, which prevents remyelination. CSF from MS patients contains antibodies that can specifically recognize Hsp90 molecules located on OPCs with consequent activation of the complement and significant reduction of the OPCs [106]. These features indicate that the use of Hsp90 inhibitors could be beneficial in EAE and probably also in MS.

4. Conclusions

Neuroinflammation and protein misfolding and aggregation are currently recognized as important players in neurodegenerative diseases. The chaperone system, the main component of which are molecular chaperones, is critical for maintaining protein homeostasis. While protein quality control encompasses the canonical functions of chaperones, these also have noncanonical functions, and both have an impact on the nervous system in health and disease. A malfunction of a chaperone may cause disease, a chaperonopathy. Thus, while normal chaperones are typically cytoprotective, abnormal ones can be pathogenic and contribute to the initiation–progression of neuropathies. This knowledge opens the road for considering chaperonotherapy as a therapeutic resource in the field of neurodegenerative diseases. If the chaperones are working in a cytoprotective mode, namely, promoting protein folding, preventing misfolding and aggregation, and dissolving reversible aggregates, their levels ought to be enhanced if necessary, for example, when they become quantitatively insufficient because of excessive demand and/or depletion because of sequestration in the aggregates. In these chaperonopathies by defect, positive chaperonotherapy would be appropriate and would involve administration of chaperone stimulators or the chaperones themselves as proteins or via gene therapy. The same would apply in other instances of chaperonopathies by defect, for example, when a chaperone is structurally damaged by mutation or by an aberrant post-translation modification. If, on the contrary, a chaperone plays an etiologic–pathogenic role and favors development of neuroinflammation and neurodegeneration, negative chaperonotherapy would be required. The pathogenic chaperone protein or its gene should be blocked or eliminated. These are options now open for investigating novel therapeutic approaches targeting neuroinflammation and neurodegeneration. For example, arimoclomol, which induces expression of chaperone genes, is a potential agent to potentiate neuroprotection in ALS [107–109]. Progress in these kinds of therapeutic strategies centering on chaperonotherapy is desperately needed, considering the severity of most neurodegenerative diseases and the current scarcity of efficacious treatments.

Author Contributions: Conceptualization A.J.L.M. and A.M.G.; original draft preparation G.A., L.P., and A.M.V.; supervision E.C.d.M., C.C., and C.C.B., writing and editing, A.J.L.M. and A.M.G. All authors have read and agreed to the published version of the manuscript.

Funding: A.J.L.M. and E.C.d.M. were partially supported by IMET and IEMEST. This is IMET contribution number IMET 21-001.

Institutional Review Board Statement: Not applicable.

Informed Consent Statement: Not applicable.

Conflicts of Interest: The authors declare no conflict of interest.

Abbreviations

Aβ, Amyloid-β peptide; ALS, Amyotrophic Lateral Sclerosis; AD, Alzeheimer's Disease; APC, Antigen Presenting Cell; ATP, Adenosine Triphosphate; BBB, Blood-Brain Barrier; BDNF, Brain-derived Neurotrophic Factor; CD14, Cluster of Differentiation 14; CNS, Central Nervous System; CSF, Cerebrospinal Fluid; EAE, Experimental Autoimmune Encephalomyelitis; HD, Huntington's Disease; HOP, Hsp70/Hsp90 organizing protein; HSF1, Heat Shock Factor 1; Hsp, Heat Shock Protein; HSR, Heat Shock Response; Htt, Huntingtin; IL-1β, Interleukin 1 beta; IL-6, Interleukin 6; lkB, Inhibitory Subunit I kappa B-alpha; IKK, inhibitor of κB kinasen; iNOS, Inducible Isoform of Nitric Oxide Synthase; MBP, Myelin Basic Protein; MHC, Major Histocompatibility Complex; mHtt, Mutant Huntingtin; MMP, Matrix Metalloproteinase; MS, Multiple Sclerosis; NF-kB, Nuclear Factor-kappa B; 6-OHDA, 6-hydroxydopamine; OPC, Oligodendrocyte Precursor Cell; PLP, Proteolipid Protein; P, Phosphate; PD, Parkinson's Disease; REST, RE1-Silencing Transcription factor; RIP, Receptor Interacting Protein Kinase; SOD1, Superoxide Dismutase-1; SR, Scavenger Receptor; TLR, Toll-like Receptor; TNFα, Tumor Necrosis Factor-α; NTF, Neurofibrillary Tangle.

References

1. Yang, Q.Q.; Zhou, J.W. Neuroinflammation in the central nervous system: Symphony of glial cells. *Glia* **2019**, *67*, 1017–1035. [CrossRef] [PubMed]
2. Bader, V.; Winklhofer, K.F. Mitochondria at the interface between neurodegeneration and neuroinflammation. *Semin. Cell Dev. Biol.* **2020**, *99*, 163–171. [CrossRef] [PubMed]
3. Guerreiro, S.; Privat, A.L.; Bressac, L.; Toulorge, D. CD38 in Neurodegeneration and Neuroinflammation. *Cells* **2020**, *9*, 471. [CrossRef] [PubMed]
4. Liyanagamage, D.S.N.K.; Martinus, R.D. Role of Mitochondrial Stress Protein HSP60 in Diabetes-Induced Neuroinflammation. *Mediators Inflamm.* **2020**, *2020*, 8073516. [CrossRef] [PubMed]
5. Webers, A.; Heneka, M.T.; Gleeson, P.A. The role of innate immune responses and neuroinflammation in amyloid accumulation and progression of Alzheimer's disease. *Immunol. Cell Biol.* **2020**, *98*, 28–41. [CrossRef] [PubMed]
6. Doyle, D. The fine structure of the nervous system: The neurons and supporting cells. *J. Neurol. Neurosurg. Psychiatry* **1978**, *41*, 191. [CrossRef]
7. Banjara, M.; Ghosh, C. Sterile Neuroinflammation and Strategies for Therapeutic Intervention. *Int. J. Inflamm.* **2017**, 8385961. [CrossRef] [PubMed]
8. Tang, Y.; Le, W. Differential Roles of M1 and M2 Microglia in Neurodegenerative Diseases. *Mol. Neurobiol.* **2016**, *53*, 1181–1194. [CrossRef] [PubMed]
9. DiSabato, D.J.; Quan, N.; Godbout, J.P. Neuroinflammation: The devil is in the details. *J. Neurochem.* **2016**, *139*, 136–153. [CrossRef] [PubMed]
10. Macario, A.J.L.; Conway de Macario, E. Sick chaperones, cellular stress, and disease. *N. Engl. J. Med.* **2005**, *353*, 1489–1501. [CrossRef]
11. Chaperone proteins and chaperonopathies. In *Stress Physiology, Biochemistry, and Pathology*; Handbook of Stress; Academic Press: Cambridge, MA, USA, 2019; Volume 3, pp. 135–152.
12. Marino Gammazza, A.; Caruso Bavisotto, C.; Barone, R.; de Macario, E.C.; Macario, A.J.L. Alzheimer's disease and molecular chaperones: Current knowledge and the future of chaperonotherapy. *Curr. Pharm. Des.* **2016**, *22*, 4040–4049. [CrossRef] [PubMed]
13. Didelot, C.; Schmitt, E.; Brunet, M.; Maingret, L.; Parcellier, A.; Garrido, C. Heat shock proteins: Endogenous modulators of apoptotic cell death. In *Molecular Chaperones in Health and Disease*; Springer: Berlin/Heidelberg, Germany, 2006; Volume 172, pp. 171–198.
14. Macario, A.J.L.; Conway de Macario, E.; Cappello, F. The Chaperonopathies. In *Diseases with Defective Molecular Chaperones*; Springer: New York, NY, USA; London, UK, 2013; pp. 1–116.
15. Cappello, F.; Marino Gammazza, A.; Palumbo Piccionello, A.; Campanella, C.; Pace, A.; Conway de Macario, E.; Macario, A.J.L. Hsp60 chaperonopathies and chaperonotherapy: Targets and agents. *Expert. Opin. Ther. Targets* **2014**, *18*, 185–208. [CrossRef] [PubMed]

16. Calderwood, S.K.; Gong, J.; Murshid, A. Extracellular HSPs: The Complicated Roles of Extracellular HSPs in Immunity. *Front. Immunol.* **2016**, *7*, 159. [CrossRef] [PubMed]
17. Murshid, A.; Theriault, J.; Gong, J.; Calderwood, S.K. Investigating receptors for extracellular heat shock proteins. *Methods Mol. Biol.* **2011**, *787*, 289–302. [PubMed]
18. Thériault, J.R.; Adachi, H.; Calderwood, S.K. Role of scavenger receptors in the binding and internalization of heat shock protein 70. *J. Immunol.* **2006**, *177*, 8604–8611. [CrossRef]
19. Zininga, T.; Ramatsui, L.; Shonhai, A. Heat Shock Proteins as Immunomodulants. *Molecules* **2018**, *23*, 2846. [CrossRef]
20. Asea, A.; Rehli, M.; Kabingu, E.; Boch, J.A.; Bare, O.; Auron, P.E.; Stevenson, M.A.; Calderwood, S.K. Novel signal transduction pathway utilized by extracellular HSP70: Role of toll-like receptor (TLR) 2 and TLR4. *J. Biol. Chem.* **2002**, *277*, 15028–15034. [CrossRef]
21. Lehnardt, S.; Schott, E.; Trimbuch, T.; Laubisch, D.; Krueger, C.; Wulczyn, G.; Nitsch, R.; Weber, J.R. A vicious cycle involving release of heat shock protein 60 from injured cells and activation of toll-like receptor 4 mediates neurodegeneration in the CNS. *J. Neurosci.* **2008**, *28*, 2320–2331. [CrossRef]
22. Fiebich, B.L.; Batista, C.R.A.; Saliba, S.W.; Yousif, N.M.; de Oliveira, A.C.P. Role of Microglia TLRs in Neurodegeneration. *Front. Cell Neurosci.* **2018**, *12*, 329. [CrossRef]
23. Zhang, D.; Sun, L.; Zhu, H.; Wang, L.; Wu, W.; Xie, J.; Gu, J. Microglial LOX-1 reacts with extracellular HSP60 to bridge neuroinflammation and neurotoxicity. *Neurochem. Int.* **2012**, *61*, 1021–1035. [CrossRef]
24. Rosenberger, K.; Dembny, P.; Derkow, K.; Engel, O.; Krüger, C.; Wolf, S.A.; Kettenmann, H.; Schott, E.; Meisel, A.; Lehnardt, S. Intrathecal heat shock protein 60 mediates neurodegeneration and demyelination in the CNS through a TLR4- and MyD88-dependent pathway. *Mol. Neurodegener.* **2015**, *10*, 5. [CrossRef] [PubMed]
25. Giffard, R.G.; Han, R.; Emery, J.F.; Duan, M.; Pittet, J.F. Regulation of apoptotic and inflammatory cell signaling in cerebral ischemia: The complex roles of heat shock protein 70. *Anesthesiology* **2008**, *109*, 339–348. [CrossRef] [PubMed]
26. Yang, C.K.; He, S.D. Heat shock protein 90 regulates necroptosis by modulating multiple signaling effectors. *Cell Death Dis.* **2016**, *7*, e2126. [CrossRef]
27. Mifflin, L.; Ofengeim, D.; Yuan, J. Receptor-interacting protein kinase 1 (RIPK1) as a therapeutic target. *Nat. Rev. Drug Discov.* **2020**, *19*, 553–571. [PubMed]
28. Binder, R.J.; Zhou, Y.J.; Messmer, M.N.; Pawaria, S. CD91-Dependent Modulation of Immune Responses by Heat Shock Proteins: A Role in Autoimmunity. *Autoimmune Dis.* **2012**, *2012*, 863041.
29. Cohen-Sfady, M.; Nussbaum, G.; Pevsner-Fischer, M.; Mor, F.; Carmi, P.; Zanin-Zhorov, A.; Lider, O.; Cohen, I.R. Heat shock protein 60 activates B cells via the TLR4-MyD88 pathway. *J. Immunol.* **2005**, *175*, 3594–3602. [CrossRef] [PubMed]
30. Murshid, A.; Gong, J.; Calderwood, S.K. The role of heat shock proteins in antigen cross presentation. *Front. Immunol.* **2012**, *3*, 63. [CrossRef]
31. Liu, G.; Zhao, Y. Toll-like receptors and immune regulation: Their direct and indirect modulation on regulatory CD4+ CD25+ T cells. *Immunology* **2007**, *122*, 149–156.
32. Chen, J.; Graham, S.H.; Zhu, R.L.; Simon, R.P. Stress proteins and tolerance to focal cerebral ischemia. *J. Cereb. Blood Flow Metab.* **1996**, *16*, 566–577. [CrossRef]
33. Lee, J.E.; Kim, Y.J.; Kim, J.Y.; Lee, W.T.; Yenari, M.A.; Giffard, R.G. The 70 kDa heat shock protein suppresses matrix metalloproteinases in astrocytes. *Neuroreport* **2004**, *15*, 499–502. [CrossRef]
34. Zheng, Z.; Kim, J.Y.; Ma, H.; Lee, J.E.; Yenari, M.A. Anti-inflammatory effects of the 70 kDa heat shock protein in experimental stroke. *J. Cereb. Blood Flow Metab.* **2008**, *28*, 53–63. [CrossRef] [PubMed]
35. Sheppard, P.W.; Sun, X.; Khammash, M.; Giffard, R.G. Overexpression of heat shock protein 72 attenuates NF-κB activation using a combination of regulatory mechanisms in microglia. *PLoS Comput. Biol.* **2014**, *10*, e1003471. [CrossRef] [PubMed]
36. Evgen'ev, M.B.; Krasnov, G.S.; Nesterova, I.V.; Garbuz, D.G.; Karpov, V.L.; Morozov, A.V.; Snezhkina, A.V.; Samokhin, A.N.; Sergeev, A.; Kulikov, A.M.; et al. Molecular Mechanisms Underlying Neuroprotective Effect of Intranasal Administration of Human Hsp70 in Mouse Model of Alzheimer's Disease. *J. Alzheimers Dis.* **2017**, *59*, 1415–1426. [CrossRef] [PubMed]
37. Ferat-Osorio, E.; Sánchez-Anaya, A.; Gutiérrez-Mendoza, M.; Boscó-Gárate, I.; Wong-Baeza, I.; Pastelin-Palacios, R.; Pedraza-Alva, G.; Bonifaz, L.C.; Cortés-Reynosa, P.; Pérez-Salazar, E.; et al. Heat shock protein 70 down-regulates the production of toll-like receptor-induced pro-inflammatory cytokines by a heat shock factor-1/constitutive heat shock element-binding factor-dependent mechanism. *J. Inflamm.* **2014**, *11*, 19. [CrossRef]
38. Kim, N.; Kim, J.Y.; Yenari, M.A. Pharmacological induction of the 70-kDa heat shock protein protects against brain injury. *Neuroscience* **2015**, *284*, 912–919. [CrossRef] [PubMed]
39. Chen, Y.; Wang, B.; Liu, D.; Li, J.J.; Xue, Y.; Sakata, K.; Zhu, L.Q.; Heldt, S.A.; Xu, H.; Liao, F.F. Hsp90 chaperone inhibitor 17-AAG attenuates Aβ-induced synaptic toxicity and memory impairment. *J. Neurosci.* **2014**, *34*, 2464–2470. [CrossRef]
40. Chun, J.N.; Choi, B.; Lee, K.W.; Lee, D.J.; Kang, D.H.; Lee, J.Y.; Song, I.S.; Kim, H.I.; Lee, S.H.; Kim, H.S.; et al. Cytosolic Hsp60 is involved in the NF-kappaB-dependent survival of cancer cells via IKK regulation. *PLoS ONE* **2010**, *5*, e9422. [CrossRef]
41. Lewis, J.; Devin, A.; Miller, A.; Lin, Y.; Rodriguez, Y.; Neckers, L.; Liu, Z.G. Disruption of hsp90 function results in degradation of the death domain kinase, receptor-interacting protein (RIP), and blockage of tumor necrosis factor-induced nuclear factor-kappaB activation. *J. Biol. Chem.* **2000**, *275*, 10519–10526. [CrossRef]
42. Nakamura, H.; Minegishi, H. HSP60 as a drug target. *Curr. Pharm. Des.* **2013**, *19*, 441–451. [CrossRef]

43. Zuo, Y.; Wang, J.; Liao, F.; Yan, X.; Li, J.; Huang, L.; Liu, F. Inhibition of Heat Shock Protein 90 by 17-AAG Reduces Inflammation via P2X7 Receptor/NLRP3 Inflammasome Pathway and Increases Neurogenesis after Subarachnoid Hemorrhage in Mice. *Front. Mol. Neurosci.* **2018**, *11*, 401. [CrossRef]
44. Peterson, L.B.; Blagg, B.S. To fold or not to fold: Modulation and consequences of Hsp90 inhibition. *Future Med. Chem.* **2009**, *1*, 267–283. [CrossRef] [PubMed]
45. Theodoraki, M.A.; Caplan, A.J. Quality control and fate determination of Hsp90 client proteins. *Biochim. Biophys. Acta* **2012**, *1823*, 683–688. [CrossRef]
46. Luo, W.; Sun, W.; Taldone, T.; Rodina, A.; Chiosis, G. Heat shock protein 90 in neurodegenerative diseases. *Mol. Neurodegener.* **2010**, *5*, 24. [CrossRef] [PubMed]
47. Radons, J. The human HSP70 family of chaperones: Where do we stand? *Cell Stress Chaperones* **2016**, *21*, 379–404. [CrossRef] [PubMed]
48. Carson, M.J.; Thrash, J.C.; Walter, B. The cellular response in neuroinflammation: The role of leukocytes, microglia and astrocytes in neuronal death and survival. *Clin. Neurosci. Res* **2006**, *6*, 237–245. [CrossRef]
49. Sochocka, M.; Diniz, B.S.; Leszek, J. Inflammatory Response in the CNS: Friend or Foe? *Mol. Neurobiol.* **2017**, *54*, 8071–8089. [CrossRef] [PubMed]
50. Penke, B.; Bogár, F.; Crul, T.; Sántha, M.; Tóth, M.E.; Vígh, L. Heat Shock Proteins and Autophagy Pathways in Neuroprotection: From Molecular Bases to Pharmacological Interventions. *Int. J. Mol. Sci.* **2018**, *19*, 325. [CrossRef]
51. Campanella, C.; Bucchieri, F.; Merendino, A.M.; Fucarino, A.; Burgio, G.; Corona, D.F.; Barbieri, G.; David, S.; Farina, F.; Zummo, G.; et al. The odyssey of Hsp60 from tumor cells to other destinations includes plasma membrane-associated stages and Golgi and exosomal protein-trafficking modalities. *PLoS ONE* **2012**, *7*, e42008. [CrossRef] [PubMed]
52. Broquet, A.H.; Thomas, G.; Masliah, J.; Trugnan, G.; Bachelet, M. Expression of the molecular chaperone Hsp70 in detergent-resistant microdomains correlates with its membrane delivery and release. *J. Biol. Chem.* **2003**, *278*, 21601–21606. [CrossRef]
53. Mambula, S.S.; Stevenson, M.A.; Ogawa, K.; Calderwood, S.K. Mechanisms for Hsp70 secretion: Crossing membranes without a leader. *Methods* **2007**, *43*, 168–175. [CrossRef]
54. Kirkegaard, T.; Roth, A.G.; Petersen, N.H.; Mahalka, A.K.; Olsen, O.D.; Moilanen, I.; Zylicz, A.; Knudsen, J.; Sandhoff, K.; Arenz, C.; et al. Hsp70 stabilizes lysosomes and reverts Niemann-Pick disease-associated lysosomal pathology. *Nature* **2010**, *463*, 549–553. [CrossRef]
55. Lauwers, E.; Wang, Y.C.; Gallardo, R.; Van der Kant, R.; Michiels, E.; Swerts, J.; Baatsen, P.; Zaiter, S.S.; McAlpine, S.R.; Gounko, N.V.; et al. Hsp90 Mediates Membrane Deformation and Exosome Release. *Mol. Cell* **2018**, *71*, 689–702. [CrossRef] [PubMed]
56. Lachenal, G.; Pernet-Gallay, K.; Chivet, M.; Hemming, F.J.; Belly, A.; Bodon, G.; Blot, B.; Haase, G.; Goldberg, Y.; Sadoul, R. Release of exosomes from differentiated neurons and its regulation by synaptic glutamatergic activity. *Mol. Cell Neurosci.* **2011**, *46*, 409–418. [CrossRef] [PubMed]
57. Sun, Y.; Zheng, J.; Xu, Y.; Zhang, X. Paraquat-induced inflammatory response of microglia through HSP60/TLR4 signaling. *Hum. Exp. Toxicol.* **2018**, *37*, 1161–1168. [CrossRef] [PubMed]
58. Lane, C.A.; Hardy, J.; Schott, J.M. Alzheimer's disease. *Eur. J. Neurol.* **2018**, *25*, 59–70. [CrossRef] [PubMed]
59. Mangione, M.R.; Vilasi, S.; Marino, C.; Librizzi, F.; Canale, C.; Spigolon, D.; Bucchieri, F.; Fucarino, A.; Passantino, R.; Cappello, F.; et al. Hsp60, amateur chaperone in amyloid-beta fibrillogenesis. *Biochim. Biophys. Acta* **2016**, *1860*, 2474–2483. [CrossRef] [PubMed]
60. Cappello, F.; Angileri, F.; Conway de Macario, E.; Macario, A.J.L. Chaperonopathies and chaperonotherapy. Hsp60 as therapeutic target in cancer: Potential benefits and risks. *Curr. Pharm. Des.* **2013**, *19*, 452–457. [CrossRef]
61. Cappello, F.; Conway de Macario, E.; Marino Gammazza, A.; Bonaventura, G.; Carini, F.; Czarnecka, A.M.; Farina, F.; Zummo, G.; Macario, A.J.L. Hsp60 and human aging: Les liaisons dangereuses. *Front. Biosci.* **2013**, *18*, 626–637. [CrossRef]
62. Nemirovsky, A.; Fisher, Y.; Baron, R.; Cohen, I.R.; Monsonego, A. Amyloid beta-HSP60 peptide conjugate vaccine treats a mouse model of Alzheimer's disease. *Vaccine* **2011**, *29*, 4043–4050. [CrossRef]
63. Arispe, N.; De Maio, A. Memory Loss and the Onset of Alzheimer's Disease Could Be under the Control of Extracellular Heat Shock Proteins. *J. Alzheimers Dis.* **2018**, *63*, 927–934. [CrossRef]
64. De Mena, L.; Chhangani, D.; Fernandez-Funez, P.; Rincon-Limas, D.E. secHsp70 as a tool to approach amyloid-β42 and other extracellular amyloids. *Fly* **2017**, *11*, 179–184. [CrossRef] [PubMed]
65. Kakimura, J.; Kitamura, Y.; Takata, K.; Umeki, M.; Suzuki, S.; Shibagaki, K.; Taniguchi, T.; Nomura, Y.; Gebicke-Haerter, P.J.; Smith, M.A.; et al. Microglial activation and amyloid-beta clearance induced by exogenous heat-shock proteins. *FASEB J.* **2002**, *16*, 601–603. [CrossRef]
66. Maciejewski, A.; Ostapchenko, V.G.; Beraldo, F.H.; Prado, V.F.; Prado, M.A.; Choy, W.Y. Domains of STIP1 responsible for regulating PrPC-dependent amyloid-β oligomer toxicity. *Biochem. J.* **2016**, *473*, 2119–2130. [CrossRef]
67. Wirdefeldt, K.; Adami, H.O.; Cole, P.; Trichopoulos, D.; Mandel, J. Epidemiology and etiology of Parkinson's disease: A review of the evidence. *Eur. J. Epidemiol.* **2011**, *26*, S1–S58. [CrossRef]
68. Kim, S.D.; Allen, N.E.; Canning, C.G.; Fung, V.S.C. Parkinson disease. *Nat. Rev. Dis. Primers* **2018**, *159*, 173–193.
69. Spillantini, M.G.; Crowther, R.A.; Jakes, R.; Hasegawa, M.; Goedert, M. alpha-Synuclein in filamentous inclusions of Lewy bodies from Parkinson's disease and dementia with lewy bodies. *Proc. Natl. Acad. Sci. USA* **1998**, *95*, 6469–6473. [CrossRef] [PubMed]

70. Cheng, M.Y.; Hartl, F.U.; Martin, J.; Pollock, R.A.; Kalousek, F.; Neupert, W.; Hallberg, E.M.; Hallberg, R.L.; Horwich, A.L. Mitochondrial heat-shock protein hsp60 is essential for assembly of proteins imported into yeast mitochondria. *Nature* **1989**, *337*, 620–625. [CrossRef]
71. Pemberton, S.; Melki, R. The interaction of Hsc70 protein with fibrillar α-Synuclein and its therapeutic potential in Parkinson's disease. *Commun. Integr. Biol.* **2012**, *5*, 94–95. [CrossRef] [PubMed]
72. Bohush, A.; Bieganowski, P.; Filipek, A. Hsp90 and Its Co-Chaperones in Neurodegenerative Diseases. *Int. J. Mol. Sci.* **2019**, *20*, 4976. [CrossRef] [PubMed]
73. Luk, K.C.; Mills, I.P.; Trojanowski, J.Q.; Lee, V.M. Interactions between Hsp70 and the hydrophobic core of alpha-synuclein inhibit fibril assembly. *Biochemistry* **2008**, *47*, 12614–12625. [CrossRef] [PubMed]
74. Klucken, J.; Shin, Y.; Masliah, E.; Hyman, B.T.; McLean, P.J. Hsp70 Reduces alpha-Synuclein Aggregation and Toxicity. *J. Biol. Chem.* **2004**, *279*, 25497–25502. [PubMed]
75. Falsone, S.F.; Kungl, A.J.; Rek, A.; Cappai, R.; Zangger, K. The molecular chaperone Hsp90 modulates intermediate steps of amyloid assembly of the Parkinson-related protein alpha-synuclein. *J. Biol. Chem.* **2009**, *284*, 31190–31199. [PubMed]
76. Feng, M.; Zhang, L.; Liu, Z.; Zhou, P.; Lu, X. The expression and release of Hsp60 in 6-OHDA induced in vivo and in vitro models of Parkinson's disease. *Neurochem. Res.* **2013**, *38*, 2180–2189. [PubMed]
77. Lecours, C.; Bordeleau, M.; Cantin, L.; Parent, M.; Paolo, T.D.; Tremblay, M.È. Microglial Implication in Parkinson's Disease: Loss of Beneficial Physiological Roles or Gain of Inflammatory Functions? *Front. Cell Neurosci.* **2018**, *12*, 282.
78. Robberecht, W.; Philips, T. The changing scene of amyotrophic lateral sclerosis. *Nat. Rev. Neurosci.* **2013**, *14*, 248–264.
79. Rosen, D.R.; Siddique, T.; Patterson, D.; Figlewicz, D.A.; Sapp, P.; Hentati, A.; Donaldson, D.; Goto, J.; O'Regan, J.P.; Deng, H.X.; et al. Mutations in Cu/Zn superoxide dismutase gene are associated with familial amyotrophic lateral sclerosis. *Nature* **1993**, *362*, 59–62.
80. McAlary, L.; Aquilina, J.A.; Yerbury, J.J. Susceptibility of mutant SOD1 to form a destabilized monomer predicts cellular aggregation and toxicity but not in vitro aggregation propensity. *Front. Neurosci.* **2016**, *10*, 499.
81. San Gil, R.; Ooi, L.; Yerbury, J.J.; Ecroyd, H. The heat shock response in neurons and astroglia and its role in neurodegenerative diseases. *Mol. Neurodegener.* **2017**, *12*, 65.
82. Takeuchi, H.; Kobayashi, Y.; Yoshihara, T.; Niwa, J.; Doyu, M.; Ohtsuka, K.; Sobue, G. Hsp70 and Hsp40 improve neurite outgrowth and suppress intracytoplasmic aggregate formation in cultured neuronal cells expressing mutant SOD1. *Brain Res.* **2002**, *949*, 11–22.
83. Liu, J.; Shinobu, L.A.; Ward, C.M.; Young, D.; Cleveland, D.W. Elevation of the Hsp70 chaperone does not effect toxicity in mouse models of familial amyotrophic lateral sclerosis. *J. Neurochem.* **2005**, *93*, 875–882.
84. Apolloni, S.; Caputi, F.; Pignataro, A.; Amadio, S.; Fabbrizio, P.; Ammassari-Teule, M.; Volonté, C. Histamine Is an Inducer of the Heat Shock Response in SOD1-G93A Models of ALS. *Int. J. Mol. Sci.* **2019**, *20*, 3793. [CrossRef] [PubMed]
85. Gifondorwa, D.J.; Robinson, M.B.; Hayes, C.D.; Taylor, A.R.; Prevette, D.M.; Oppenheim, R.W.; Caress, J.; Milligan, C.E. Exogenous delivery of heat shock protein 70 increases lifespan in a mouse model of amyotrophic lateral sclerosis. *J. Neurosci.* **2007**, *27*, 13173–13180. [CrossRef] [PubMed]
86. Tytell, M. Release of heat shock proteins (Hsps) and the effects of extracellular Hsps on neural cells and tissues. *Int. J. Hyperth.* **2005**, *21*, 445–455. [CrossRef] [PubMed]
87. Giffard, R.G.; Yenari, M.A. Many mechanisms for hsp70 protection from cerebral ischemia. *J. Neurosurg. Anesth.* **2004**, *16*, 53–61. [CrossRef] [PubMed]
88. Taylor, A.R.; Robinson, M.B.; Gifondorwa, D.J.; Tytell, M.; Milligan, C.E. Regulation of heat shock protein 70 release in astrocytes: Role of signaling kinases. *Dev. Neurobiol.* **2007**, *67*, 1815–1829. [CrossRef] [PubMed]
89. Fontaine, S.N.; Zheng, D.; Sabbagh, J.J.; Martin, M.D.; Chaput, D.; Darling, A.; Trotter, J.H.; Stothert, A.R.; Nordhues, B.A.; Lussier, A.; et al. DnaJ/Hsc70 chaperone complexes control the extracellular release of neurodegenerative-associated proteins. *EMBO J.* **2016**, *35*, 1537–1549. [CrossRef]
90. Robinson, M.B.; Taylor, A.R.; Gifondorwa, D.J.; Tytell, M.; Milligan, C.E. Exogenous Hsc70, but not thermal preconditioning, confers protection to motoneurons subjected to oxidative stress. *Dev. Neurobiol.* **2008**, *68*, 1–17. [CrossRef]
91. Kim, S.D.; Fung, V.S. An update on Huntington's disease: From the gene to the clinic. *Curr Opin Neurol* **2014**, *27*, 477–483. [CrossRef]
92. McColgan, P.; Tabrizi, S.J. Huntington's disease: A clinical review. *Eur. J. Neurol.* **2018**, *25*, 24–34. [CrossRef]
93. Wang, H.Q.; Xu, Y.X.; Zhao, X.Y.; Zhao, H.; Yan, J.; Sun, X.B.; Guo, J.C.; Zhu, C.Q. Overexpression of F(0)F(1)-ATP synthase alpha suppresses mutant huntingtin aggregation and toxicity in vitro. *Biochem. Biophys. Res. Commun.* **2009**, *390*, 1294–1298. [CrossRef]
94. Novoselova, T.V.; Margulis, B.A.; Novoselov, S.S.; Sapozhnikov, A.M.; van der Spuy, J.; Cheetham, M.E.; Guzhova, I.V. Treatment with extracellular HSP70/HSC70 protein can reduce polyglutamine toxicity and aggregation. *J. Neurochem.* **2005**, *94*, 597–606. [CrossRef] [PubMed]
95. Baldo, B.; Weiss, A.; Parker, C.N.; Bibel, M.; Paganetti, P.; Kaupmann, K. A screen for enhancers of clearance identifies huntingtin as a heat shock protein 90 (Hsp90) client protein. *J. Biol. Chem.* **2012**, *287*, 1406–1414. [CrossRef] [PubMed]
96. Orozco-Díaz, R.; Sánchez-Álvarez, A.; Hernández-Hernández, J.M.; Tapia-Ramírez, J. The interaction between RE1-silencing transcription factor (REST) and heat shock protein 90 as new therapeutic target against Huntington's disease. *PLoS ONE* **2019**, *14*, e0220393. [CrossRef] [PubMed]

97. Noh, K.M.; Hwang, J.Y.; Follenzi, A.; Athanasiadou, R.; Miyawaki, T.; Greally, J.M.; Bennett, M.V.; Zukin, R.S. Repressor element-1 silencing transcription factor (REST)-dependent epigenetic remodeling is critical to ischemia-induced neuronal death. *Proc. Natl. Acad. Sci. USA* **2012**, *109*, E962–E971. [CrossRef] [PubMed]
98. Ekimova, I.V.; Nitsinskaya, L.E.; Romanova, I.V.; Pastukhov, Y.F.; Margulis, B.A.; Guzhova, I.V. Exogenous protein Hsp70/Hsc70 can penetrate into brain structures and attenuate the severity of chemically-induced seizures. *J. Neurochem.* **2010**, *115*, 1035–1044. [CrossRef] [PubMed]
99. Behrends, C.; Langer, C.A.; Boteva, R.; Böttcher, U.M.; Stemp, M.J.; Schaffar, G.; Rao, B.V.; Giese, A.; Kretzschmar, H.; Siegers, K.; et al. Chaperonin TRiC promotes the assembly of polyQ expansion proteins into nontoxic oligomers. *Mol. Cell* **2006**, *23*, 887–897. [CrossRef] [PubMed]
100. Levin, M.C.; Douglas, J.N.; Meyers, L.; Lee, S.; Shin, Y.; Gardner, L.A. Neurodegeneration in multiple sclerosis involves multiple pathogenic mechanisms. *Degener. Neurol. Neuromuscul. Dis.* **2014**, *4*, 49–63. [CrossRef]
101. Quintana, F.J.; Farez, M.F.; Izquierdo, G.; Lucas, M.; Cohen, I.R.; Weiner, H.L. Antigen microarrays identify CNS-produced autoantibodies in RRMS. *Neurology* **2012**, *78*, 532–539. [CrossRef]
102. Cwiklinska, H.; Mycko, M.P.; Luvsannorov, O.; Walkowiak, B.; Brosnan, C.F.; Raine, C.S.; Selmaj, K.W. Heat shock protein 70 associations with myelin basic protein and proteolipid protein in multiple sclerosis brains. *Int. Immunol.* **2003**, *15*, 241–249. [CrossRef]
103. Mansilla, M.J.; Costa, C.; Eixarch, H.; Tepavcevic, V.; Castillo, M.; Martin, R.; Lubetzki, C.; Aigrot, M.S.; Montalban, X.; Espejo, C. Hsp70 regulates immune response in experimental autoimmune encephalomyelitis. *PLoS ONE* **2014**, *9*, e105737. [CrossRef]
104. Mycko, M.P.; Cwiklinska, H.; Walczak, A.; Libert, C.; Raine, C.S.; Selmaj, K.W. A heat shock protein gene (Hsp70.1) is critically involved in the generation of the immune response to myelin antigen. *Eur. J. Immunol.* **2008**, *38*, 1999–2013. [CrossRef] [PubMed]
105. Aquino, D.A.; Capello, E.; Weisstein, J.; Sanders, V.; Lopez, C.; Tourtellotte, W.W.; Brosnan, C.F.; Raine, C.S.; Norton, W.T. Multiple sclerosis: Altered expression of 70- and 27-kDa heat shock proteins in lesions and myelin. *J. Neuropathol. Exp. Neurol.* **1997**, *56*, 664–672. [CrossRef]
106. Cid, C.; Alvarez-Cermeño, J.C.; Camafeita, E.; Salinas, M.; Alcázar, A. Antibodies reactive to heat shock protein 90 induce oligodendrocyte precursor cell death in culture. Implications for demyelination in multiple sclerosis. *FASEB J.* **2004**, *18*, 409–411. [CrossRef] [PubMed]
107. Kalmar, B.; Lu, C.H.; Greensmith, L. The role of heat shock proteins in Amyotrophic Lateral Sclerosis: The therapeutic potential of Arimoclomol. *Pharmacol. Ther.* **2014**, *141*, 40–54. [CrossRef] [PubMed]
108. Benatar, M.; Wuu, J.; Andersen, P.M.; Atassi, N.; David, W.; Cudkowicz, M.; Schoenfeld, D. Randomized, double-blind, placebo-controlled trial of arimoclomol in rapidly progressive *SOD1* ALS. *Neurology* **2018**, *90*, e565–e574. [CrossRef] [PubMed]
109. Kuta, R.; Larochelle, N.; Fernandez, M.; Pal, A.; Minotti, S.; Tibshirani, M.; St Louis, K.; Gentil, B.J.; Nalbantoglu, J.N.; Hermann, A.; et al. Depending on the stress, histone deacetylase inhibitors act as heat shock protein co-inducers in motor neurons and potentiate arimoclomol, exerting neuroprotection through multiple mechanisms in ALS models. *Cell Stress Chaperones* **2020**, *25*, 173–191. [CrossRef] [PubMed]

applied sciences

MDPI

Brief Report

A Pilot Longitudinal Evaluation of MicroRNAs for Monitoring the Cognitive Impairment in Pediatric Multiple Sclerosis

Nicoletta Nuzziello [1], Arianna Consiglio [1], Rosa Gemma Viterbo [2], Flavio Licciulli [1], Sabino Liuni [1], Maria Trojano [2] and Maria Liguori [1,*]

1 National Research Council of Italy, Institute of Biomedical Technologies, Bari Section, 70125 Bari, Italy; nicoletta.nuzziello@gmail.com (N.N.); arianna.consiglio@ba.itb.cnr.it (A.C.); flavio.licciulli@ba.itb.cnr.it (F.L.); sabino.liuni@ba.itb.cnr.it (S.L.)

2 Department of Basic Sciences, Neurosciences and Sense Organs, University of Bari, 70121 Bari, Italy; rgviterbo@uniba.it (R.G.V.); mstojano@uniba.it (M.T.)

* Correspondence: maria.liguori@cnr.it; Tel.: +39-080-592-9663; Fax: +39-080-592-9690

Received: 30 September 2020; Accepted: 20 November 2020; Published: 22 November 2020

check for updates

Abstract: MicroRNAs (miRNAs), a class of non-coding RNAs, seem to play a key role in complex diseases like multiple sclerosis (MS), as well as in many cognitive functions associated with the disease. In a previous cross-sectional evaluation on pediatric MS (PedMS) patients, the expression of some miRNAs and their target genes were found to be associated with the scores of some neuropsychiatric tests, thus suggesting that they may be involved in early processes of cognitive impairment. To verify these data, we asked the same patients to be re-evaluated after a 1-year interval; unfortunately, only nine of them agreed to this further clinical and molecular analysis. The main results showed that 13 differentially expressed miRNAs discriminated the two time-points. Among them, the expression of miR-182-5p, miR-320a-3p, miR-744-5p and miR-192-5p significantly correlated with the attention and information processing speed performances, whereas the expression of miR-182-5p, miR-451a, miR-4742-3p and miR-320a-3p correlated with the expressive language performances. The analysis of mRNA expression uncovered 58 predicted and/or validated miRNA-target pairs, including 23 target genes, some of them already associated with cognitive impairment, such as the transducing beta like 1 X-linked receptor-1 gene (*TBL1XR1*), correlated to disorders of neurodevelopment; the Snf2 related CREBBP activator protein gene (*SRCAP*) that was found implicated in a rare form of dementia; and the glia maturation factor beta gene (*GMFB*), which has been reported to be implicated in neurodegeneration and neuroinflammation. No molecular pathways involving the most targeted genes survived the adjustment for multiple data. Although preliminary, these findings showed the feasibility of the methods also applied to longitudinal investigations, as well as the reliability of the obtained results. These findings should be confirmed in larger PedMS cohorts in order to identify early markers of cognitive impairment, towards which more efficient therapeutic efforts can be addressed.

Keywords: pediatric multiple sclerosis; microRNA; gene target; high-throughput next-generation sequencing

1. Introduction

Multiple sclerosis (MS) is a demyelinating autoimmune disease of the central nervous system (CNS) that usually affects young adults [1]. As in other complex multifactorial diseases, in MS the possibility to identify reliable biomarkers that are predictive of the disease course has so far been elusive due to the multifactorial nature of MS, which involves environmental factors, several genes and their epigenetic interactions [2]. Lately, the evidence that up to 5% of patients with MS experience their first clinical symptoms/signs before the age of 18 years [3] has offered the opportunity to investigate the very early steps of this heterogeneous disease, also helping to disentangle the complex molecular scenario of the latest phases and possibly leading to more targeted therapeutic efforts. To this aim, integrated approaches using deep sequencing technologies of next-generation sequencing (HT-NGS) provide valuable support, e.g., to assess the differential expression of microRNAs (miRNAs), which are epigenetic factors, the functions of which have been associated with MS and its phenotypic features [4].

Following these suggestions, in a previous study performed on a selected group of 19 pediatric MS (PedMS), 12 upregulated and 1 downregulated miRNAs resulted, compared to 20 age-matched controls [5]; six miRNAs were confirmed in an adult MS population (n = 58 patients versus n = 20 controls) suggesting that they may be considered as distinctive markers of the disease independently of their age [6]. It is also worth noting that the expression of 11 miRNAs correlated with the scores obtained in different cognitive tests; although none of them passed the 0.05 threshold of statistical significance, they seemed to point in the direction of a possible impact of miRNAs in PedMS cognitive performance [7].

To verify these findings, we asked the same PedMS population to be re-evaluated after 1-year by following the same clinical, neuropsychological and molecular protocols.

2. Subjects and Methods

The criteria for the recruitment of PedMS patients were detailed previously [5]. The study was approved by the Ethical Committee of Azienda Ospedaliera Policlinico, University of Bari. Since the PedMS patients were under the age of 18, their legal guardians signed written informed consent forms (according to the Declaration of Helsinki) at the time of the evaluations.

For the clinical, molecular, bioinformatic and biostatistical analyses, we followed the same methods used for the first study [5]. Briefly, the clinical disability of each patient was scored by expanded disability status scale (EDSS) [8], whereas their neuropsychological performances were tested by a validated battery exploring several cognitive domains [9]. Peripheral blood samples of PedMS patients were collected in PAXgene blood RNA tubes, coded, anonymized and frozen at −20 °C until use. Total RNA was isolated using a PAXgene blood RNA kit (PreAnalytiX Qiagen/BD, Hilden, Germany) and sequenced by an Illumina HiSeq2500 platform, after the opportune processing of smallRNA (sRNA) and mRNA libraries. The HT-NGS data were analyzed with an integrated bioinformatics pipeline, developed by our group, that includes both miRNAs and genes expression estimation using mirDeep2, Bowtie, STAR and RSEM; differential expression analysis using edgeR and DESeq2; miRNA-target gene interaction evaluation using both experimentally validated microRNA–target interaction databases (DIANA–Tarbase and miRTarBase) and miRNA-target genes prediction algorithms (miRanda, RNAhybrid, RNA22, miRDB and TargetScan); pathway enrichment analysis using Database for Annotation, Visualization and Integrated Discovery (DAVID v6.8 https://david.ncifcrf.gov) tool that provides a large series of functional annotation tools and pathways databases, such as KEGG.

The EDSS score and the individual scores obtained by each administered neuropsychiatric test, at the study entry and after the follow-up period, were analyzed by a non-parametric Wilcoxon signed rank test for paired samples. Spearman's correlation test was also applied to miRNA expressions and to the scores of the cognitive tests.

3. Results

Only 9 out of 19 PedMS patients (3 males and 6 females) agreed to this further evaluation after a mean interval of 1.3 ± 0.6 SD years from the first exams (baseline demographic and clinical characteristics are summarized in Table 1A); five of them had started Interferon beta-1a (IFNβ-1a) treatment in an average of 140 days (range 21–576 days) from the study entry.

The neuropsychiatric scores registered at the second time point suggested a trend to deterioration particularly in the domains of attention and information processing speed, visual–spatial memory, executive functioning and expressive language (Table 1B). However, no significant changes of clinical and cognitive abilities were recorded within the observational interval.

Table 1. A: demographic and clinical features of the investigated pediatric multiple sclerosis (PedMS) patients (at baseline). **B:** Mean scores (± SD) obtained at the neuropsychiatric tests (baseline and follow-up).

A.		
PedMS (No. 9)	**Mean ± SD**	
Age at onset (years)	14.1 ± 2.1	
Disease duration (years)	2.2 ± 2.0	
EDSS	2.7 ± 0.9 (range 1.5–4.5)	
FSS	3.1 ± 2.0	
CDI	5.4 ± 1.8	
Schooling (years)	10.2 ± 2.0	
B.		
Cognitive Domains and Tests	**Baseline (Mean ± SD)**	**Follow-Up (Mean ± SD)**
Verbal memory		
SRT-LTS	30.4 ± 12	30.8 ± 10.8
SRT-CLTR	22 ± 12.6	23.6 ± 11.2
SRT-D	6.7 ± 2.2	6.8 ± 2.1
Visual-spatial memory		
SPART	21 ± 4.9	20.6 ± 3.2
SPART-D	7.3 ± 2.2	7.1 ± 1.6
Attention, IPS		
SDMT	40.9 ± 7.9	42.9 ± 8.9
TMT-A	41.8 ± 12.7	41.8 ± 11.8
TMT-B	97.4 ± 46.3	93.6 ± 45.7
Executive functioning		
TOL	25.4 ± 5.2	25 ± 4.7
Expressive language		
SVFT	24.6 ± 10.6	22.6 ± 8.6
PVFT	15.6 ± 5.7	17.4 ± 4.9

Abbreviations: EDSS = expanded disability status scale; FSS = fatigue severity scale; CDI = children depression inventory; IPS = information processing speed; SRT-LTS = selective reminding test long-term storage; SRT-CLTR = selective reminding test consistent long-term retrieval; SRT-D = selective reminding test—delayed; SPART = spatial recall test; SPART-D= spatial recall test—delayed; SDMT = symbol digit modalities test; TMT = trail making test (TMT-A and TMT-B); TOL = Tower of London test; SVFT = semantic verbal fluency test; PVFT= phonemic verbal fluency test.

The multidimensional scaling (MSD) plot of dispersion index between the miRNA expressions of all PedMS subjects at baseline and at follow-up reported a definite separation between T0 and T12 that was due to 13 significantly differentially expressed (DE) miRNAs (adjusted p-value < 0.05, absolute log2 fold change >1) (Table 2; Supplementary Figure S1). Since the clinical features of the examined PedMS patients did not significantly change during the interval, we could not verify whether any of these miRNAs might be considered as suggestive markers of the disease progression. However, it is interesting to note that the expression of miR-182-5p, miR-320a-3p, miR-744-5p and miR-192-5p significantly correlated with the scores at follow-up of the Trailing Making Test A (TMT-A) whereas the expression of miR-182-5p, miR-451a, miR-4742-3p and miR-320a-3p correlated with the scores obtained at the SVFT test (Spearman's rho correlation test, $p < 0.05$; Supplementary Table S1 for details). Unfortunately, no adjustments for variables such as sex and concomitant Disease-Modifying Treatments (DMTs) (type and duration of treatment) were allowed due to the small sample size.

Table 2. miRNAs whose expression (logFC) results were statistically significantly different between the observational interval (adjusted *p*-value< 0.05) and their published associations with MS and/or other neurodegenerative disorders.

TRANSCRIPT ID	LogFC	Adjusted *p*-Value	Reported Associations (References, See Supplementary Files)
hsa-miR-26b-5p	2.254118	0.009	Upregulated in AD [10] Upregulated in AD vs. FTD vs. HC [11] Upregulated in ADHD [12] Downregulated in SPMS [13] Downregulated in AD [14] Downregulated in ALS [15] *Related to A-beta expression in cortical neurons animal model* [16]
hsa-miR-127-3p	−6.144506	<0.0001	Downregulated in FTD [17] Upregulated in PPMS exosomes [4]
hsa-miR-182-5p	1.341916	0.01	Upregulated in PedMS [5] Downregulated in prion disease and AD [18] Downregulated in ALS [19] Inhibition of oxidative stress and apoptosis in inflammatory disease [20]
hsa-miR-192-5p	1.14595	0.03	Upregulated in ALS vs. MS [21] Regulatory factor in AD [22] *Related to A-beta expression in cortical neurons animal model* [16]
hsa-miR-320a-3p	−0.862674	0.03	Downregulated in schizophrenia [23]
hsa-miR-451a	1.435193	0.02	Upregulated in RR [4] Upregulated in YOAD [24] Upregulated in depression [25] Downregulated in ALS [15] Downregulated in MDD [26] Downregulated in AD [27]
hsa-miR-486-5p	−1.332208	0.03	Upregulated (NC) in MS with low BPV [19] Upregulated in HD [28]
hsa-miR-501-3p	−1.562031	0.002	Downregulated in sera AD [29] Upregulated in brain tissue—progression [29]
hsa-miR-576-5p	1.74505	0.004	Upregulated in Relapse MS and ON vs. HC [30] Downregulated in NMOSD [30] Downregulated in active inflammation [31]
hsa-miR-744-5p	−0.817595	0.04	
hsa-miR-1275	−1.241061	0.01	
hsa-miR-4742-3p	1.418434	0.03	Downregulated in ASD [32]
hsa-miR-548o-3p	1.391054	0.03	

Abbreviations: FC = fold change; AD = Alzheimer's disease; FTD = frontotemporal dementia; HC = healthy controls; ADHD = attention deficit hyperactive disorders; SPMS = secondary progressive multiple sclerosis; ALS = amyotrophic lateral sclerosis; PPMS = primary progressive multiple sclerosis; PedMS = pediatric multiple sclerosis; BD = bipolar disorder; RRMS = relapsing remitting multiple sclerosis; ASD = autism spectrum disorders; MDD = major depressive disorders; HD = Huntington's disease; NMOSD = neuromyelitis optica spectrum disorders.

The analysis of mRNA expression from RNA-Seq data identified 64 DE genes (adjusted *p*-value < 0.05, absolute log2 fold change >1). The DE miRNAs and genes were used for the miRNA-target analysis, under the assumption that changes in miRNA expression may influence the expression of their target/s. This analysis uncovered 58 predicted and/or validated miRNA-target pairs, including 23 target genes (Figure 1, Supplementary Table S2); most of them were genes already reported associated with several neuropsychiatric and neurodegenerative disorders, such as *TBL1XR1* (targeted by miR-576-5p, miR-26b-5p, miR-182-5p, miR-451a, miR-4742-3p, miR-548o-3p), *SRCAP* (by miR-127-3p, miR-1275, miR-320a-3p, miR-486-5p, miR-744-5p) and *GMFB* (targeted by miR-182-5p, miR-548o-3p), as discussed below.

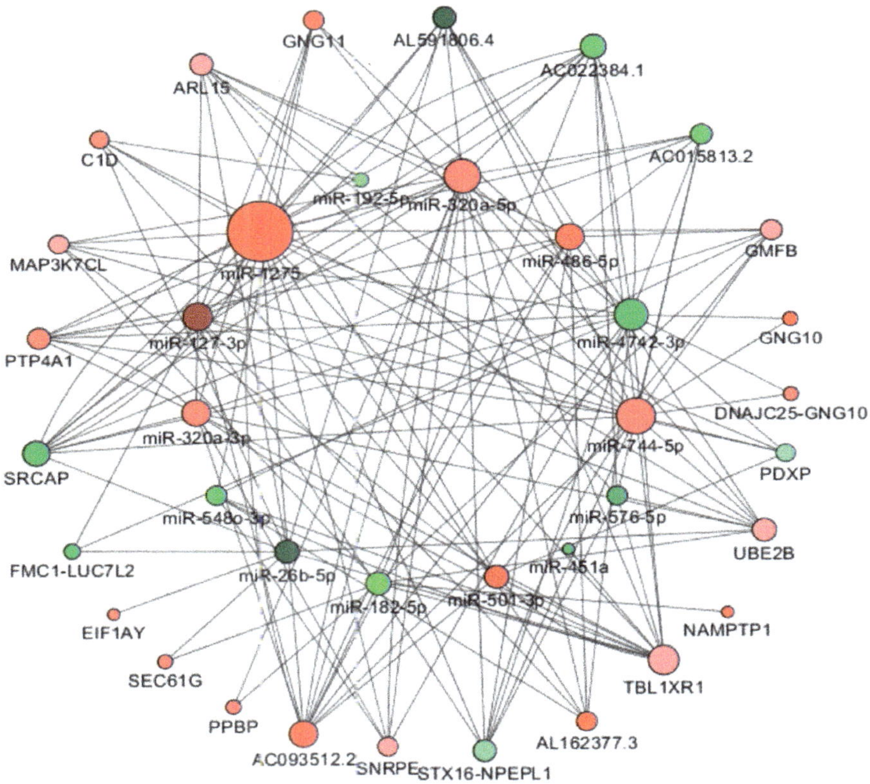

Figure 1. Graphical representation of miRNA-target genes interaction network using Cytoscape v3.7.2. The node intensity color is proportional to the log2 fold change values in the DE analysis (red: under-regulated; green: upregulated); the node size is proportional to the number of miRNA/mRNA connections. Please note that only downregulated genes were selected as targets of upregulated miRNAs (in the lower semicircle) and vice versa for downregulated miRNAs (in the upper semicircle).

No molecular pathways involving the most targeted genes survived the adjustment for multiple data; the only trend to significance was obtained by the KEGG chemokine signaling pathway that enclosed the genes *GNG10*, *GNG11* and *PPBB* ($p = 0.034$, adjusted $p > 0.05$).

4. Discussion

The possibility that miRNA pathways (including their mRNA gene targets) are involved in the modulation of MS during the developmental age carries the chance to actively interfere with novel and selective therapeutic strategies in order to prevent detrimental effects due to the progression of the disease or its disabling symptoms like the cognitive impairment. In the same direction, genetic variants (SNPs) within the genes coding for suggestive miRNAs or located in a miRNA-binding site of a target gene were evaluated as well, following the suggestion that they may impact the "normal" miRNA functions [33].

The hypothesis that there may be a molecular signature associated with the cognitive dysfunctions also in PedMS is more than plausible [7]. Several pieces of evidence also reported significant changes of selective miRNA/mRNA expression under the treatment available for MS, such as IFNβ, one of the first approved DMTs since 1996, in the adult MS population [34]. These data, together with the numerous clinical evidences showing significant improvement of the MS cognitive performances under the IFNβ therapy [35], suggested that novel therapeutic strategies should be considered, with the final aim to target the dysregulated molecular pathways possibly identified during the MS cognitive performances.

Unfortunately, the small number of PedMS patients available for this further evaluation did not allow us to reach conclusive findings on this hypothesis, even in the evaluation of possible interference of the concomitant IFNβ in both the transcriptomic and the cognitive profiles of the studied PedMS cohort. However, in our view, several useful pieces of information were provided as well.

First of all, the neuropsychiatric evaluation showed that over an interval of 1-year, in the domains of attention and information processing speed, visual–spatial memory, executive functioning and expressive language, results were more compromised in the PedMS subjects than in the general MS population [36,37]. It is also worthy to note that for some other cognitive tests, the follow-ups registered slight improvements in the correspondent functions, without any concomitant cognitive rehabilitation, confirming once more that the interval for performing longitudinal investigation should be longer to avoid any interferences due to learning curve processing, especially in younger MS patients, as reported in other investigations [38].

Looking at the molecular results, our data demonstrated that the two time points differed significantly in the expression of 13 miRNAs. Among them, it is worthwhile to note that the expressions of miR-182-5p and miR-320a(-3p) were preliminarily associated with cognitive dysfunctions in the original cohort of 19 PedMS patients [7], whereas those of miR-26b-5p, miR-192-5p, miR-451a, miR-486-5p and miR-576-5p were already mentioned in adult MS populations [4,13,19,21]. Some of these significant miRNAs were reported jointly dysregulated in previous analyses, as they were part of common molecular pathways; this was observed, e.g., for miR-26b-5p, miR-182-5p and miR-451a in ALS [15]. Further investigations in larger sample sizes and possibly over a reasonably longer interval, as well as functional studies in MS and other neurodegenerative diseases, will shed light on the possibility that these significant miRNAs may be considered markers of some clinical changes over time.

Other intriguing tips derived from the analysis of mRNA expressions that led to several target genes of interest, some of which were already reported as associated with diseases that impair the cognitive performances. We cannot speculate on these preliminary results either, since they also suffer for the same weaknesses of the small-RNA data, as summarized below. However, in our view, they might be considered as key actors of a more complex molecular pathway underlying the cognitive functions that, especially in MS, still needs to be disentangled. For example, the most influenced gene (targeted by 6 miRNAs) is the transducing beta-like 1X-linked receptor 1 (*TBL1XR1*), which has been found associated with autism [39], mental retardation and disorders of neurodevelopment [40]. Five miRNAs targeted the Snf2 related CREBBP activator protein gene (*SRCAP*) that was found implicated in a rare form of AD [41]. Interestingly, *SRCAP* impacts the transcription of CREB, involved with its transcription factors in memory retention and consolidation by hippocampal neurogenesis. Finally, although targeted only by two miRNAs, we would direct attention to the glia maturation factor beta gene (*GMFB*), a growth and differentiation factor for both glia and neurons, since it has been considered as a candidate therapeutic target for both neuroinflammatory and neurodegenerative diseases, given its role in mediating apoptosis and neuroinflammation [42].

Limitations of the study have been mostly already mentioned; despite the efforts of the investigators and the high motivations of the PedMS patients and their families during the first time period, only some of them were available for the follow-up. A longer interval would had been preferred for other inclusions, as well as for revealing further changes in the cognitive performances; however, this time extension was not possible due to several technical reasons (related to the incoming deadline of the financed project). Another unfortunate event was that we were not able to examine the second set of MRI scans, so we could not verify the associations between the miRNA expressions and the regional volumetric measurements of the brain, together with the molecular data, as we did at the baseline [7].

On the other hand, this strategy provided a unique opportunity to analyze simultaneously for the first time the miRNAs and their targeted mRNAs in a selected pediatric population of MS patients, which was confirmed as an amazing and reliable source of data and information. In fact, the study of miRNome and Targetome in pediatric patients enabled us to understand how genetic dysregulation during the developmental age could lead to autoimmune disorders and how a differential genetic expression may represent a substrate of the phenotype heterogeneity observed during the MS lifespan. Since children with MS can be considered environmentally naïve, we believe that their genetic load may provide more significant information than adult patients, especially about those factors that modulate their phenotypes.

Further investigations and collaborative efforts will be critical, as well as comparisons with other cognitive disorders of developmental ages, in order to test the possibility that they may share some of the molecular pathways implicated in cognitive dysfunctions.

Supplementary Materials: The following are available online at http://www.mdpi.com/2076-3417/10/22/8274/s1, Figure S1: Multidimensional scaling plot of Dispersion Index in longitudinal miRNA expressions, Table S1: Spearman's correlations between miRNA expressions and scores obtained at the individual cognitive tests, Table S2: Significant miRNAs/mRNAs pairs resulted from the analysis.

Author Contributions: M.L. conceived and designed the experiments, performed the neurological exams of the patients, supervised the study and wrote the manuscript; N.N. performed the molecular experiments and the preliminary analysis of the data, and significantly contribute to the manuscript; R.G.V. performed the neuropsychological exams of the patients; A.C. and F.L. performed the bioinformatics/biostatistics analysis of the data with the contribution of S.L.; as Chief of the MS Outpatients Clinic, M.T. supervised the clinical data. All authors have read and agreed to the published version of the manuscript.

Funding: This research was funded by Fondazione Italiana Sclerosi Multipla (FISM), Grant no. 2014/R/10.

Acknowledgments: The Authors are grateful to the MS patients and their families who kindly agreed to be part of the project.

Conflicts of Interest: Authors declare no conflicts of interests.

References

1. Compston, A.; Coles, A.J. Multiple sclerosis. *Lancet* **2008**, *372*, 1502–1517. [CrossRef]
2. Olsson, T.; Barcellos, L.F.; Alfredsson, L. Interactions between genetic, lifestyle and environmental risk factors for multiple sclerosis. *Nat. Rev. Neurol.* **2017**, *13*, 25–36. [CrossRef] [PubMed]
3. Banwell, B. Multiple sclerosis in children. *Handb. Clin. Neurol.* **2014**, *122*, 427–441. [CrossRef] [PubMed]
4. Ebrahimkhani, S.; Vafaee, F.; Young, P.E.; Hur, S.S.J.; Hawke, S.; Devenney, E.; Beadnall, H.; Barnett, M.H.; Suter, C.M.; Buckland, M.E. Exosomal microRNA signatures in multiple sclerosis reflect disease status. *Sci. Rep.* **2017**, *7*, 1–10. [CrossRef]
5. Liguori, M.; Nuzziello, N.; Licciulli, F.; Consiglio, A.; Simone, M.; Viterbo, R.G.; Creanza, T.M.; Ancona, N.; Tortorella, C.; Margari, L.; et al. Combined microRNA and mRNA expression analysis in pediatric multiple sclerosis: An integrated approach to uncover novel pathogenic mechanisms of the disease. *Hum. Mol. Genet.* **2018**, *27*, 66–79. [CrossRef]
6. Nuzziello, N.; Vilardo, L.; Pelucchi, P.; Consiglio, A.; Liuni, S.; Trojano, M.; Liguori, M. Investigating the Role of MicroRNA and Transcription Factor Co-regulatory Networks in Multiple Sclerosis Pathogenesis. *Int. J. Mol. Sci.* **2018**, *19*, 3652. [CrossRef]

7. Liguori, M.; Nuzziello, N.; Simone, M.; Amoroso, N.; Viterbo, R.G.; Tangaro, S.; Consiglio, A.; Giordano, P.; Bellotti, R.; Trojano, M. Association between miRNAs expression and cognitive performances of Pediatric Multiple Sclerosis patients: A pilot study. *Brain Behav.* **2019**, *9*, e01199. [CrossRef]
8. Kurtzke, J.F. Rating neurologic impairment in multiple sclerosis: An expanded disability status scale (EDSS). *Neurology* **1983**, *33*, 1444. [CrossRef]
9. Amato, M.P.; Goretti, B.; Ghezzi, A.; Lori, S.; Zipoli, V.; Moiola, L.; Falautano, M.; De Caro, M.F.; Viterbo, R.; Patti, F.; et al. Cognitive and psychosocial features in childhood and juvenile MS: Two-year follow-up. *Neurology* **2010**, *75*, 1134–1140. [CrossRef]
10. Chang, W.-S.; Wang, Y.-H.; Zhu, X.-T.; Wu, C.-J. Genome-Wide Profiling of miRNA and mRNA Expression in Alzheimer's Disease. *Med. Sci. Monit.* **2017**, *23*, 2721–2731. [CrossRef]
11. Denk, J.; Oberhauser, F.; Kornhuber, J.; Wiltfang, J.; Fassbender, K.; Schroeter, M.L.; Volk, A.E.; Diehl-Schmid, J.; Prudlo, J.; Danek, A.; et al. Specific serum and CSF microRNA profiles distinguish sporadic behavioural variant of frontotemporal dementia compared with Alzheimer patients and cognitively healthy controls. *PLoS ONE* **2018**, *13*, e0197329. [CrossRef] [PubMed]
12. Sánchez-Mora, C.; Artigas, M.S.; Garcia-Martínez, I.; Pagerols, M.; Rovira, P.; Richarte, V.; Corrales, M.; Fadeuilhe, C.; Padilla, N.; De La Cruz, X.; et al. Epigenetic signature for attention-deficit/hyperactivity disorder: Identification of miR-26b-5p, miR-185-5p, and miR-191-5p as potential biomarkers in peripheral blood mononuclear cells. *Neuropsychopharmacology* **2019**, *44*, 890–897. [CrossRef] [PubMed]
13. Sanders, K.A.; Benton, M.C.; Lea, R.A.; Maltby, V.E.; Agland, S.; Griffin, N.; Scott, R.J.; Tajouri, L.; Lechner-Scott, J. Next-generation sequencing reveals broad down-regulation of microRNAs in secondary progressive multiple sclerosis CD4+ T cells. *Clin. Epigenet.* **2016**, *8*, 87. [CrossRef] [PubMed]
14. Chen, J.; Qi, Y.; Liu, C.-F.; Lu, J.-M.; Shi, J.; Shi, Y. MicroRNA expression data analysis to identify key miRNAs associated with Alzheimer's disease. *J. Gene Med.* **2018**, *20*, e3014. [CrossRef] [PubMed]
15. Liguori, M.; Nuzziello, N.; Introna, A.; Consiglio, A.; Licciulli, F.; D'Errico, E.; Scarafino, A.; Distaso, E.; Simone, I.L. Dysregulation of MicroRNAs and Target Genes Networks in Peripheral Blood of Patients With Sporadic Amyotrophic Lateral Sclerosis. *Front. Mol. Neurosci.* **2018**, *11*, 288. [CrossRef] [PubMed]
16. Dursun, E.; Candaş, E.; Yılmazer, S.; Gezen-Ak, D. Amyloid Beta 1–42 Alters the Expression of miRNAs in Cortical Neurons. *J. Mol. Neurosci.* **2019**, *67*, 181–192. [CrossRef]
17. Piscopo, P.; Grasso, M.; Puopolo, M.; D'Acunto, E.; Talarico, G.; Crestini, A.; Gasparini, M.; Campopiano, R.; Gambardella, S.; Castellano, A.E.; et al. Circulating miR-127-3p as a Potential Biomarker for Differential Diagnosis in Frontotemporal Dementia. *J. Alzheimer's Dis.* **2018**, *65*, 455–464. [CrossRef]
18. Boese, A.S.; Saba, R.; Campbell, K.; Majer, A.; Medina, S.; Burton, L.; Booth, T.F.; Chong, P.; Westmacott, G.R.; Dutta, S.M.; et al. MicroRNA abundance is altered in synaptoneurosomes during prion disease. *Mol. Cell. Neurosci.* **2016**, *71*, 13–24. [CrossRef]
19. Regev, K.; Healy, B.C.; Khalid, F.; Paul, A.; Chu, R.; Tauhid, S.; Tummala, S.; Diaz-Cruz, C.; Raheja, R.; Mazzola, M.A.; et al. Association Between Serum MicroRNAs and Magnetic Resonance Imaging Measures of Multiple Sclerosis Severity. *JAMA Neurol.* **2017**, *74*, 275–285. [CrossRef]
20. Qin, S.-B.; Peng, D.-Y.; Lu, J.-M.; Ke, Z.-P. MiR-182-5p inhibited oxidative stress and apoptosis triggered by oxidized low-density lipoprotein via targeting toll-like receptor 4. *J. Cell. Physiol.* **2018**, *233*, 6630–6637. [CrossRef]
21. Raheja, R.; Regev, K.; Healy, B.C.; Mazzola, M.A.; Beynon, V.; Von Glehn, F.; Paul, A.; Diaz-Cruz, C.; Gholipour, T.; Glanz, B.I.; et al. Correlating serum micrornas and clinical parameters in amyotrophic lateral sclerosis. *Muscle Nerve* **2018**, *58*, 261–269. [CrossRef] [PubMed]
22. Rahman, R.; Islam, T.; Turanli, B.; Zaman, T.; Faruquee, H.M.; Rahman, M.; Mollah, N.H.; Nanda, R.K.; Arga, K.Y.; Gov, E.; et al. Network-based approach to identify molecular signatures and therapeutic agents in Alzheimer's disease. *Comput. Biol. Chem.* **2019**, *78*, 431–439. [CrossRef] [PubMed]
23. Wang, Y.; Wang, J.; Guo, T.; Peng, Y.; Wang, K.; Bai, K.; Huang, Y. Screening of schizophrenia associated miRNAs and the regulation of miR-320a-3p on integrin β1. *Medicine* **2019**, *98*, e14332. [CrossRef] [PubMed]

24. McKeever, P.M.; Schneider, R.; Taghdiri, F.; Weichert, A.; Multani, N.; Brown, R.A.; Boxer, A.L.; Karydas, A.; Miller, B.; Robertson, J.; et al. MicroRNA Expression Levels Are Altered in the Cerebrospinal Fluid of Patients with Young-Onset Alzheimer's Disease. *Mol. Neurobiol.* **2018**, *55*, 8826–8841. [CrossRef]
25. Camkurt, M.A.; Acar Şenel, T.; Coşkun, S.; Gunes, M.L.; Güneş, S.; Yılmaz, M.F.; Görür, A.; Tamer, L. Comparison of plasma MicroRNA levels in drug naive, first episode depressed patients and healthy controls. *J. Psychiatr. Res.* **2015**, *69*, 67–71. [CrossRef] [PubMed]
26. Wan, Y.; Liu, Y.; Wang, X.; Wu, J.; Liu, K.; Zhou, J.; Liu, L.; Zhang, C. Identification of Differential MicroRNAs in Cerebrospinal Fluid and Serum of Patients with Major Depressive Disorder. *PLoS ONE* **2015**, *10*, e0121975. [CrossRef]
27. Gámez-Valero, A.; Campdelacreu, J.; Vilas, D.; Ispierto, L.; Reñé, R.; Álvarez, R.; Armengol, M.P.; Borràs, F.E.; Beyer, K. Exploratory study on microRNA profiles from plasma-derived extracellular vesicles in Alzheimer's disease and dementia with Lewy bodies. *Transl. Neurodegener.* **2019**, *8*, 1–17. [CrossRef]
28. Hoss, A.G.; Bs, V.N.L.; Frank, S.; Hadzi, T.C.; Myers, R.H.; Dsc, J.C.L. Study of plasma-derived miRNAs mimic differences in Huntington's disease brain. *Mov. Disord.* **2015**, *30*, 1961–1964. [CrossRef]
29. Hara, N.; Kikuchi, M.; Miyashita, A.; Hatsuta, H.; Saito, Y.; Kasuga, K.; Murayama, S.; Ikeuchi, T.; Kuwano, R. Serum microRNA miR-501-3p as a potential biomarker related to the progression of Alzheimer's disease. *Acta Neuropathol. Commun.* **2017**, *5*, 1–9. [CrossRef]
30. Chen, J.; Zhu, J.; Wang, Z.; Yao, X.; Wu, X.; Liu, F.; Zheng, W.; Li, Z.; Lin, A. MicroRNAs Correlate with Multiple Sclerosis and Neuromyelitis Optica Spectrum Disorder in a Chinese Population. *Med. Sci. Monit.* **2017**, *23*, 2565–2583. [CrossRef]
31. Mansouri, L.; Lundwall, K.; Moshfegh, A.; Jacobson, S.H.; Lundahl, J.; Spaak, J. Vitamin D receptor activation reduces inflammatory cytokines and plasma MicroRNAs in moderate chronic kidney disease—A randomized trial. *BMC Nephrol.* **2017**, *18*, 1–7. [CrossRef] [PubMed]
32. Ander, B.P.; Barger, N.; Stamova, B.; Sharp, F.R.; Schumann, C.M. Atypical miRNA expression in temporal cortex associated with dysregulation of immune, cell cycle, and other pathways in autism spectrum disorders. *Mol. Autism* **2015**, *6*, 37. [CrossRef] [PubMed]
33. Rhead, B.; Shao, X.; Graves, J.S.; Chitnis, T.; Waldman, A.T.; Lotze, T.; Schreiner, T.; Belman, A.; Krupp, L.; Greenberg, B.M.; et al. miRNA contributions to pediatric-onset multiple sclerosis inferred from GWAS. *Ann. Clin. Transl. Neurol.* **2019**, *6*, 1053–1061. [CrossRef] [PubMed]
34. Fattahi, M.; Rezaei, N.; Nematalahi, F.S.; Shaygannejad, V.; Fouladi, S.; Karimi, L.; Fathi, F.; Dehghani, L.; Mirmosayyeb, O.; Eskandari, N. MicroRNA-29b variants and MxA expression change during interferon beta therapy in patients with relapsing-remitting multiple sclerosis. *Mult. Scler. Relat. Disord.* **2019**, *35*, 241–245. [CrossRef]
35. Benešová, Y.; Tvaroh, A. Cognition and fatigue in patients with relapsing multiple sclerosis treated by subcutaneous interferon β-1a: An observational study SKORE. *Ther. Adv. Neurol. Disord.* **2017**, *10*, 18–32. [CrossRef]
36. Nunan-Saah, J.H.; Paulraj, S.R.; Waubant, E.; Krupp, L.B.; Gomez, R.G. Neuropsychological correlates of multiple sclerosis across the lifespan. *Mult. Scler. J.* **2015**, *21*, 1355–1364. [CrossRef]
37. Sokolov, A.A.; Grivaz, P.; Bove, R. Cognitive Deficits in Multiple Sclerosis: Recent Advances in Treatment and Neurorehabilitation. *Curr. Treat. Options Neurol.* **2018**, *20*, 53. [CrossRef]
38. Öztürk, Z.; Gucuyener, K.; Acar, A.; Şebnem, S.; Konuşkan, G.D.; Konuşkan, B.; Dikmen, A.U.; Anlar, B. Cognitive functions in pediatric multiple sclerosis: 2-years follow-up. *Neurol. Res.* **2020**, *42*, 159–163. [CrossRef]
39. O'Roak, B.J.; Vives, L.; Fu, W.; Egertson, J.D.; Stanaway, I.B.; Phelps, I.G.; Carvill, G.; Kumar, A.; Lee, C.; Ankenman, K.; et al. Multiplex Targeted Sequencing Identifies Recurrently Mutated Genes in Autism Spectrum Disorders. *Science* **2012**, *338*, 1519–1622. [CrossRef]
40. Yamamoto, T.; Imaizumi, T.; Yamamoto-Shimojima, K.; Lu, Y.; Yanagishita, T.; Shimada, S.; Chong, P.F.; Kira, R.; Ueda, R.; Ishiyama, A.; et al. Genomic backgrounds of Japanese patients with undiagnosed neurodevelopmental disorders. *Brain Dev.* **2019**, *41*, 776–782. [CrossRef]

41. Vardarajan, B.N.; Tosto, G.; Lefort, R.; Yu, L.; Bennett, D.A.; De Jager, P.L.; Barral, S.; Reyes-Dumeyer, D.; Nagy, P.L.; Lee, J.H.; et al. Ultra-rare mutations in SRCAP segregate in Caribbean Hispanic families with Alzheimer disease. *Neurol. Genet.* **2017**, *3*, e178. [CrossRef] [PubMed]
42. Fan, J.; Fong, T.; Chen, X.; Chen, C.; Luo, P.; Xie, H. Glia maturation factor-β: A potential therapeutic target in neurodegeneration and neuroinflammation. *Neuropsychiatr. Dis. Treat.* **2018**, *14*, 495–504. [CrossRef] [PubMed]

Publisher's Note: MDPI stays neutral with regard to jurisdictional claims in published maps and institutional affiliations.

applied sciences

MDPI

Review

Brain Tumor-Derived Extracellular Vesicles as Carriers of Disease Markers: Molecular Chaperones and MicroRNAs

Alessandra Maria Vitale [1,2,†], Radha Santonocito [1,†], Giuseppe Vergilio [1], Antonella Marino Gammazza [1], Claudia Campanella [1], Everly Conway de Macario [3], Fabio Bucchieri [1], Alberto J. L. Macario [2,3] and Celeste Caruso Bavisotto [1,2,*]

1 Department of Biomedicine, Neuroscience and Advanced Diagnostics (BIND), Section of Human Anatomy, University of Palermo, 90127 Palermo, Italy; alessandramaria.vitale@unipa.it (A.M.V.); radha.santonocito@unipa.it (R.S.); peppe04@tiscali.it (G.V.); antonella.marinogammazza@unipa.it (A.M.G.); claudia.campanella@unipa.it (C.C.); fabio.bucchieri@unipa.it (F.B.)

2 Euro-Mediterranean Institute of Science and Technology (IEMEST), 90139 Palermo, Italy; Ajlmacario@som.umaryland.edu

3 Department of Microbiology and Immunology, School of Medicine, University of Maryland at Baltimore-Institute of Marine and Environmental Technology (IMET), Baltimore, MD 21202, USA; econwaydemacario@som.umaryland.edu

* Correspondence: celestebavisotto@gmail.com or celeste.carusobavisotto@unipa.it; Tel.: +39-091-2386-5700

† These authors contributed equally to this work.

Received: 6 September 2020; Accepted: 1 October 2020; Published: 5 October 2020

check for updates

Abstract: Primary and metastatic brain tumors are usually serious conditions with poor prognosis, which reveal the urgent need of developing rapid diagnostic tools and efficacious treatments. To achieve these objectives, progress must be made in the understanding of brain tumor biology, for example, how they resist natural defenses and therapeutic intervention. One resistance mechanism involves extracellular vesicles that are released by tumors to meet target cells nearby or distant via circulation and reprogram them by introducing their cargo. This consists of different molecules among which are microRNAs (miRNAs) and molecular chaperones, the focus of this article. miRNAs modify target cells in the immune system to avoid antitumor reaction and chaperones are key survival molecules for the tumor cell. Extracellular vesicles cargo reflects the composition and metabolism of the original tumor cell; therefore, it is a source of markers, including the miRNAs and chaperones discussed in this article, with potential diagnostic and prognostic value. This and their relatively easy availability by minimally invasive procedures (e.g., drawing venous blood) illustrate the potential of extracellular vesicles as useful materials to manage brain tumor patients. Furthermore, understanding extracellular vesicles circulation and interaction with target cells will provide the basis for using this vesicle for delivering therapeutic compounds to selected tumor cells.

Keywords: brain tumors; extracellular vesicles; miRNA; molecular chaperones; diagnostic tools; drug delivery

1. Introduction

Brain tumors entail high mortality and morbidity worldwide, with 296,851 new cases and 241,037 deaths in 2018, according to the Global Cancer Observatory [1]. Currently, the diagnosis of brain tumors is based on neuroimaging techniques complemented by biopsy. However, these methods are not always applicable because of the type and/or the localization of the tumor. Therapeutic choices include surgery, and radio-, chemo-, and immunotherapy. However, several complications can occur, such as venous thrombosis, pulmonary embolism, intracranial bleeding, wound and systemic infection, seizures, depression, deteriorating neurologic conditions, and adverse drug reaction [2]. This worrying scenario points to the need for novel methods that will allow earlier and more accurate diagnosis and better patient monitoring and treatment than those currently available.

Remarkable achievements in the molecular and genetic fields have opened new frontiers in the management of brain tumors, including the identification of new diagnostic and prognostic molecular biomarkers released by the tumor. Noteworthy examples of these biomarkers are carried by extracellular vesicles (EVs), released by tumors. EVs, which are considered mediators of intercellular communication, carry bioactive molecules between cells close by or distant, affecting physiological and pathological processes in recipient cells [3–9]. MicroRNAs (miRNAs) and molecular chaperones are among the bioactive molecules carried by EVs that affect, either directly or indirectly, cancer initiation, cell proliferation and growth, and metastasization [10–14]. Consequently, specific miRNAs and molecular chaperones have been suggested as targets for analysis that can provide useful information for differential diagnosis, assessing prognosis and response to treatment, and for developing novel therapies [15]. One added advantage is that EVs can be obtained from biological fluids with minimally invasive procedures.

Here, we discuss the role of EVs in the pathogenesis of brain tumors, summarizing current knowledge regarding their miRNAs and molecular chaperones cargo. We examine their role in the development of brain tumors, and their impact on the tumor microenvironment.

2. Intercellular Communication: EVs

Cells can share between them biological information using lipids, proteins, or nucleic acids as mediators, which are carried within small, nano-to-micrometer lipid-membraned EVs released by them [16]. EVs are present in biological fluids, including blood, urine, milk, saliva, and cerebrospinal, amniotic, and seminal fluids [17].

EVs play a key role in intercellular communication in physiological and pathological cellular processes [18,19] and are considered a valuable source of useful biomarkers [20–22].

The International Society for Extracellular Vesicles (ISEV) encourages the use of the term "extracellular vesicles (EVs)" as a generic term for all secreted vesicles, considering the lack of consensus for the identification of specific markers to distinguish between the different subtypes of EVs [23]. Formerly, the nomenclature assigned to EVs subgroups was based on differences in the size and formation mechanism [23], and classified them into three main groups: 1. shedding microvesicles, with a size range of 100–1000 nm [24]; 2. apoptotic bodies (1–5 μm diameter), released into the extracellular environment by dying cells [25]; and 3. exosomes, small vesicles between 30–150 nm in size produced from the endosomal compartment [23].

However, distinguishing between the various groups is still problematic, so in this review, we refer to EVs, regardless of the classification used in the works cited, which often do not report specific data that would allow a precise identification of EVs subtypes.

It is known that EVs are involved in numerous physiological and pathological processes, including immune response, signal transduction, tumor progression, and inflammation [4]. Consequently, EVs have potential as carriers of molecular biomarkers for diagnosis and prognosis in a range of conditions, including cardiovascular, renal, and neurodegenerative diseases, and cancer [26], or as transporters of therapeutic agents [27,28].

EVs carry proteins, lipids, mRNA, and miRNAs, and their contents depend on the type and function of the cell in which they originate [29].

EVs contain substantial amounts of different RNA species, such as miRNA, mRNA, ribosomal RNA, long noncoding RNA, transfer RNA, and small nuclear RNA [30]. Although mRNAs are the most abundant class of RNAs in EVs, many studies have focused on miRNAs because of their apparent role in cancer progression. Vesicular miRNAs were suggested as novel diagnostic, prognostic, and predictive biomarkers in several common cancers [31].

Exosomes are highly enriched in proteins with various functions, such as proteins associated with cell membrane interaction, invasion, and fusion, e.g., the tetraspanins CD9, CD63, CD81, and CD82. The tetraspanins CD9, CD63, and CD81 are used as specific markers [23]. Furthermore, EVs contain proteins involved in maintenance of cell homeostasis and protection of cells against stress/apoptosis, e.g., molecular chaperones, some of which are called heat shock proteins (Hsp). Hsp60, Hsp70, and Hsp90 are the most commonly present Hsps in EVs and are among the most used vesicular markers [32].

3. miRNAs and Molecular Chaperones in Brain Tumors

Brain tumors are a heterogeneous group of neoplasms that differ in etiology, morphology, clinical manifestations, prognosis, and treatment. They were initially classified by the World Health Organization (WHO) according to their histologic features and presumed cellular origin. This type of characterization was the main tool used for many years in the diagnosis and management of patients, including decision on treatment strategy [33–36] (Table 1). However, increasing genetic and epigenetic discoveries have shown that a classification only based on histopathological findings fell short of the mark. For this reason, in 2016, the WHO formulated an updated classification, which went beyond the old principle of diagnosis based only on microscopy and incorporated molecular parameters to define brain tumors entities. This new classification allows a more objective and accurate diagnosis, ensuring a more accurate assessment of prognosis and treatment response than the old one [37].

Brain tumors can be either benign (noncancerous), or malignant (cancerous). The latter can be distinguished into primary, arising directly within the brain, and secondary, namely metastatic brain tumors derived from other parts of the body that have metastasized to the brain [2,38–40] (Table 1). Up to half of the metastatic brain tumors derive from lung cancer. Other types of tumors that commonly spread to the brain include melanoma, and breast, kidney, and colon cancer, although determining the site of the primary tumor is often difficult [38,41–44].

Genetic and environmental factors are implicated in the onset of primary brain tumors. The former are exemplified by neurofibromatosis types 1 and 2, adenomatous polyposis syndrome, tuberous sclerosis, nevoid basal cell carcinoma syndrome, Turcot syndrome, Li–Fraumeni syndrome, and von Hippel–Lindau syndrome, whereas those linked to environmental factors are exemplified by those caused by ionizing radiation, and they are all considerably less frequent than secondary tumors (Table 1) [45–51] (Table 1).

Table 1. Characteristics of the most common primary brain tumors.

Tumor	Cell of Origin	Molecular Features	Clinical Features	Ref.
Astrocytomas Diffuse astrocytoma (WHO Grade II); Anaplastic astrocytoma (WHO Grade III); Pilocytic astrocytoma (WHO Grade I); Glioblastoma multiforme (WHO Grade IV); Gliosarcoma (WHO Grade IV) [1].	Astrocyte	p16 deletion p53 mutation PTEN mutation EGFR amplification IDH1 mutation	Astrocytomas, oligodendrogliomas, and ependimomas are gliomas that originate in the glia and are extremely invasive and malignant. Main symptoms are headache, seizures, nausea, vomiting, disturbed vision, tingling sensations, weakness, difficult ambulation.	[52–64]
Oligodendrogliomas Oligodendroglioma (WHO Grade II); -Anaplastic Oligodendroglioma; (WHO Grade III).	Oligodendrocyte	PEG3 deletion EGFR amplification p53 mutation p16 deletion IDH1, IDH2 mutation		
Ependymomas Subependymoma (WHO Grade I); -Myxopapillary ependymoma (WHO Grade I); Ependymoma (WHO Grade II); Ependymoma, RELA fusion-positive (WHO Grade II or III); Anaplastic ependymoma (WHO Grade III).	Ependymal cell	NF2 mutation MT3 underexpression hTERT overexpression miR-485-5p downregulation IGF1 upregulation p16 deletion EGFR amplification		

Table 1. *Cont.*

Tumor	Cell of Origin	Molecular Features	Clinical Features	Ref.
Meningiomas Meningioma (WHO Grade I); Atypical meningioma (WHO Grade II); Anaplastic meningioma (WHO Grade III).	Meningeal cell	NF2 mutation DAL1 loss PTEN mutation p16 deletion EGFR overexpression	Meningiomas are tumors of the meninges. Main symptoms are headache, seizures, psychotic-motor disabilities, mental weakening, personality changes, visual disorders, language dysfunction.	[37,65–69]
Medulloblastomas Medulloblastoma (WHO Grade IV); Desmoplastic/Nodular Medulloblastoma (WHO Grade IV); Medulloblastoma with Extensive Nodularity (WHO Grade IV); Anaplastic Medulloblastoma (WHO Grade IV).	Neuron	p53 mutation TRKC, ERBB2, FSTL5 overexpression PTCH1, CTNNB1 mutation MYC amplification DDX3X mutation	Medulloblastomas are tumors of the cerebellum. Main symptoms are headache, morning vomiting, ataxia.	[70–78]

[1] WHO, World Health Organization.

Current diagnostic approaches are based on imaging methods with subsequent histological examination of a biopsy. However, these approaches are limited by tumor localization and heterogeneity. Treatment choices can vary depending on tumor type and location, malignancy potential and patient's conditions, and include surgery, radiotherapy, chemotherapy, or a combination. Complete safe surgical resection, followed by radio/chemotherapy, represents the most common initial treatment for many primary brain tumors. The main aims are to achieve an accurate histological diagnosis, define the tumor's molecular genotype, reduce the mass effect and tumor burden, improve patient's quality of life, and prolong survival time [2,79,80].

Despite efforts to develop new therapeutic strategies, including surgical procedures, and radio-, chemo-, and immunotherapies, brain tumors continue to be a substantial source of morbidity and mortality worldwide, a situation compounded by late diagnosis and the development of resistance to anticancer agents [81–83].

For this reason, repeated attempts have been made over the last few years to identify specific biomarkers that could be detected/measured using noninvasive methods and that would allow early diagnosis and disease monitoring, including controlling the response to treatment [84].

Among the diagnostic and prognostic markers currently under investigation, miRNAs and molecular chaperones released by EVs hold promise, considering their roles in physiology and pathology. For example, they are implicated in the regulation of the cellular proteome at transcriptional and post-transcriptional levels. Thus, it is likely that cancer cells may use miRNAs and chaperones by delivering them to other cells and influence them in ways favoring tumorigenesis.

3.1. miRNAs in Brain Tumors

MicroRNAs are short noncoding single stranded RNA (ssRNA) molecules, 19–25 nucleotides long, which regulate the expression of target genes post-transcriptionally by affecting either the stability or the translation of their mRNA [85].

The biogenesis of miRNAs consists of two highly regulated cleavage events [86]. The first one, occurring within the nucleus, generates a long hairpin-shaped RNA molecule called pre-miRNA [87–92] which is exported to the cytoplasm, in which the second step occurs, producing a shorter double stranded RNA [93–97]. One of the strands, the star (*) strand or passenger strand, is degraded. The other strand (the guide strand or mature miRNA) forms the miRNA-induced silencing complex (miRISC) that specifically recognizes a target mRNA, and downregulates gene expression by repression of translation or by mRNA cleavage [98–104]. However, in recent years it became clear that the passenger strand may not be degraded and could act as miRNA. Thus, according to a more recent nomenclature proposed by the miRBase registry, the two miRNA strands produced after the second cleavage are referred to as predominant product (indicated without *) and the strand from the opposite arm of the precursor (indicated with *). When the data are not sufficient to determine which sequence is predominant, the strands are indicated as 5p, i.e., present in the forward (5′–3′) position, and 3p, i.e., located in the reverse (3′–5′) position [105].

Since their discovery over 30 years ago in the nematode *Caenorhabditis elegans* [106,107], a myriad of new miRNAs have been identified and annotated in the miRbase registry, and their number continuously increases thanks to the development of new high-throughput sequencing technologies and computational and bioinformatics prediction methods, which facilitate identification of miRNAs targets and their biological functions [108–114].

Currently, in humans over 2000 miRNAs have been annotated and validated, which regulate the vast majority of protein encoding genes, and thus most if not all biological events [115–121]. For this reason, any alteration of a miRNA normal expression profile was often related to pathology, including cancer as firstly suggested by the depletion or downregulation of miR-15a and miR-16a genes in the majority of B-cell chronic lymphocytic leukemia [122]. Typically, when miRNAs expression is amplified in cancer cells, they function as oncogenes and promote cancer development by negatively regulating tumor suppressor genes and/or genes that control cell differentiation or apoptosis. Conversely, when downregulated, miRNAs act as tumor suppressors and may inhibit tumorigenesis by regulating oncogenes and/or genes that control cell differentiation or apoptosis [123,124]. There are several mechanisms inducing miRNAs dysregulation in cancer, such as (i) miRNAs genes amplification or deletion [122,125]; (ii) abnormal transcriptional regulation of miRNA genes [126,127]; (iii) epigenetic alterations, such as aberrant DNA methylation and histone acetylation of miRNAs genes, which, in turn, affect miRNAs levels [128,129]; and (iv) defects in miRNAs biogenesis and maturation pathways, which can alter their expression [130,131].

MicroRNAs with expression levels different in the tumoral tissues as compared with the normal tissue counterparts were identified for several tumor types, for instance glioblastoma [132–134].

More than 70 percent of all brain tumors are gliomas that are classified according to the type of the glial cell involved, and include astrocytoma (astrocytoma, anaplastic astrocytoma, and glioblastoma), ependymomas (anaplastic ependymoma, myxopapillary ependymoma, and subependymoma), and oligodendrogliomas (oligodendroglioma, anaplastic oligodendroglioma, and anaplastic oligoastrocytoma) (Table 2). Microarray studies on miRNAs have shown significant changes of their expression profile in gliomas, both in children and in adults [135,136] (Table 2).

Compared to normal brain tissues, miRNA-155 (miR-155) was found overexpressed in glioma tissues, and its overexpression was associated with poor overall survival rates, suggesting that monitoring its expression levels could be a way to assess prognosis [137]. The positive correlation between miR-155 expression level and glioma malignancy was further established by in vitro and in vivo studies [138]. In vitro, miR-155 promoted tumor cells proliferation, invasion, and migration by downregulating two components of the mitogen-activated protein kinase (MAPK) signaling pathway and, in turn, enhancing secretion of matrix metalloproteinases 2 (MMP2) and MMP9 [138]. Moreover, it competed with miR-185 to induce ANXA2 (annexin A2), which exhibited oncogenic functions in glioblastoma multiforme (GBM) [139]. In vivo studies showed that miR-155 facilitated the progression of glioblastoma and confers drug resistance by modulating Six1 expression [140]. These results confirm the role of miR-155 as oncomiR and suggest its potential use as biomarker and as an anticancer drug target. In fact, its knockdown sensitized glioma cells to temozolomide, a common anticancer drug, through the induction of MAPK13, MAPK14, and Six1, and mediated oxidative stress and apoptosis [138,140] (Table 2).

The other two miRNAs proposed as oncomiRs, and potential prognostic biomarkers, are miR-221 and miR-222, whose high levels of expression are positively correlated with glioma aggressiveness and poor prognosis [141,142]. Among miR-221 and miR-222 targets that are involved in these protumorigenic effects are the tissue inhibitor of metalloproteinase (TIMP3), which is downregulated concomitantly with miR-221 and miR-222 overexpression, promoting glioma cell invasion [141], and the Akt pathway, which is activated, promoting cancer cell proliferation [143]. Conversely, cosuppression of miR-221 and miR-222 suppresses human glioma-cell growth and proliferation by a mechanism involving the upregulation of the cell cycle inhibitor p27Kip1 both in vitro and in vivo [144]. Moreover, as reported for the oncomiR-155, downregulation of miR-221 and miR222 sensitizes glioma cells to temozolomide by increasing the expression of proapoptotic factors [145] (Table 2).

There are also several examples of miRNAs considered tumor suppressors of malignant gliomas. Low plasma level of miR-185 is a signature of a glioma that correlates with poor survival [146]. Its inhibition after miR-155 overexpression in GBM promoted ANXA2 expression and tumor growth and progression [139]. Similarly, serum miR-205 expression was significantly lower in patients with glioma than in healthy controls, as well as in other brain tumor cohorts, and its serum level appeared inversely correlated with pathological grades and overall survival, since patients with glioma at an advanced pathological grade (grade III or IV) and a higher miR-205 serum level had a longer overall survival than those with a lower miR-205 serum concentration [147]. Thus, miR-185 and miR-205 were identified as tumor suppressors and were proposed as biomarkers with predictive prognostic potential to be used with noninvasive tools for monitoring cancer progression and response to treatment [146–149] (Table 2).

Glioblastoma multiforme (WHO grade IV astrocytoma) is characterized by poorly differentiated glial cells with polymorphism, nuclear atypia, and high mitotic activity, and is the most common malignant primary brain tumor, with an incidence of 3.19 cases per 100,000 person/year and a remarkably poor prognosis due to the still limited therapeutic options [150]. Therefore, the identification of new diagnostic and prognostic biomarkers is necessary to develop novel and personalized therapeutic treatments. Also, in this case, great interest was elicited by miRNAs, which often show an altered expression level in GMB patients compared to healthy controls, as already discussed for miR-155 and miR-185. It was observed that the serum level of miR-203 is decreased in GBM patients compared with low grade glioma (LGG) patients and healthy controls, and it positively correlates with poor overall survival [151]. These results confirm the tumor-suppressor activity of miR-203 in GMB, whose low expression level in two human GBM cell lines was previously shown to induce their epithelial–mesenchymal transition and confer them chemoresistance [152] (Table 2).

Similarly, miR-605 has a reduced expression level in GMB tissues and cell lines, and this was correlated with patients' poor survival [153]. Conversely, increased levels of miR-605 inhibited cancer cell proliferation and growth in vitro and in vivo, by directly targeting SOX9 (SRY-box 9) and by inhibiting the activation of the PI3K (phosphatidylinositol 3-kinase)/Akt (protein kinase B) pathway [153] (Table 2).

Compared to GBM, much less is known about circulating miRNAs as useful indicators for diagnosis and prognosis of meningiomas. In addition, for these tumors, miRNAs have been identified that act as oncogenes or tumor suppressors. For instance, miR-200a has been implicated in the pathogenesis of meningiomas and was found downregulated in sporadic benign human meningioma tumors (WHO grade I), compared to the arachnoid tissues from which these tumors arise [154]. This miRNA functions as a potential tumor suppressor since its upregulation inhibits Wnt/β-catenin signaling, involved in cell proliferation, through two complementary mechanisms: a) direct targeting of the β-catenin mRNA, which reduces the levels of β-catenin, acting as the main activator of the Wnt signaling; and b) targeting the mRNAs for ZEB1 (Zinc Finger E-Box Binding Homeobox 1) and SIP1 (Smad Interacting Protein 1), which negatively regulate the expression of the E-cadherin gene, with a consequent upregulation of E-cadherin levels and sequestration of β-catenin [154]. Another mir-200a target is the nonmuscle myosin heavy chain IIb (NMHCIIb), involved in regulation of cells motility. NMHCIIb downregulation concomitantly with miR-200a overexpression in malignant meningioma cells significantly reduced the rate of cancer cells migration, and thus tumor invasiveness [155].

The expression level of miR-145 appears significantly reduced in atypical and anaplastic tumors as compared with benign meningiomas [156]. In vitro, the overexpression of miR-145 reduced meningioma cells proliferation and motility, thanks to the associated downregulation of collagen type V alpha (COL5A1) and induced apoptotic cell death [156]. These effects translated into a decreased growth of an orthotopic tumor in a nude mice model, with reduction in tumor cell infiltration upon overexpression of miR-145 [156] (Table 2).

The microRNA miR-335 is overexpressed in meningiomas and acts as an oncomiR. It has been shown that elevated levels of miR-335 in vitro increased tumor cell growth by directly targeting the signaling pathway of the tumor suppressor Rb1, whereas reduction of the miR-335 levels had the opposite effect on tumor growth and progression, leading to cell cycle arrest in the G0/G1 phase [157].

The oncogenic role of miR-335 was observed also in astrocytoma, a type of glioma deriving from astrocytes or astroglial precursors; it was overexpressed in astrocytoma cells, promoting their growth and invasiveness by targeting the tumor suppressor disheveled-associated activator of morphogenesis 1 (Daam1), a member of the formin protein family acting downstream of Wnt signaling, and responsible for the regulation of cell polarization, migration, proliferation, and tissue morphogenesis during embryonic development [158]. On the contrary, miR-335 inhibition suppressed growth and induced apoptosis of astrocytoma cells in vitro and in vivo, suggesting its potential use as therapeutic target [158] (Table 2).

The microRNA miR-224 level was found higher in meningioma tissues compared to normal brain and was positively correlated with advanced pathological grade [159]. miR-224 could be a promising therapeutic target for treating malignant meningiomas, since its downregulation in vitro suppressed cell growth and increased apoptosis through the activation of the ERG2-BAK-induced apoptosis pathway [159] (Table 2). Six miRNAs (miR-106a-5p, miR-219-5p, miR-375, miR-409-3p miR-197, and miR-224) were identified in serum from patients with meningioma [160]. The serum levels of miR-106a-5p, miR-219-5p, miR-375, and miR-409-3p were increased in meningioma patients compared to healthy controls and decrease after tumor removal. On the contrary, the serum levels of miR-197 and miR-224 were markedly decreased in meningioma patients but significantly increased in the postoperative samples of the same patients. Therefore, the use of this panel of miRNAs was suggested as potentially useful for the diagnosis and the evaluation of clinical outcomes during management of meningioma patients [160] (Table 2).

Similarly, very recently the use of miRNAs profiling was proposed as novel tool to predict meningiomas recurrence, and improve patients' clinical management [161].

MicroRNA profiling in brain tumor biology has been conducted considering not only the miRNAs expressed in malignant tissues and in blood from patients, but also the miRNAs released in the cerebrospinal fluid (CSF). CSF is considered the ideal source of nervous tissue-specific miRNAs to use as diagnostic biomarkers for brain tumors, since it is in direct contact with the entire central nervous system, and has the advantage of containing fewer miRNAs than blood plasma or serum, which, flowing throughout the body, collect miRNAs generated by all tissues [162–167]. Recently, a screening was conducted, using CSF samples from patients with glioblastoma, low-grade glioma, meningioma, and brain metastasis, and from nontumor patients as controls, with the aim of identifying specific CSF miRNA patterns that could differentiate brain tumors from one another [166]. CSF miRNA signatures were identified for all the cancer types studied, which revealed the potential of CSF miRNAs for diagnosis of brain tumors, especially in cases with borderline or uncertain imaging results [166] (Figure 1).

Table 2. MicroRNAs in brain tumors.

Tumor	miRNA	Function	Quantity		Ref.
			Level	Tissue/Cell	
Glioma	miR-21	OncomiR	Increased	GBM cells and derived EVs [1]	[168–174]
			Increased	Blood; GBM cells derived EVs	[175]
	miR-148a	OncomiR	Increased	Glioma tissues; glioma cell line; GBM specimens and cell line	[137–140]
	miR-155	OncomiR	Increased	Blood; Glioma cell line	[141–145]
	miR-221/222	OncomiR	Increased	Blood	[176]
	miR-301a	OncomiR	Increased	GBM patient's serum	[177]
	miR-222, miR-124-3p, miR-221, miR-320, miR-574-3p, and miR-301a	Positive diagnostic biomarkers	Increased	Astrocytoma cells	[158]

Table 2. *Cont.*

Tumor	miRNA	Function	Quantity		Ref.
			Level	Tissue/Cell	
Glioma	miR-335	OncomiR	Increased	GBM cells and derived EVs	[168]
	miR-451	OncomiR	Increased in cells and released in EVs	GBM cells and derived EVs	[178]
	miR-1238	OncomiR	Low levels in cells	GBM cells and derived EVs	[179]
	miR-1	Tumor suppressor	Released in EVs	GBM cells derived EVs	[180]
	miR-151a	Tumor suppressor	Low levels	Blood; GBM specimens and cell line	[139,146,149]
	miR-185	Tumor suppressor	Decreased	GBM cell lines and blood	[151,152]
	miR-203	Tumor suppressor	Low levels	Blood	[147,148]
	miR-205	Tumor suppressor	Low levels	GBM tissues and cell lines	[181,182]
	miR-454	Tumor suppressor	Low levels	GMB tissues and cell lines	[153]
	miR-605	Tumor suppressor	Increased	Meningioma tissues	[159]
Meningioma	miR-224	OncomiR	Increased	Meningioma tissues; meningioma cell	[157]
	miR-335	OncomiR	Low levels	Atypical and anaplastic meningiomas; meningioma cell	[156]
	miR-145	Tumor suppressor	Low levels	Benign meningioma tissue	[154,155]
	miR-200a	Tumor suppressor	Increased	Malignant meningioma cells	[154,155]
			High levels	Blood	[160]
	miR-106a-5p, miR-219-5p, miR-375, and miR-409-3p	Diagnostic and prognostic biomarkers	Low levels	Blood	[160]
	miR-197 and miR-224	Diagnostic and prognostic biomarkers	Increased following recurrence	Tumor samples	[161]
	miR-15a-5p, miR-146a-5p, and miR-331-3p	Prognostic biomarkers			[161]

[1] GBM, glioblastoma multiforme.

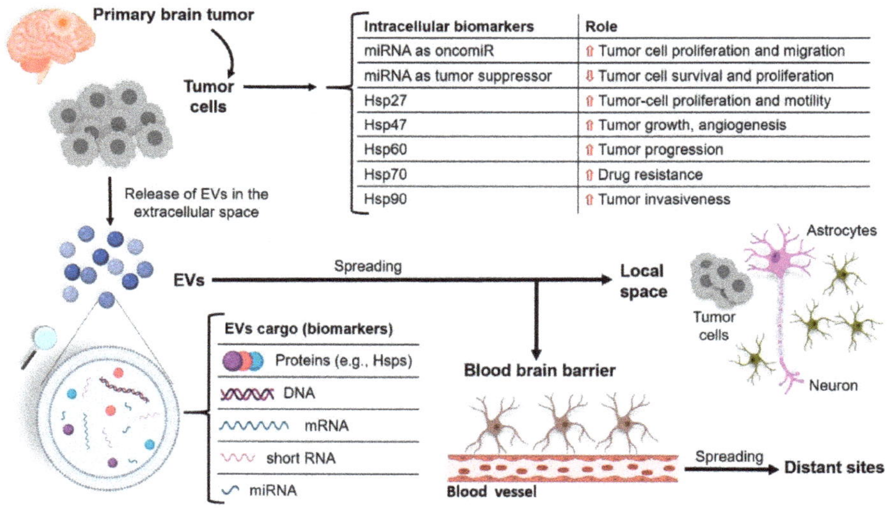

Figure 1. Schematics of the production and migration of extracellular vesicles (EVs) released by brain tumors and their local and distant cell targets (top half of the figure) with the components of their cargo discussed in this article (bottom half).

3.2. Molecular Chaperones in Brain Tumors

Molecular chaperones are the main components of the chaperoning (chaperone) system (CS) of an organism, which is also constituted of co-chaperones, chaperone cofactors and chaperone receptors and interactors, forming different functional networks [183,184]. Some chaperones are called Hsp (from heat shock protein), since they increase in response to heat shock and other stressors [185–188]. Although not all chaperones are Hsps and vice versa not all Hsps are chaperones, both terms have been used indistinctly for years, as if they were true synonyms. This unfortunate confusion is practically impossible to eradicate from the literature and continues to thrive. Therefore, we use the terms chaperone and Hsp interchangeably in this work.

The canonical functions of chaperones pertain to protein homeostasis and quality control [189–198]. However, chaperones have also noncanonical functions unrelated to the maintenance of protein homeostasis, including participation in immune and inflammatory reactions [12,199–201].

The CS interacts with the immune system and when malfunctional, it becomes a pathogenic factor in autoimmune and inflammatory diseases. The diseases in which components of the chaperone system play an etiological–pathogenic role are the chaperonopathies [202].

As for most diseases, chaperonopathies can be genetic or acquired, with the former being the result of a gene variant, e.g., mutation, while the acquired chaperonopathies are characterized by structural and functional abnormalities in the chaperone protein, but its gene is normal. In addition, chaperonopathies can be classified according to its main feature as by defect, by excess, and by mistake [203]. Typically, chaperonopathies associated with cancer are by mistake, namely the pathogenic chaperone helps the tumor cell rather than defend the human host against it. Since chaperones are classically considered cytoprotective and guardians of protein homeostasis, their helping malignant cells to grow, proliferate, and disseminate appear as mistaken activities, so to speak. This concept has played a key role in alerting physicians and pathologists to the fact that chaperones may be determinant pathogenic factors and should be looked upon as tumor biomarkers and targets for treatment, with negative chaperonotherapy being the modality of choice most often, which consists of inhibiting, blocking, or eliminating the "mistaken" chaperone.

Molecular chaperones can confer resistance against chemo- and radiotherapies, and support glial tumor growth and invasion, i.e., the typical role of "mistaken" chaperones that characterize the chaperonopathies by mistake, underpinning carcinogenic mechanisms in certain types of tumors [13,204–208] (Table 3).

A positive correlation between Hsp27 expression level and the growth rate of different types of high-grade astrocytoma, including glioblastoma, has been reported, suggesting its involvement in promoting tumor growth [209,210]. Hsp27 is an important regulator of F-actin polymerization and it was shown that p38MAPK activation, followed by Hsp27 phosphorylation, was required for Phorbol 12-myristate 13-acetate (PMA)-induced migration of glioblastoma cells, suggesting that this chaperone is a potential target of negative chaperonotherapy to inhibit cancer invasion and progression [211]. This hypothesis was further supported by findings showing that Hsp27 downregulation synergizes the anticancer effects of different drugs and treatments, reducing GBM cell proliferation and promoting caspase 3-mediated apoptosis [212–214].

Another Hsp involved in glioblastoma tumorigenesis is Hsp47. Hsp47 was found overexpressed in glioma tissues and cell lines and associated with glioma tumor grade [215]. Moreover, its expression level was positively correlated with tumor vascularization, since its silencing consistently decreased VEGF expression in glioma cells, and reduced glioma vasculature [215]. Hsp47 knockdown also inhibited glioma cell growth, migration, and invasion in vitro and in vivo [216]. Contrarily, in vivo Hsp47 overexpression promoted primary glioma cell tumor formation and stemlike properties maintenance, as well as tumor invasion and angiogenesis, thanks to the upregulation of extracellular matrix related genes, such as CD44, LAMC1m COL4A2, ITGB1, FN1, and MMP9, through the TGF-β pathway [217]. These data indicate a key role of Hsp47 in glioma angiogenesis, suggesting its potential use as therapeutic target to treat glioma tumors (Table 3).

In addition, the involvement of the mitochondrial chaperonin Hsp60 in glioblastoma tumorigenesis and progression has been studied. The Hsp60 expression level was found higher in glioblastoma patients and cell lines, with an antiapoptotic and a prosurvival role [218,219]. Hsp60 through its interaction with cyclophilin D, Hsp90, and other cofactors, modulates tumor growth and prevents apoptosis in vivo [218]. Moreover, the chaperonin downregulation in glioblastoma cells leads to epithelial–mesenchymal transition and increases production of reactive oxygen species (ROS), ultimately suppressing cell growth and proliferation through the ROS/AMPK/mTOR pathway [219]. The protumorigenic role of Hsp60 was suggested also by immunohistochemical analysis on a subset of human brain neoplasms by comparing the levels of Hsp60 and Hsp70, another Hsp commonly implicated in carcinogenesis [220–222]. The results showed a significant difference between Hsp60 and Hsp70 levels in neuroepithelial tumors, while levels of both molecules did not differ among each other in meningeal neoplasms. It was suggested that Hsp60 is not increased by a passive phenomenon, but may play an active role in tumor progression, although other studies are needed to fully understand this issue [220,223,224].

Hsp70 is another chaperone that is thought to play a role in carcinogenesis, since it was found abundantly expressed in malignant cells, and with different roles [225]. For instance, the cytosolic, membrane-bound, and extracellular forms of Hsp70 are augmented in primary glioblastomas [226], and its increase was associated with increased proliferation, migration, and invasion rates, as well as with acquisition of radio resistance by human glioblastoma cell lines [227,228]. It has been shown that Hsp70 promotes survival of C6 and U87 glioma cells, by protecting ATF5 from proteasome and caspase-dependent proteolytic degradation [229]. Moreover, in a rat model of GBM (C6 cells), Hsp70 elicited cytoprotective activity and rescued glioblastoma cells from oxidative stress and death by sequestrating the aggregation-prone GAPDH, which is usually responsible for pathogenic aggregation of proteins after cell exposure to oxidative stress. The protective power of the chaperone could be abolished by specific inhibitors of Hsp70 expression [230] (Table 3).

Another Hsp with a protumorigenic role in glioma is Hsp90, which through different signaling pathways, promotes cancer cell motility and invasion [231–234]. In addition to migration, Hsp90 regulates other protumorigenic processes in GBM cells, such as cell-survival mechanisms and apoptosis. Consequently, a combined treatment with Hsp90 and PI-3 kinase inhibitors has been shown to increase the apoptotic death of GBM cells, likely by disrupting AKT signaling and promoting G2/M arrest [235]. In agreement with these results, several works demonstrated the efficacy of Hsp90 inhibitors in counteracting malignant gliomas, proposing their use, alone or in combination with other traditional anticancer drugs, as potential therapeutic agents in gliomas treatment—all examples of negative chaperonotherapy [208,236–238] (Table 3).

Compared to malignant gliomas, less is known about the role played by Hsps in other brain tumors. Two different studies investigated different Hsps on paraffin-embedded sections from medulloblastoma patients using immunohistochemistry and found substantial amounts of them [239,240]. However, these data are still preliminary and further studies, involving a larger series of patients, are necessary to clarify the relationship of Hsps with tumor aggressiveness and prognosis [239,240] (Figure 1).

Table 3. Reported roles of Hsp27, Hsp47, Hsp60, Hsp70, and Hsp90 in gliomas.

Chaperone/Hsp	Role in Gliomas	Ref.
Hsp27	Promotes tumor growth, and cancer-cell proliferation and motility	[209–214]
Hsp47	Promotes tumor growth, invasiveness, and angiogenesis	[215–217]
Hsp60	Promotes tumor progression by enhancing cancer-cell proliferation, preventing apoptosis, and inhibiting the antitumor immune response	[218–220]
Hsp70	Promotes cancer cell proliferation, migration, and invasion, and protects cancer cells from apoptosis and anticancer drugs	[227–230]
Hsp90	Promotes cancer-cell motility, tumor invasiveness, and drug resistance	[231–235]
CCT6	Promotes GBM cell invasion and has a negative association with patient survival [1]	[241]

[1] GBM, glioblastoma multiforme.

4. The Release of miRNAs and Chaperones through EVs as Molecular Signaling and Source of Biomarkers

Intercellular communication between tumor cells and their neighboring structures, including other cells, is vital for cancer growth and progression, and EVs are key elements in this crosstalk. However, the mechanisms involved are still poorly understood, which stands in the way of progress in cancer treatment. The involvement of brain tumor-derived EVs in the modulation of the tumor microenvironment has been suggested by studies that revealed in those vesicles functionally active molecules that can play a role in cancer progression.

The release of Hsps inside vesicles has been reported for different brain tumor cell lines [242]. Moreover, it has been found that Hsp27, Hsp70, and Hsp90 can also be present on the surface of brain tumor-derived EVs [243,244], which indicate their potential as tumor biomarkers [245,246]. However, although molecular chaperones have been described as key players in brain tumor biology, in the current literature there are still very few data about the possible role of extracellular Hsps in these tumors. On the contrary, the role of miRNAs carried by EVs released by brain tumor cells is currently under active investigation and debate. MicroRNAs in EVs are abundant by comparison with other cargo molecules [247] and their expression patterns reflect their source, thus providing information about their cells of origin [248]. For instance, gliomas secrete EVs that transport receptors and signaling molecules for oncogenes [16,249,250]. The EV-mediated transfer of miRNAs appears to be a way for the tumor to communicate with distinct sets of surrounding nontumor cells, including neurons and glial and vascular cells, which are thus reprogrammed to modify the tumor microenvironment to make it suitable to tumor growth and dissemination [251]. For example, it has been shown that glioma-derived EV saltered synaptic activity in neurons, contributing to tumor growth [251]. In addition, several studies have pinpointed the specific role of certain vesicular miRNAs on target cells, which in essence consisted in mediating the aggressive properties of gliomas [178–180,252]. The microRNAs miR-451 and miR-21 are present at very high levels in the EVs produced by primary GBM cells, and are uptaken by microglia, which is followed by a phenotype change of the recipient cell accompanied by upregulation of cytokines, chemokines, and matrix metallopeptidases (MMP), all of which promote growth and invasion of GBM cells while lessening of the immune response [168]. MiR-21 transferred by GBM–EVs to microglia regulates the expression of the Btg2 gene, involved in the control of cellular proliferation and differentiation [169]. The dysregulation of c-Myc and Btg2 attests that the GBM–EVs have a functional activity on neighboring target cells. Several studies suggest the oncogenic effect of miR-21, whose overexpression occurring in many tumor types favors the establishment of tumor-permissive pathways in glioblastoma [170,171]. It has been observed that malignant brain tumor-derived EVs support neoangiogenesis through miR-21/VEGF signaling [172].

MicroRNAs can cross the blood–brain barrier (BBB) and reach far target sites through the circulation. For instance, the vesicular miR-21 has been identified in the blood [250] and CSF [173,174], suggesting that these EVs carrying miRNAs could have diagnostic and prognostic usefulness.

Both miR-105 [253] and miR-181c [254] have the ability to disrupt the blood–brain barrier and function as pro-oncogenes by downregulating tumor suppressor genes, and thereby favor invasiveness and metastasization. It has also been reported that miR-148a delivered by EVs promotes glioma-cell proliferation and metastasis via targeting CADM1 to activate the STAT3 pathway [175].

The content in miRNAs within EVs released by a tumor can vary depending on the stage of tumor progression, thus providing key diagnostic indicators, which is particularly convenient when it is possible to isolate the EVs from biological fluids such as plasma [163]. For instance, miR-222, miR-124-3p, miR-221, miR-320, miR-574-3p, and miR-301a were found to be increased in EVs derived from serum of high-grade glioma patients [176,177] and their distinctive quantitative patterns allowed for distinguishing between tumoral and control samples [177].

The pathogenic role of some miRNAs depends on their tissue localization. For instance, miR-454 may act as an oncogene in gastric cancer [255], but is a tumor suppressor in gliomas [181]. The microRNA miR-454 is elevated in EVs from preoperative sera when compared with its levels after surgery, indicating that cancer cells can differentially secrete specific miRNAs into the circulation in EVs, which support the idea that these vesicles and their contents have potential as useful biomarkers accessible with minimally invasive procedures [182].

Despite their fundamental role in cell homeostasis and the knowledge that molecular chaperones are differentially expressed in cancer, little is known about the secretion of these proteins by brain tumors via EVs. The stress response of tumor cells, included those of brain tumors, consists of the overexpression of Hsps and their release in the extracellular environment also by EVs. However, little is known on the role of molecular chaperones in the brain tumor microenvironment; one working hypothesis postulates that the release of chaperones via EVs could be a defense mechanism from injury [242,256].

CCT6A, a subunit of the molecular chaperone CCT (Chaperonin-containing Tcp-1) is present in EVs from glioma cells derived from surgical specimens [241]. Secretion of CCT6A via EVs could be linked to GBM cell invasion and has a negative association with patient survival [241] (Figure 1).

After their release in the extracellular space, EVs can either perform a local autocrine/paracrine signalling, targeting other nearby tumor cells and nervous-system cells (neurons and astrocytes), or cross the blood–brain barrier and reach distant cells through the systemic circulation. In both cases, EVs release their cargo into the cytoplasm of recipient cells by fusing with the target-cell membrane. At this point, the delivered molecules can mediate many physiological and pathological processes. EVs cargo reflects the characteristics of the cells from which they originate. Therefore, it is likely that both miRNAs and Hsps normally produced by tumor cells may be selected for the sorting into the released EVs and affect the tumor microenvironment and other distant tissues. As indicated in the bracket at the top right half of Figure 1, miRNAs may act either as oncomiR, favoring tumor cells proliferation and migration, or as tumor suppressor, by inhibiting cancer cells survival and proliferation. In both cases, the role is mediated by the post-transcriptional regulation of factors involved in the control of the cell life cycle and is correlated to the level of the pertinent miRNA. On the contrary, Hsps act mainly as protumorigenic factors (i.e., chaperonopathies by mistake), since Hsps promote cancer cell proliferation and motility (Hsp27); tumor growth, invasiveness, and angiogenesis (Hsp47); prevent tumor cell apoptosis and inhibit antitumor immune response (Hsp60); and confer resistance to anticancer treatment (Hsp70 and Hsp90).

5. Conclusions and Perspectives

Tumor-derived EVs with their miRNA and molecular chaperone contents have potential applications in medicine for diagnostic purposes, for monitoring patients, and for designing and implementing novel therapeutic procedures. Methods for isolating EVs and for characterizing their cargo have been developed over the last few years and interest has increased in their possible role as intercellular messengers. Tumor-derived EVs are considered a potential source of biomarkers, reflecting the status and metabolism of the cells from which they originate. Because of this and of their availability via minimally invasive sampling procedures, EVs have a promising future in brain tumor management, including early diagnosis and patient follow up. It is hoped that as knowledge about these vesicles and their physiological and pathogenic roles progresses, mechanistic insights will be gained that will pave the way for developing novel treatment strategies and drugs. Here, we have discussed two components of the EVs cargo, miRNAs and molecular chaperones, the key functions of which in carcinogenesis and in the fight of the organism against malignancies, and vice versa, are becoming increasingly clear. Therefore, measurement and characterization of miRNA and chaperones in EVs from brain tumors in liquid biopsies are promising endeavors in research and medical practice, particularly in neurosurgery. Furthermore, learning about EVs from brain tumors and their migrations within the brain and the rest of the body will tell whether these vesicles have potential as vectors for delivering anticancer drugs to specific cell targets.

The efficacy of EV-carried noncoding RNAs has been examined in neurodegenerative diseases [257] and in tumors [258]. Systemically administered engineered EVs, targeting the transferrin receptor, were found to bind glioblastoma cells and enhance the action of the antisense miRNA oligonucleotide produced to inhibit endogenous miR-21 [259].

Natural or artificially constructed EVs carrying chaperones have been tested for their ability to enhance the antitumoral immune response. For example, Hsp70 contained in EVs from dendritic cells activated T lymphocytes toward becoming glioma-specialized cytotoxic T lymphocytes [260]. Chaperone-enriched EVs have been tested in glioma immunotherapy in an in vivo model and have shown strong induction of the CD4+ and CD8+ T cell activity and enhancement of T cell infiltration in intracranial glioma tissues, causing inhibition of tumor growth [261].

Other advances in EVs application in tumor therapy are their use as target-specific carriers to deliver tumor suppressor miRNAs or miRNA-mimic molecules. Some miRNA-mimic molecules have the capability of targeting and reducing the protumoral effect mediated by Hsps. This has been demonstrated in a triple-negative breast cancer model, in which miR-134-enriched EVs reduced the levels of Hsp90, a chaperone that favors the survival of cancer cells by stabilizing oncogenic proteins [262]. This treatment resulted in reduced cell migration and invasion, and enhanced sensitivity to anti-Hsp90 drugs in breast cancer cells [262]. Delivery of specific miRNAs or anti-miRNA molecules played important roles in the modulation of the expression of Hsps, with therapeutic effects in cancer [263]. In the vascular endothelial cells of gliomas, miR-144 targeted directly Heat Shock Factor 2, which regulates Hsps expression, modifying the permeability of the blood–tumor barrier, which opens a potential new way in glioma treatment centered on the regulation of Hsp expression by Heat Shock Factors [264].

These findings point the way for future research aiming at finding and producing miRNAs and miRNA-mimic molecules targeting Hsps in brain tumors. Thus, EVs and miRNA and chaperones in their cargo have potential not only in diagnosis and patient monitoring but also for brain cancer treatment.

Author Contributions: Conceptualization, A.M.V., R.S., C.C., E.C.d.M., F.B., A.J.L.M., and C.C.B.; original draft preparation, A.M.V., R.S., G.V., A.M.G., and C.C.B.; supervision, F.B., A.J.L.M., and C.C.B. All authors have read and agreed to the published version of the manuscript.

Funding: This work was funded in part by the Italian National Operational Programme (PON) «Imprese e Competitività» 2014-2020 FESR, grant awarded by the Italian Ministry of Economic Development to the project titled «Gestione di un servizio integrato multicentrico di diagnostica e terapia personalizzata in oncologia» (Project code: F/090012/01-02/X36). A.J.L.M and E.C.d.M. were partially supported by IMET and IEMEST. This is IMET contribution number IMET 20-019.

Conflicts of Interest: The authors declare no conflict of interest.

References

1. Bray, F.; Ferlay, J.; Soerjomataram, I.; Siegel, R.L.; Torre, L.A.; Jemal, A. Global cancer statistics 2018: GLOBOCAN estimates of incidence and mortality worldwide for 36 cancers in 185 countries. *CA Cancer J. Clin.* **2018**, *68*, 394–424. [CrossRef] [PubMed]
2. Perkins, A.; Liu, G. Primary brain tumors in adults: Diagnosis and treatment. *Am. Fam. Physician* **2016**, *93*, 211–217.
3. Tai, Y.L.; Chen, K.C.; Hsieh, J.T.; Shen, T.L. Exosomes in cancer development and clinical applications. *Cancer Sci.* **2018**, *109*, 2364–2374. [CrossRef] [PubMed]
4. Zhang, Y.; Liu, Y.; Liu, H.; Tang, W.H. Exosomes: Biogenesis, biologic function and clinical potential. *Cell Biosci.* **2019**, *9*, 19. [CrossRef] [PubMed]
5. Zhang, L.; Yu, D. Exosomes in cancer development, metastasis, and immunity. *Biochim. Biophys. Acta Rev. Cancer* **2019**, *1871*, 455–468. [CrossRef] [PubMed]
6. Caruso Bavisotto, C.; Marino Gammazza, A.; Rappa, F.; Fucarino, A.; Pitruzzella, A.; David, S.; Campanella, C. Exosomes: Can doctors still ignore their existence? *Euromediterranean Biomed. J.* **2013**, *8*, 136–139.
7. Caruso Bavisotto, C.; Cappello, F.; Macario, A.J.L.; Conway de Macario, E.; Logozzi, M.; Fais, S.; Campanella, C. Exosomal HSP60: A potentially useful biomarker for diagnosis, assessing prognosis, and monitoring response to treatment. *Expert Rev. Mol Diagn.* **2017**, *17*, 815–822. [CrossRef]
8. Campanella, C.; Rappa, F.; Sciumè, C.; Marino Gammazza, A.; Barone, R.; Bucchieri, F.; David, S.; Curcurù, G.; Caruso Bavisotto, C.; Pitruzzella, A.; et al. Heat shock protein 60 levels in tissue and circulating exosomes in human large bowel cancer before and after ablative surgery. *Cancer* **2015**, *121*, 3230–3239. [CrossRef]
9. Caruso Bavisotto, C.; Cipolla, C.; Graceffa, G.; Barone, R.; Bucchieri, F.; Bulone, D.; Cabibi, D.; Campanella, C.; Marino Gammazza, A.; Pitruzzella, A.; et al. Immunomorphological pattern of molecular chaperones in normal and pathological thyroid tissues and circulating exosomes: Potential use in clinics. *Int. J. Mol. Sci.* **2019**, *20*, 4496. [CrossRef]
10. Peng, Y.; Croce, C.M. The role of microRNAs in human cancer. *Signal Transduct. Target. Ther.* **2016**, *1*, 15004. [CrossRef]
11. Takahashi, R.U.; Prieto-Vila, M.; Hironaka, A.; Ochiya, T. The role of extracellular vesicle microRNAs in cancer biology. *Clin. Chem. Lab. Med.* **2017**, *55*, 648–656. [CrossRef] [PubMed]

12. Lang, B.J.; Guerrero-Giménez, M.E.; Prince, T.L.; Ackerman, A.; Bonorino, C.; Calderwood, S.K. Heat shock proteins are essential components in transformation and tumor progression: Cancer cell intrinsic pathways and beyond. *Int. J. Mol. Sci.* **2019**, *20*, 4507. [CrossRef] [PubMed]
13. Graziano, F.; Caruso Bavisotto, C.; Marino Gammazza, A.; Rappa, F.; Conway de Macario, E.; Macario, A.J.L.; Cappello, F.; Campanella, C.; Maugeri, R.; Iacopino, D.G. Chaperonology: The third eye on brain gliomas. *Brain Sci.* **2018**, *8*, 110. [CrossRef] [PubMed]
14. Caruso Bavisotto, C.; Graziano, F.; Rappa, F.; Marino Gammazza, A.; Logozzi, M.; Fais, S.; Maugeri, R.; Bucchieri, F.; Conway de Macario, E.; Macario, A.J.L.; et al. Exosomal chaperones and miRNAs in gliomagenesis: State-of-art and theranostics perspectives. *Int. J. Mol. Sci.* **2018**, *19*, 2626. [CrossRef]
15. Marino Gammazza, A.; Caruso Bavisotto, C.; Barone, R.; Conway de Macario, E.; Macario, A.J.L. Alzheimer's disease and molecular chaperones: Current knowledge and the future of chaperonotherapy. *Curr. Pharm. Des.* **2016**, *22*, 4040–4049. [CrossRef]
16. Abels, E.R.; Breakefield, X.O. Introduction to extracellular vesicles: Biogenesis, rna cargo selection, content, release, and uptake. *Cell. Mol. Neurobiol.* **2016**, *36*, 301–312. [CrossRef]
17. Urbanelli, L.; Magini, A.; Buratta, S.; Brozzi, A.; Sagini, K.; Polchi, A.; Tancini, B.; Emiliani, C. Signaling pathways in exosomes biogenesis, secretion and fate. *Genes (Basel)* **2013**, *4*, 152–170. [CrossRef]
18. Perut, F.; Roncuzzi, L.; Baldini, N. The emerging roles of extracellular vesicles in osteosarcoma. *Front. Oncol.* **2019**, *9*, 1342. [CrossRef]
19. Caruso Bavisotto, C.; Scalia, F.; Marino Gammazza, A.; Carlisi, D.; Bucchieri, F.; Conway de Macario, E.; Macario, A.J.L.; Cappello, F.; Campanella, C. Extracellular vesicle-mediated Cell⁻Cell communication in the nervous system: Focus on neurological diseases. *Int. J. Mol. Sci.* **2019**, *20*, 434. [CrossRef]
20. Cufaro, M.C.; Pieragostino, D.; Lanuti, P.; Rossi, C.; Cicalini, I.; Federici, L.; De Laurenzi, V.; Del Boccio, P. Extracellular vesicles and their potential use in monitoring cancer progression and therapy: The contribution of proteomics. *J. Oncol.* **2019**, *2019*, 1639854. [CrossRef]
21. Cappello, F.; Logozzi, M.; Campanella, C.; Caruso Bavisotto, C.; Marcilla, A.; Properzi, F.; Fais, S. Exosome levels in human body fluids: A tumor marker by themselves? *Eur. J. Pharm. Sci.* **2017**, *96*, 93–98. [CrossRef] [PubMed]
22. Logozzi, M.; Mizzoni, D.; Bocca, B.; Di Raimo, R.; Petrucci, F.; Caimi, S.; Alimonti, A.; Falchi, M.; Cappello, F.; Campanella, C.; et al. Human primary macrophages scavenge AuNPs and eliminate it through exosomes. A natural shuttling for nanomaterials. *Eur. J. Pharm. Biopharm.* **2019**, *137*, 23–36. [CrossRef] [PubMed]
23. Théry, C.; Witwer, K.W.; Aikawa, E.; Alcaraz, M.J.; Anderson, J.D.; Andriantsitohaina, R.; Antoniou, A.; Arab, T.; Archer, F.; Atkin-Smith, G.K.; et al. Minimal information for studies of extracellular vesicles 2018 (MISEV2018): A position statement of the International Society for Extracellular Vesicles and update of the MISEV2014 guidelines. *J. Extracell. Vesicles* **2018**, *7*, 1535750. [CrossRef] [PubMed]
24. Cocucci, E.; Racchetti, G.; Meldolesi, J. Shedding microvesicles: Artefacts no more. *Trends Cell Biol.* **2009**, *19*, 43–51. [CrossRef]
25. Kakarla, R.; Hur, J.; Kim, Y.J.; Kim, J.; Chwae, Y.J. Apoptotic cell-derived exosomes: Messages from dying cells. *Exp. Mol. Med.* **2020**, *52*, 1–6. [CrossRef]
26. Xu, L.; Wu, L.-F.; Deng, F.-Y. Exosome: An emerging source of biomarkers for human diseases. *Curr. Mol. Med.* **2019**, *19*, 387–394. [CrossRef]
27. Reiner, A.T.; Witwer, K.W.; Van Balkom, B.W.M.; De Beer, J.; Brodie, C.; Corteling, R.L.; Gabrielsson, S.; Gimona, M.; Ibrahim, A.G.; De Kleijn, D.; et al. Concise review: Developing best-practice models for the therapeutic use of extracellular vesicles. *Stem Cells Transl. Med.* **2017**, *6*, 1730–1739. [CrossRef]
28. Burgio, S.; Noori, L.; Marino Gammazza, A.; Campanella, C.; Logozzi, M.; Fais, S.; Bucchieri, F.; Cappello, F.; Caruso Bavisotto, C. Extracellular vesicles-based drug delivery systems: A new challenge and the exemplum of malignant pleural mesothelioma. *Int. J. Mol. Sci.* **2020**, *21*, 5432. [CrossRef]
29. Nawaz, M.; Malik, M.I.; Hameed, M.; Zhou, J. Research progress on the composition and function of parasite-derived exosomes. *Acta Trop.* **2019**, *196*, 30–36. [CrossRef]
30. Van Dommelen, S.M.; Vader, P.; Lakhal, S.; Kooijmans, S.A.A.; Van Solinge, W.W.; Wood, M.J.A.; Schiffelers, R.M. Microvesicles and exosomes: Opportunities for cell-derived membrane vesicles in drug delivery. *J. Control. Release* **2012**, *161*, 635–644. [CrossRef]

31. Bhome, R.; Del Vecchio, F.; Lee, G.H.; Bullock, M.D.; Primrose, J.N.; Sayan, A.E.; Mirnezami, A.H. Exosomal microRNAs (exomiRs): Small molecules with a big role in cancer. *Cancer Lett.* **2018**, *420*, 228–235. [CrossRef] [PubMed]
32. Théry, C.; Zitvogel, L.; Amigorena, S. Exosomes: Composition, biogenesis and function. *Nat. Rev. Immunol.* **2002**, *2*, 569–579. [CrossRef]
33. Kleihues, P.; Burger, P.C.; Scheithauer, B.W. The new WHO classification of brain tumours. *Brain Pathol.* **1993**, *3*, 255–268. [CrossRef] [PubMed]
34. Biernat, W. World health organization classification of tumors, of the nervous system. *Pol. J. Pathol.* **2000**, *51*, 107–114. [PubMed]
35. Kleihues, P.; Louis, D.N.; Scheithauer, B.W.; Rorke, L.B.; Reifenberger, G.; Burger, P.C.; Cavenee, W.K. The WHO classification of tumors of the nervous system. *J. Neuropathol. Exp. Neurol.* **2002**, *61*, 215–225. [CrossRef]
36. Louis, D.N.; Ohgaki, H.; Wiestler, O.D.; Cavenee, W.K.; Burger, P.C.; Jouvet, A.; Scheithauer, B.W.; Kleihues, P. The 2007 WHO classification of tumours of the central nervous system. *Acta Neuropathol.* **2007**, *114*, 97–109. [CrossRef] [PubMed]
37. Louis, D.N.; Perry, A.; Reifenberger, G.; von Deimling, A.; Figarella-Branger, D.; Cavenee, W.K.; Ohgaki, H.; Wiestler, O.D.; Kleihues, P.; Ellison, D.W. The 2016 world health organization classification of tumors of the central nervous system: A summary. *Acta Neuropathol.* **2016**, *131*, 803–820. [CrossRef]
38. Posner, J.B.; Chernik, N.L. Intracranial metastases from systemic cancer. *Adv. Neurol.* **1978**, *19*, 579–592.
39. Lapointe, S.; Perry, A.; Butowski, N.A. Primary brain tumours in adults. *Lancet* **2018**, *392*, 432–446. [CrossRef]
40. Achrol, A.S.; Rennert, R.C.; Anders, C.; Soffietti, R.; Ahluwalia, M.S.; Nayak, L.; Peters, S.; Arvold, N.D.; Harsh, G.R.; Steeg, P.S.; et al. Brain metastases. *Nat. Rev. Dis. Prim.* **2019**, *5*, 5. [CrossRef]
41. Gaspar, L.E. Brain metastases in lung cancer. *Expert Rev. Anticancer* **2004**, *4*, 259–270. [CrossRef] [PubMed]
42. Gállego Pérez-Larraya, J.; Hildebrand, J. Brain metastases. In *Handbook of Clinical Neurology*; Elsevier: Amsterdam, The Netherlands, 2014; Volume 121, pp. 1143–1157.
43. Michl, M.; Thurmaier, J.; Schubert-Fritschle, G.; Wiedemann, M.; Laubender, R.P.; Nüssler, N.C.; Ruppert, R.; Kleeff, J.; Schepp, W.; Reuter, C.; et al. Brain metastasis in colorectal cancer patients: Survival and analysis of prognostic factors. *Clin. Colorectal Cancer* **2015**, *14*, 281–290. [CrossRef] [PubMed]
44. Witzel, I.; Oliveira-Ferrer, L.; Pantel, K.; Müller, V.; Wikman, H. Breast cancer brain metastases: Biology and new clinical perspectives. *Breast Cancer Res.* **2016**, *18*, 8. [CrossRef] [PubMed]
45. Hemminki, K.; Li, X. Familial risks in nervous system tumors. *Cancer Epidemiol. Prev. Biomark.* **2003**, *12*, 1137–1142.
46. Goh, S.; Butler, W.; Thiele, E.A. Subependymal giant cell tumors in tuberous sclerosis complex. *Neurology* **2004**, *63*, 1457–1461. [CrossRef]
47. Farrell, C.J.; Plotkin, S.R. Genetic causes of brain tumors: Neurofibromatosis, tuberous sclerosis, von hippel-lindau, and other syndromes *Neurol. Clin.* **2007**, *25*, 925–946. [CrossRef]
48. Hottinger, A.F.; Khakoo, Y. Neuro-oncology of neurofibromatosis type 1. *Curr. Treat. Options Neurol.* **2009**, *11*, 306–314. [CrossRef]
49. Kijima, C.; Miyashita, T.; Suzuki, M.; Oka, H.; Fujii, K. Two cases of nevoid basal cell carcinoma syndrome associated with meningioma caused by a PTCH1 or SUFU germline mutation. *Fam. Cancer* **2012**, *11*, 565–570. [CrossRef]
50. Braganza, M.Z.; Kitahara, C.M.; Berrington De González, A.; Inskip, P.D.; Johnson, K.J.; Rajaraman, P. Ionizing radiation and the risk of brain and central nervous system tumors: A systematic review. *Neuro Oncol.* **2012**, *14*, 1316–1324. [CrossRef]
51. Smoll, N.R.; Brady, Z.; Scurrah, K.; Mathews, J.D. Exposure to ionizing radiation and brain cancer incidence: The life span study cohort. *Cancer Epidemiol.* **2016**, *42*, 60–65. [CrossRef]
52. Libermann, T.A.; Nusbaum, H.R.; Razon, N.; Kris, R.; Lax, I.; Soreq, H.; Whittle, N.; Waterfield, M.D.; Ullrich, A.; Schlessinger, J. Amplification, enhanced expression and possible rearrangement of EGF receptor gene in primary human brain tumours of glial origin. *Nature* **1985**, *313*, 144–147. [CrossRef] [PubMed]
53. Kamb, A.; Gruis, N.A.; Weaver-Feldhaus, J.; Liu, Q.; Harshman, K.; Tavtigian, S.V.; Stockert, E.; Day, R.S.; Johnson, B.E.; Skolnick, M.H. A cell cycle regulator potentially involved in genesis of many tumor types. *Science* **1994**, *264*, 436–440. [CrossRef] [PubMed]

54. Ebert, C.; Von Haken, M.; Meyer-Puttlitz, B.; Wiestler, O.D.; Reifenberger, G.; Pietsch, T.; Von Deimling, A. Molecular genetic analysis of ependymal tumors: NF2 mutations and chromosome 22q loss occur preferentially in intramedullary spinal ependymomas. *Am. J. Pathol.* **1999**, *155*, 627–632. [CrossRef]
55. Wani, K.; Armstrong, T.S.; Vera-Bolanos, E.; Raghunathan, A.; Ellison, D.; Gilbertson, R.; Vaillant, B.; Goldman, S.; Packer, R.J.; Fouladi, M.; et al. A prognostic gene expression signature in infratentorial ependymoma. *Acta Neuropathol.* **2012**, *123*, 727–738. [CrossRef]
56. Gupta, K.; Salunke, P. Understanding Ependymoma oncogenesis: An Update on recent molecular advances and current perspectives. *Mol. Neurobiol.* **2017**, *54*, 15–21. [CrossRef]
57. Li, J.; Yen, C.; Liaw, D.; Podsypanina, K.; Bose, S.; Wang, S.I.; Puc, J.; Miliaresis, C.; Rodgers, L.; McCombie, R.; et al. PTEN, a putative protein tyrosine phosphatase gene mutated in human brain, breast, and prostate cancer. *Science* **1997**, *275*, 1943–1947. [CrossRef]
58. Nigro, J.M.; Baker, S.J.; Preisinger, A.C.; Jessup, J.M.; Hosteller, R.; Cleary, K.; Bigner, S.H.; Davidson, N.; Baylin, S.; Devilee, P.; et al. Mutations in the p53 gene occur in diverse human tumour types. *Nature* **1989**, *342*, 705–708. [CrossRef]
59. Wolter, M.; Reifenberger, J.; Blaschke, B.; Ichimura, K.; Schmidt, E.E.; Collins, V.P.; Reifenberger, G. Oligodendroglial tumors frequently demonstrate hypermethylation of the CDKN2A (MTS1, p16INK4a), p14ARF, and CDKN2B (MTS2, p15INK4b) tumor suppressor genes. *J. Neuropathol. Exp. Neurol.* **2001**, *60*, 1170–1180. [CrossRef]
60. Bortolotto, S.; Chiadò-Piat, L.; Cavalla, P.; Bosone, I.; Chiò, A.; Mauro, A.; Schiffer, D. CDKN2A/p16 inactivation in the prognosis of oligodendrogliomas. *Int. J. Cancer* **2000**, *88*, 554–557. [CrossRef]
61. Watanabe, T.; Nobusawa, S.; Kleihues, P.; Ohgaki, H. IDH1 mutations are early events in the development of astrocytomas and oligodendrogliomas. *Am. J. Pathol.* **2009**, *174*, 1149–1153. [CrossRef]
62. Costa, F.F.; Bischof, J.M.; Vanin, E.F.; Lulla, R.R.; Wang, M.; Sredni, S.T.; Rajaram, V.; de Fátima Bonaldo, M.; Wang, D.; Goldman, S.; et al. Identification of micrornas as potential prognostic markers in ependymoma. *PloS ONE* **2011**, *6*, e25114. [CrossRef] [PubMed]
63. Taylor, M.D.; Poppleton, H.; Fuller, C.; Su, X.; Liu, Y.; Jensen, P.; Magdaleno, S.; Dalton, J.; Calabrese, C.; Board, J.; et al. Radial glia cells are candidate stem cells of ependymoma. *Cancer Cell* **2005**, *8*, 323–335. [CrossRef]
64. Peyre, M.; Commo, F.; Dantas-Barbosa, C.; Andreiuolo, F.; Puget, S.; Lacroix, L.; Drusch, F.; Scott, V.; Varlet, P.; Mauguen, A.; et al. Portrait of ependymoma recurrence in children: Biomarkers of tumor progression identified by dual- color microarray-based gene expression analysis. *PloS ONE* **2010**, *5*, e12932. [CrossRef] [PubMed]
65. Hartmann, C.; Sieberns, J.; Gehlhaar, C.; Simon, M.; Paulus, W.; von Deimling, A. NF2 mutations in secretory and other rare variants of meningiomas. *Brain Pathol.* **2006**, *16*, 15–19. [CrossRef] [PubMed]
66. Gutmann, D.H.; Donahoe, J.; Perry, A.; Lemke, N.; Gorse, K.; Kittiniyom, K.; Rempel, S.A.; Gutierrez, J.A.; Newsham, I.F. Loss of DAL-1, a protein 4.1-related tumor suppressor, is an important early event in the pathogenesis of meningiomas. *Hum. Mol. Genet.* **2000**, *9*, 1495–1500. [CrossRef]
67. Peters, N.; Wellenreuther, R.; Rollbrocker, B.; Hayashi, Y.; Meyer-Puttlitz, B.; Duerr, E.M.; Lenartz, D.; Marsh, D.J.; Schramm, J.; Wiestler, O.D.; et al. Analysis of the PTEN gene in human meningiomas. *Neuropathol. Appl. Neurobiol.* **1998**, *24*, 3–8. [CrossRef] [PubMed]
68. Carvalho, L.H.; Smirnov, I.; Baia, G.S.; Modrusan, Z.; Smith, J.S.; Jun, P.; Costello, J.F.; McDermott, M.W.; VandenBerg, S.R.; Lal, A. Molecular signatures define two main classes of meningiomas. *Mol. Cancer* **2007**, *6*, 64. [CrossRef]
69. Arnli, M.B.; Backer-Grøndahl, T.; Ytterhus, B.; Granli, U.S.; Lydersen, S.; Gulati, S.; Torp, S.H. Expression and clinical value of EGFR in human meningiomas. *PeerJ* **2017**, *2017*, e3140. [CrossRef]
70. Pfaff, E.; Remke, M.; Sturm, D.; Benner, A.; Witt, H.; Milde, T.; Von Bueren, A.O.; Wittmann, A.; Schöttler, A.; Jorch, N.; et al. TP53 mutation is frequently associated with CTNNB1 mutation or MYCN amplification and is compatible with long-term survival in medulloblastoma. *J. Clin. Oncol.* **2010**, *28*, 5188–5196. [CrossRef]
71. Hernan, R.; Fasheh, R.; Calabrese, C.; Frank, A.J.; Maclean, K.H.; Allard, D.; Barraclough, R.; Gilbertson, R.J. ERBB2 up-regulates S100A4 and several other prometastatic genes in medulloblastoma. *Cancer Res.* **2003**, *63*, 140–148. [PubMed]

72. Grotzer, M.A.; Janss, A.J.; Phillips, P.C.; Trojanowski, J.Q. Neurotrophin receptor TrkC predicts good clinical outcome in medulloblastoma and other primitive neuroectodermal brain tumors. *Klin. Padiatr.* **2000**, *212*, 196–199. [CrossRef] [PubMed]
73. Eberhart, C.G.; Kratz, J.; Wang, Y.; Summers, K.; Stearns, D.; Cohen, K.; Dang, C.V.; Burger, P.C. Histopathological and molecular prognostic markers in medulloblastoma: C-myc, N-myc, TrkC, and Anaplasia. *J. Neuropathol. Exp. Neurol.* **2004**, *63*, 441–449. [CrossRef] [PubMed]
74. Raffel, C.; Jenkins, R.B.; Frederick, L.; Hebrink, D.; Alderete, B.; Fults, D.W.; David James, C. Sporadic medulloblastomas contain PTCH mutations. *Cancer Res.* **1997**, *57*, 842–845. [PubMed]
75. Eberhart, C.G.; Tihan, T.; Burger, P.C. Nuclear localization and mutation of β-catenin in medulloblastomas. *J. Neuropathol. Exp. Neurol.* **2000**, *59*, 333–337. [CrossRef] [PubMed]
76. Northcott, P.A.; Buchhalter, I.; Morrissy, A.S.; Hovestadt, V.; Weischenfeldt, J.; Ehrenberger, T.; Gröbner, S.; Segura-Wang, M.; Zichner, T.; Rudneva, V.A.; et al. The whole-genome landscape of medulloblastoma subtypes. *Nature* **2017**, *547*, 311–317. [CrossRef]
77. Northcott, P.A.; Shih, D.J.H.; Peacock, J.; Garzia, L.; Sorana Morrissy, A.; Zichner, T.; Stútz, A.M.; Korshunov, A.; Reimand, J.; Schumacher, S.E.; et al. Subgroup-specific structural variation across 1,000 medulloblastoma genomes. *Nature* **2012**, *487*, 49–56. [CrossRef] [PubMed]
78. Tantravedi, S.; Vesuna, F.; Winnard, P.T.; Martin, A.; Lim, M.; Eberhart, C.G.; Berlinicke, C.; Raabe, E.; van Diest, P.J.; Raman, V. Targeting DDX3 in medulloblastoma using the small molecule inhibitor RK-33. *Transl. Oncol.* **2019**, *12*, 96–105. [CrossRef]
79. Sanai, N.; Polley, M.Y.; Berger, M.S. Insular glioma resection: Assessment of patient morbidity, survival, and tumor progression—Clinical article. *J. Neurosurg.* **2010**, *112*, 1–9. [CrossRef]
80. Ius, T.; Isola, M.; Budai, R.; Pauletto, G.; Tomasino, B.; Fadiga, L.; Skrap, M. Low-grade glioma surgery in eloquent areas: Volumetric analysis of extent of resection and its impact on overall survival. A single-institution experience in 190 patients—Clinical article. *J. Neurosurg.* **2012**, *117*, 1039–1052. [CrossRef]
81. Mansoori, B.; Mohammadi, A.; Davudian, S.; Shirjang, S.; Baradaran, B. The different mechanisms of cancer drug resistance: A brief review. *Adv. Pharm. Bull.* **2017**, *7*, 339–348. [CrossRef]
82. Ostrom, Q.T.; Cioffi, G.; Gittleman, H.; Patil, N.; Waite, K.; Kruchko, C.; Barnholtz-Sloan, J.S. CBTRUS statistical report: Primary brain and other central nervous system tumors diagnosed in the United States in 2012–2016. *Neuro Oncol.* **2019**, *21*, V1–V100. [CrossRef] [PubMed]
83. Patel, A.P.; Fisher, J.L.; Nichols, E.; Abd-Allah, F.; Abdela, J.; Abdelalim, A.; Abraha, H.N.; Agius, D.; Alahdab, F.; Alam, T.; et al. Global, regional, and national burden of brain and other CNS cancer, 1990–2016: A systematic analysis for the Global Burden of Disease Study 2016. *Lancet Neurol.* **2019**, *18*, 376–393. [CrossRef]
84. Osei, E.; Walters, P.; Masella, O.; Tennant, Q.; Fishwick, A.; Dadzie, E.; Bhangu, A.; Darko, J. A review of predictive, prognostic and diagnostic biomarkers for brain tumours: Towards personalised and targeted cancer therapy. *J. Radiother. Pract.* **2019**, 1–16. [CrossRef]
85. Bartel, D.P. MicroRNAs: Genomics, biogenesis, mechanism, and function. *Cell* **2004**, *116*, 281–297. [CrossRef]
86. Lee, Y.; Jeon, K.; Lee, J.T.; Kim, S.; Kim, V.N. MicroRNA maturation: Stepwise processing and subcellular localization. *EMBO J.* **2002**, *21*, 4663–4670. [CrossRef] [PubMed]
87. Lee, Y.; Kim, M.; Han, J.; Yeom, K.H.; Lee, S.; Baek, S.H.; Kim, V.N. MicroRNA genes are transcribed by RNA polymerase II. *EMBO J.* **2004**, *23*, 4051–4060. [CrossRef]
88. Lee, Y.; Ahn, C.; Han, J.; Choi, H.; Kim, J.; Yim, J.; Lee, J.; Provost, P.; Rådmark, O.; Kim, S.; et al. The nuclear RNase III Drosha initiates microRNA processing. *Nature* **2003**, *425*, 415–419. [CrossRef]
89. Han, J.; Lee, Y.; Yeom, K.H.; Kim, Y.K.; Jin, H.; Kim, V.N. The Drosha-DGCR8 complex in primary microRNA processing. *Genes Dev.* **2004**, *18*, 3016–3027. [CrossRef]
90. Cai, X.; Hagedorn, C.H.; Cullen, B.R. Human microRNAs are processed from capped, polyadenylated transcripts that can also function as mRNAs. *RNA* **2004**, *10*, 1957–1966. [CrossRef]
91. Morlando, M.; Ballarino, M.; Gromak, N.; Pagano, F.; Bozzoni, I.; Proudfoot, N.J. Primary microRNA transcripts are processed co-transcriptionally. *Nat. Struct. Mol. Biol.* **2008**, *15*, 902–909. [CrossRef]
92. Denli, A.M.; Tops, B.B.J.; Plasterk, R.H.A.; Ketting, R.F.; Hannon, G.J. Processing of primary microRNAs by the Microprocessor complex. *Nature* **2004**, *432*, 231–235. [CrossRef] [PubMed]
93. Bohnsack, M.T.; Czaplinski, K.; Görlich, D. Exportin 5 is a RanGTP-dependent dsRNA-binding protein that mediates nuclear export of pre-miRNAs. *RNA* **2004**, *10*, 185–191. [CrossRef] [PubMed]

94. Bernstein, E.; Caudy, A.A.; Hammond, S.M.; Hannon, G.J. Role for a bidentate ribonuclease in the initiation step of RNA interference. *Nature* **2001**, *409*, 363–366. [CrossRef]
95. Hutvágner, G.; McLachlan, J.; Pasquinelli, A.E.; Bálint, É.; Tuschl, T.; Zamore, P.D. A cellular function for the RNA-interference enzyme dicer in the maturation of the let-7 small temporal RNA. *Science* **2001**, *293*, 834–838. [CrossRef] [PubMed]
96. Ketting, R.F.; Fischer, S.E.J.; Bernstein, E.; Sijen, T.; Hannon, G.J.; Plasterk, R.H.A. Dicer functions in RNA interference and in synthesis of small RNA involved in developmental timing in C. elegans. *Genes Dev.* **2001**, *15*, 2654–2659. [CrossRef]
97. Park, J.E.; Heo, I.; Tian, Y.; Simanshu, D.K.; Chang, H.; Jee, D.; Patel, D.J.; Kim, V.N. Dicer recognizes the 5′ end of RNA for efficient and accurate processing. *Nature* **2011**, *475*, 201–205. [CrossRef]
98. Olsen, P.H.; Ambros, V. The lin-4 regulatory RNA controls developmental timing in Caenorhabditis elegans by blocking LIN-14 protein synthesis after the initiation of translation. *Dev. Biol.* **1999**, *216*, 671–680. [CrossRef]
99. Liu, J.; Carmell, M.A.; Rivas, F.V.; Marsden, C.G.; Thomson, J.M.; Song, J.J.; Hammond, S.M.; Joshua-Tor, L.; Hannon, G.J. Argonaute2 is the catalytic engine of mammalian RNAi. *Science* **2004**, *305*, 1437–1441. [CrossRef]
100. Gregory, R.I.; Chendrimada, T.P.; Cooch, N.; Shiekhattar, R. Human RISC couples microRNA biogenesis and posttranscriptional gene silencing. *Cell* **2005**, *123*, 631–640. [CrossRef]
101. Lytle, J.R.; Yario, T.A.; Steitz, J.A. Target mRNAs are repressed as efficiently by microRNA-binding sites in the 5′ UTR as in the 3′ UTR. *Proc. Natl. Acad. Sci. USA* **2007**, *104*, 9667–9672. [CrossRef]
102. Guo, H.; Ingolia, N.T.; Weissman, J.S.; Bartel, D.P. Mammalian microRNAs predominantly act to decrease target mRNA levels. *Nature* **2010**, *466*, 835–840. [CrossRef] [PubMed]
103. Meister, G.; Landthaler, M.; Patkaniowska, A.; Dorsett, Y.; Teng, G.; Tuschl, T. Human Argonaute2 mediates RNA cleavage targeted by miRNAs and siRNAs. *Mol. Cell* **2004**, *15*, 185–197. [CrossRef] [PubMed]
104. Seggerson, K.; Tang, L.; Moss, E.G. Two genetic circuits repress the Caenorhabditis elegans heterochronic gene lin-28 after translation initiation. *Dev. Biol.* **2002**, *243*, 215–225. [CrossRef] [PubMed]
105. miRbase. Available online: http://www.mirbase.org/help/nomenclature.shtml (accessed on 23 September 2020).
106. Lee, R.C.; Feinbaum, R.L.; Ambros, V. The C. elegans heterochronic gene lin-4 encodes small RNAs with antisense complementarity to lin-14. *Cell* **1993**, *75*, 843–854. [CrossRef]
107. Wightman, B.; Ha, I.; Ruvkun, G. Posttranscriptional regulation of the heterochronic gene lin-14 by lin-4 mediates temporal pattern formation in C. elegans. *Cell* **1993**, *75*, 855–862. [CrossRef]
108. Grad, Y.; Aach, J.; Hayes, G.D.; Reinhart, B.J.; Church, G.M.; Ruvkun, G.; Kim, J. Computational and experimental identification of C. elegans microRNAs. *Mol. Cell* **2003**, *11*, 1253–1263. [CrossRef]
109. Griffiths-Jones, S. The microRNA registry. *Nucleic Acids Res.* **2004**, *32*, D109–D111. [CrossRef]
110. Huang, T.H.; Fan, B.; Rothschild, M.F.; Hu, Z.L.; Li, K.; Zhao, S.H. MiRFinder: An improved approach and software implementation for genome-wide fast microRNA precursor scans. *BMC Bioinform.* **2007**, *8*, 341. [CrossRef]
111. Friedländer, M.R.; Chen, W.; Adamidi, C.; Maaskola, J.; Einspanier, R.; Knespel, S.; Rajewsky, N. Discovering microRNAs from deep sequencing data using miRDeep. *Nat. Biotechnol.* **2008**, *26*, 407–415. [CrossRef]
112. Lewis, B.P.; Shih, I.H.; Jones-Rhoades, M.W.; Bartel, D.P.; Burge, C.B. Prediction of mammalian microRNA targets. *Cell* **2003**, *115*, 787–798. [CrossRef]
113. Sethupathy, P.; Megraw, M.; Hatzigeorgiou, A.G. A guide through present computational approaches for the identification of mammalian microRNA targets. *Nat. Methods* **2006**, *3*, 881–886. [CrossRef] [PubMed]
114. Ritchie, W. MicroRNA target prediction. In *Methods in Molecular Biology*; Humana Press Inc.: Totowa, NJ, USA, 2017; Volume 1513, pp. 193–200.
115. Alles, J.; Fehlmann, T.; Fischer, U.; Backes, C.; Galata, V.; Minet, M.; Hart, M.; Abu-Halima, M.; Grässer, F.A.; Lenhof, H.P.; et al. An estimate of the total number of true human miRNAs. *Nucleic Acids Res.* **2019**, *47*, 3353–3364. [CrossRef] [PubMed]
116. Friedländer, M.R.; Lizano, E.; Houben, A.J.S.; Bezdan, D.; Báñez-Coronel, M.; Kudla, G.; Mateu-Huertas, E.; Kagerbauer, B.; González, J.; Chen, K.C.; et al. Evidence for the biogenesis of more than 1000 novel human microRNAs. *Genome Biol.* **2014**, *15*, R57. [CrossRef] [PubMed]
117. Lewis, B.P.; Burge, C.B.; Bartel, D.P. Conserved seed pairing, often flanked by adenosines, indicates that thousands of human genes are microRNA targets. *Cell* **2005**, *120*, 15–20. [CrossRef] [PubMed]

118. Chen, J.F.; Mandel, E.M.; Thomson, J.M.; Wu, Q.; Callis, T.E.; Hammond, S.M.; Conlon, F.L.; Wang, D.Z. The role of microRNA-1 and microRNA-133 in skeletal muscle proliferation and differentiation. *Nat. Genet.* **2006**, *38*, 228–233. [CrossRef] [PubMed]
119. Sugatani, T.; Hruska, K.A. Impaired micro-RNA pathways diminish osteoclast differentiation and function. *J. Biol. Chem.* **2009**, *284*, 4667–4678. [CrossRef]
120. Shi, Y.; Jin, Y.X. MicroRNA in cell differentiation and development. *Sci. China Ser. C Life Sci.* **2009**, *52*, 205–211. [CrossRef]
121. Tzur, G.; Israel, A.; Levy, A.; Benjamin, H.; Meiri, E.; Shufaro, Y.; Meir, K.; Khvalevsky, E.; Spector, Y.; Rojansky, N.; et al. Comprehensive gene and microRNA expression profiling reveals a role for microRNAs in human liver development. *PloS ONE* **2009**, *4*, e7511. [CrossRef]
122. Calin, G.A.; Dumitru, C.D.; Shimizu, M.; Bichi, R.; Zupo, S.; Noch, E.; Aldler, H.; Rattan, S.; Keating, M.; Rai, K.; et al. Frequent deletions and down-regulation of micro- RNA genes miR15 and miR16 at 13q14 in chronic lymphocytic leukemia. *Proc. Natl. Acad. Sci. USA* **2002**, *99*, 15524–15529. [CrossRef]
123. Zhang, B.; Pan, X.; Cobb, G.P.; Anderson, T.A. MicroRNAs as oncogenes and tumor suppressors. *Dev. Biol.* **2007**, *302*, 1–12. [CrossRef]
124. Svoronos, A.A.; Engelman, D.M.; Slack, F.J. OncomiR or tumor suppressor? The duplicity of MicroRNAs in cancer. *Cancer Res.* **2016**, *76*, 3666–3670. [CrossRef] [PubMed]
125. Hayashita, Y.; Osada, H.; Tatematsu, Y.; Yamada, H.; Yanagisawa, K.; Tomida, S.; Yatabe, Y.; Kawahara, K.; Sekido, Y.; Takahashi, T. A polycistronic MicroRNA cluster, miR-17-92, is overexpressed in human lung cancers and enhances cell proliferation. *Cancer Res.* **2005**, *65*, 9628–9632. [CrossRef] [PubMed]
126. Chang, T.C.; Yu, D.; Lee, Y.S.; Wentzel, E.A.; Arking, D.E.; West, K.M.; Dang, C.V.; Thomas-Tikhonenko, A.; Mendell, J.T. Widespread microRNA repression by Myc contributes to tumorigenesis. *Nat. Genet.* **2008**, *40*, 43–50. [CrossRef]
127. Wong, Q.W.L.; Lung, R.W.M.; Law, P.T.Y.; Lai, P.B.S.; Chan, K.Y.Y.; To, K.F.; Wong, N. MicroRNA-223 is commonly repressed in hepatocellular carcinoma and potentiates expression of stathmin1. *Gastroenterology* **2008**, *135*, 257–269. [CrossRef] [PubMed]
128. Lehmann, U.; Hasemeier, B.; Christgen, M.; Müller, M.; Römermann, D.; Länger, F.; Kreipe, H. Epigenetic inactivation of microRNA gene hsa-mir-9-1 in human breast cancer. *J. Pathol.* **2008**, *214*, 17–24. [CrossRef] [PubMed]
129. Saito, Y.; Liang, G.; Egger, G.; Friedman, J.M.; Chuang, J.C.; Coetzee, G.A.; Jones, P.A. Specific activation of microRNA-127 with downregulation of the proto-oncogene BCL6 by chromatin-modifying drugs in human cancer cells. *Cancer Cell* **2006**, *9*, 435–443. [CrossRef] [PubMed]
130. Walz, A.L.; Ooms, A.; Gadd, S.; Gerhard, D.S.; Smith, M.A.; Guidry Auvil, J.M.; Meerzaman, D.; Chen, Q.-R.; Hsu, C.H.; Yan, C.; et al. Recurrent DGCR8, DROSHA, and SIX homeodomain mutations in favorable histology Wilms tumors. *Cancer Cell* **2015**, *27*, 286–297. [CrossRef] [PubMed]
131. Iliou, M.S.; Da Silva-Diz, V.; Carmona, F.J.; Ramalho-Carvalho, J.; Heyn, H.; Villanueva, A.; Muñoz, P.; Esteller, M. Impaired DICER1 function promotes stemness and metastasis in colon cancer. *Oncogene* **2014**, *33*, 4003–4015. [CrossRef]
132. Ciafrè, S.A.; Galardi, S.; Mangiola, A.; Ferracin, M.; Liu, C.G.; Sabatino, G.; Negrini, M.; Maira, G.; Croce, C.M.; Farace, M.G. Extensive modulation of a set of microRNAs in primary glioblastoma. *Biochem. Biophys. Res. Commun.* **2005**, *334*, 1351–1358. [CrossRef]
133. Shi, L.; Cheng, Z.; Zhang, J.; Li, R.; Zhao, P.; Fu, Z.; You, Y. Hsa-mir-181a and hsa-mir-181b function as tumor suppressors in human glioma cells. *Brain Res.* **2008**, *1236*, 185–193. [CrossRef]
134. Li, W.B.; Ma, M.W.; Dong, L.J.; Wang, F.; Chen, L.X.; Li, X.R. MicroRNA-34a targets notch1 and inhibits cell proliferation in glioblastoma multiforme. *Cancer Biol.* **2011**, *12*, 477–483. [CrossRef] [PubMed]
135. Liu, F.; Xiong, Y.; Zhao, Y.; Tao, L.; Zhang, Z.; Zhang, H.; Liu, Y.; Feng, G.; Li, B.; He, L.; et al. Identification of aberrant microRNA expression pattern in pediatric gliomas by microarray. *Diagn. Pathol.* **2013**, *8*, 158. [CrossRef] [PubMed]
136. Zhu, X.P.; Mou, K.J.; Xu, Q.F.; Tang, J.H.; Huang, G.H.; Xu, J.P.; Li, G.H.; Ai, S.J.; Hugnot, J.P.; Zhou, Z.; et al. Microarray analysis of the aberrant microRNA expression pattern in gliomas of different grades. *Oncol. Rep.* **2015**, *34*, 318–324. [CrossRef] [PubMed]
137. Sun, J.; Shi, H.; Lai, N.; Liao, K.; Zhang, S.; Lu, X. Overexpression of microRNA-155 predicts poor prognosis in glioma patients. *Med. Oncol.* **2014**, *31*, 911. [CrossRef] [PubMed]

138. Liu, Q.; Zou, R.; Zhou, R.; Gong, C.; Wang, Z.; Cai, T.; Tan, C.; Fang, J. MiR-155 regulates glioma cells invasion and chemosensitivity by p38 isforms in vitro. *J. Cell. Biochem.* **2015**, *116*, 1213–1221. [CrossRef]
139. Wu, W.; Yu, T.; Wu, Y.; Tian, W.; Zhang, J.; Wang, Y. The miR155HG/miR-185/ANXA2 loop contributes to glioblastoma growth and progression. *J. Exp. Clin. Cancer Res.* **2019**, *38*, 133. [CrossRef]
140. Chen, G.; Chen, Z.; Zhao, H. MicroRNA-155-3p promotes glioma progression and temozolomide resistance by targeting Six1. *J. Cell. Mol. Med.* **2020**, *24*, 5363–5374. [CrossRef]
141. Zhang, C.; Zhang, J.; Hao, J.; Shi, Z.; Wang, Y.; Han, L.; Yu, S.; You, Y.; Jiang, T.; Wang, J.; et al. High level of miR-221/222 confers increased cell invasion and poor prognosis in glioma. *J. Transl. Med.* **2012**, *10*, 119. [CrossRef]
142. Zhang, R.; Pang, B.; Xin, T.; Guo, H.; Xing, Y.; Xu, S.; Feng, B.; Liu, B.; Pang, Q. Plasma miR-221/222 Family as novel descriptive and prognostic biomarkers for glioma. *Mol. Neurobiol.* **2016**, *53*, 1452–1460. [CrossRef]
143. Zhang, J.; Han, L.; Ge, Y.; Zhou, X.; Zhang, A.; Zhang, C.; Zhong, Y.; You, Y.; Pu, P.; Kang, C. MiR-221/222 promote malignant progression of glioma through activation of the Akt pathway. *Int. J. Oncol.* **2010**, *36*, 913–920.
144. Zhang, C.; Kang, C.; You, Y.; Pu, P.; Yang, W.; Zhao, P.; Wang, G.; Wang, A.; Jia, Z.; Han, L.; et al. Co-suppression of miR-221/222 cluster suppresses human glioma cell growth by targeting p27kip1 in vitro and in vivo. *Int. J. Oncol.* **2009**, *34*, 1653–1660. [PubMed]
145. Chen, L.; Zhang, J.; Han, L.; Zhang, A.; Zhang, C.; Zheng, Y.; Jiang, T.; Pu, P.; Jiang, C.; Kang, C. Downregulation of miR-221/222 sensitizes glioma cells to temozolomide by regulating apoptosis independently of p53 status. *Oncol. Rep.* **2012**, *27*, 854–860. [PubMed]
146. Tang, H.; Liu, Q.; Liu, X.; Ye, F.; Xie, X.; Xie, X.; Wu, M. Plasma miR-185 as a predictive biomarker for prognosis of malignant glioma. *J. Cancer Res.* **2015**, *11*, 630–634.
147. Yue, X.; Lan, F.; Hu, M.; Pan, Q.; Wang, Q.; Wang, J. Downregulation of serum microRNA-205 as a potential diagnostic and prognostic biomarker for human glioma. *J. Neurosurg.* **2016**, *124*, 122–128. [CrossRef]
148. Yue, X.; Wang, P.; Xu, J.; Zhu, Y.; Sun, G.; Pang, Q.; Tao, R. MicroRNA-205 functions as a tumor suppressor in human glioblastoma cells by targeting VEGF-A. *Oncol. Rep.* **2012**, *27*, 1200–1206. [CrossRef]
149. Tang, H.; Wang, Z.; Liu, X.; Liu, Q.; Xu, G.; Li, G.; Wu, M. LRRC4 inhibits glioma cell growth and invasion through a miR-185- dependent pathway. *Curr. Cancer Drug Targets* **2012**, *12*, 1032–1042. [CrossRef]
150. Batash, R.; Asna, N.; Schaffer, P.; Francis, N.; Schaffer, M. Glioblastoma multiforme, diagnosis and treatment; Recent literature review. *Curr. Med. Chem.* **2017**, *24*, 3002–3009. [CrossRef]
151. Chen, J.; Yang, L.; Wang, X. Reduced circulating microRNA-203 predicts poor prognosis for glioblastoma. *Cancer Biomark.* **2017**, *20*, 521–526. [CrossRef]
152. Liao, H.; Bai, Y.; Qiu, S.; Zheng, L.; Huang, L.; Liu, T.; Wang, X.; Liu, Y.; Xu, N.; Yan, X.; et al. MiR-203 downregulation is responsible for chemoresistance in human glioblastoma by promoting epithelial-mesenchymal transition via SNAI2. *Oncotarget* **2015**, *6*, 8914–8928. [CrossRef]
153. Jia, J.; Wang, J.; Yin, M.; Liu, Y. MicroRNA-605 directly targets sox9 to alleviate the aggressive phenotypes of glioblastoma multiforme cell lines by deactivating the PI3K/Akt pathway. *Onco Targets Ther.* **2019**, *12*, 5437–5448. [CrossRef]
154. Saydam, O.; Shen, Y.; Würdinger, T.; Senol, O.; Boke, E.; James, M.F.; Tannous, B.A.; Stemmer-Rachamimov, A.O.; Yi, M.; Stephens, R.M.; et al. Downregulated microRNA-200a in meningiomas promotes tumor growth by reducing E-cadherin and activating the Wnt/β-catenin signaling pathway. *Mol. Cell. Biol.* **2009**, *29*, 5923–5940. [CrossRef] [PubMed]
155. Senol, O.; Schaaij-Visser, T.B.M.; Erkan, E.P.; Dorfer, C.; Lewandrowski, G.; Pham, T.V.; Piersma, S.R.; Peerdeman, S.M.; Ströbel, T.; Tannous, B.; et al. MiR-200a-mediated suppression of non-muscle heavy chain IIb inhibits meningioma cell migration and tumor growth in vivo. *Oncogene* **2014**, *34*, 1790–1798. [CrossRef] [PubMed]
156. Kliese, N.; Gobrecht, P.; Pachow, D.; Andrae, N.; Wilisch-Neumann, A.; Kirches, E.; Riek-Burchardt, M.; Angenstein, F.; Reifenberger, G.; Riemenschneider, M.; et al. MiRNA-145 is downregulated in atypical and anaplastic meningiomas and negatively regulates motility and proliferation of meningioma cells. *Oncogene* **2013**, *32*, 4712–4720. [CrossRef]
157. Shi, L.; Jiang, D.; Sun, G.; Wan, Y.; Zhang, S.; Zeng, Y.; Pan, T.; Wang, Z. MiR-335 promotes cell proliferation by directly targeting Rb1 in meningiomas. *J. Neurooncol.* **2012**, *110*, 155–162. [CrossRef] [PubMed]

158. Shu, M.; Zheng, X.; Wu, S.; Lu, H.; Leng, T.; Zhu, W.; Zhou, Y.; Ou, Y.; Lin, X.; Lin, Y.; et al. Targeting oncogenic miR-335 inhibits growth and invasion of malignant astrocytoma cells. *Mol. Cancer* **2011**, *10*, 59. [CrossRef]
159. Wang, M.; Deng, X.; Ying, Q.; Jin, T.; Li, M.; Liang, C. MicroRNA-224 targets ERG2 and contributes to malignant progressions of meningioma. *Biochem. Biophys. Res. Commun.* **2015**, *460*, 354–361. [CrossRef]
160. Zhi, F.; Shao, N.; Li, B.; Xue, L.; Deng, D.; Xu, Y.; Lan, Q.; Peng, Y.; Yang, Y. A serum 6-miRNA panel as a novel non-invasive biomarker for meningioma. *Sci. Rep.* **2016**, *6*, 32067. [CrossRef]
161. Slavik, H.; Balik, V.; Vrbkova, J.; Rehulkova, A.; Vaverka, M.; Hrabalek, L.; Ehrmann, J.; Vidlarova, M.; Gurska, S.; Hajduch, M.; et al. Identification of meningioma patients at high risk of tumor recurrence using microRNA profiling. *Neurosurgery* **2020**, 3, nyaa009. [CrossRef]
162. Xiao, F.; Lv, S.; Zong, Z.; Wu, L.; Tang, X.; Kuang, W.; Zhang, P.; Li, X.; Fu, J.; Xiao, M.; et al. Cerebrospinal fluid biomarkers for brain tumor detection: Clinical roles and current progress. *Am. J. Transl. Res.* **2020**, *12*, 1379–1396.
163. Drusco, A.; Bottoni, A.; Lagana', A.; Acunzo, M.; Fassan, M.; Cascione, L.; Antenucci, A.; Kumchala, P.; Vicentini, C.; Gardiman, M.P.; et al. A differentially expressed set of microRNAs in cerebro-spinal fluid (CSF) can diagnose CNS malignancies. *Oncotarget* **2015**, *6*, 20829–20839. [CrossRef]
164. Petrescu, G.E.D.; Sabo, A.A.; Torsin, L.I.; Calin, G.A.; Dragomir, M.P. MicroRNA based theranostics for brain cancer: Basic principles. *J. Exp. Clin. Cancer Res.* **2019**, *38*, 231. [CrossRef] [PubMed]
165. Baraniskin, A.; Kuhnhenn, J.; Schlegel, U.; Maghnouj, A.; Zöllner, H.; Schmiegel, W.; Hahn, S.; Schroers, R. Identification of microRNAs in the cerebrospinal fluid as biomarker for the diagnosis of glioma. *Neuro Oncol.* **2012**, *14*, 29–33. [CrossRef] [PubMed]
166. Kopkova, A.; Sana, J.; Machackova, T.; Vecera, M.; Radova, L.; Trachtova, K.; Vybihal, V.; Smrcka, M.; Kazda, T.; Slaby, O.; et al. Cerebrospinal fluid microRNA signatures as diagnostic biomarkers in brain tumors. *Cancers* **2019**, *11*, 1546. [CrossRef]
167. Teplyuk, N.M.; Mollenhauer, B.; Gabriely, G.; Giese, A.; Kim, E.; Smolsky, M.; Kim, R.Y.; Saria, M.G.; Pastorino, S.; Kesari, S.; et al. MicroRNAs in cerebrospinal fluid identify glioblastoma and metastatic brain cancers and reflect disease activity. *Neuro Oncol.* **2012**, *14*, 689–700. [CrossRef] [PubMed]
168. van der Vos, K.E.; Abels, E.R.; Zhang, X.; Lai, C.; Carrizosa, E.; Oakley, D.; Prabhakar, S.; Mardini, O.; Crommentuijn, M.H.W.; Skog, J.; et al. Directly visualized glioblastoma-derived extracellular vesicles transfer RNA to microglia/macrophages in the brain. *Neuro Oncol.* **2016**, *18*, 58–69. [CrossRef] [PubMed]
169. Abels, E.R.; Maas, S.L.N.; Nieland, L.; Wei, Z.; Cheah, P.S.; Tai, E.; Kolsteeg, C.J.; Dusoswa, S.A.; Ting, D.T.; Hickman, S.; et al. Glioblastoma-associated microglia reprogramming is mediated by functional transfer of extracellular miR-21. *Cell Rep.* **2019**, *28*, 3105–3119.e7. [CrossRef] [PubMed]
170. Gabriely, G.; Wurdinger, T.; Kesari, S.; Esau, C.C.; Burchard, J.; Linsley, P.S.; Krichevsky, A.M. MicroRNA 21 promotes glioma invasion by targeting matrix metalloproteinase regulators. *Mol. Cell. Biol.* **2008**, *28*, 5369–5380. [CrossRef] [PubMed]
171. Papagiannakopoulos, T.; Shapiro, A.; Kosik, K.S. MicroRNA-21 targets a network of key tumor-suppressive Pathways in glioblastoma cells. *Cancer Res.* **2008**, *68*, 8164–8172. [CrossRef]
172. Sun, X.; Ma, X.; Wang, J.; Zhao, Y.; Wang, Y.; Bihl, J.C.; Chen, Y.; Jiang, C. Glioma stem cells-derived exosomes promote the angiogenic ability of endothelial cells through miR-21/VEGF signal. *Oncotarget* **2017**, *8*, 36137–36148. [CrossRef]
173. Akers, J.C.; Ramakrishnan, V.; Kim, R.; Skog, J.; Nakano, I.; Pingle, S.; Kalinina, J.; Hua, W.; Kesari, S.; Mao, Y.; et al. MiR-21 in the extracellular vesicles (EVs) of cerebrospinal fluid (CSF): A platform for glioblastoma biomarker development. *PloS ONE* **2013**, *8*, e78115. [CrossRef]
174. Shi, R.; Wang, P.Y.; Li, X.Y.; Chen, J.X.; Li, Y.; Zhang, X.Z.; Zhang, C.G.; Jiang, T.; Li, W.B.; Ding, W.; et al. Exosomal levels of miRNA-21 from cerebrospinal fluids associated with poor prognosis and tumor recurrence of glioma patients. *Oncotarget* **2015**, *6*, 26971–26981. [CrossRef] [PubMed]
175. Cai, Q.; Zhu, A.; Gong, L. Exosomes of glioma cells deliver miR-148a to promote proliferation and metastasis of glioblastoma via targeting CADM1. *Bull. Cancer* **2018**, *105*, 643–651. [CrossRef] [PubMed]
176. Lan, F.; Qing, Q.; Pan, Q.; Hu, M.; Yu, H.; Yue, X. Serum exosomal miR-301a as a potential diagnostic and prognostic biomarker for human glioma. *Cell. Oncol.* **2018**, *41*, 25–33. [CrossRef] [PubMed]

177. Manterola, L.; Guruceaga, E.; Gállego Pérez-Larraya, J.; González-Huarriz, M.; Jauregui, P.; Tejada, S.; Diez-Valle, R.; Segura, V.; Samprón, N.; Barrena, C.; et al. A small noncoding RNA signature found in exosomes of GBM patient serum as a diagnostic tool. *Neuro Oncol.* **2014**, *16*, 520–527. [CrossRef] [PubMed]
178. Yin, J.; Zeng, A.; Zhang, Z.; Shi, Z.; Yan, W.; You, Y. Exosomal transfer of miR-1238 contributes to temozolomide-resistance in glioblastoma. *EBioMedicine* **2019**, *42*, 238–251. [CrossRef] [PubMed]
179. Bronisz, A.; Wang, Y.; Nowicki, M.O.; Peruzzi, P.; Ansari, K.I.; Ogawa, D.; Balaj, L.; De Rienzo, G.; Mineo, M.; Nakano, I.; et al. Extracellular vesicles modulate the glioblastoma microenvironment via a tumor suppression signaling network directed by miR-1. *Cancer Res.* **2014**, *74*, 738–750. [CrossRef]
180. Zeng, A.; Wei, Z.; Yan, W.; Yin, J.; Huang, X.; Zhou, X.; Li, R.; Shen, F.; Wu, W.; Wang, X.; et al. Exosomal transfer of miR-151a enhances chemosensitivity to temozolomide in drug-resistant glioblastoma. *Cancer Lett.* **2018**, *436*, 10–21. [CrossRef]
181. Fang, B.; Zhu, J.; Wang, Y.; Geng, F.; Li, G. MiR-454 inhibited cell proliferation of human glioblastoma cells by suppressing PDK1 expression. *Biomed. Pharm.* **2015**, *75*, 148–152. [CrossRef]
182. Shao, N.; Xue, L.; Wang, R.; Luo, K.; Zhi, F.; Lan, Q. MiR-454-3p is an exosomal biomarker and functions as a tumor suppressor in glioma. *Mol. Cancer* **2019**, *18*, 459–469. [CrossRef]
183. Macario, A.J.L.; Conway de Macario, E. Chaperone proteins and chaperonopathies. In *Stress: Physiology, Biochemistry, and Pathology*; Elsevier: Amsterdam, The Netherlands, 2019; pp. 135–152.
184. Dahiya, V.; Buchner, J. Functional principles and regulation of molecular chaperones. In *Advances in Protein Chemistry and Structural Biology*; Academic Press Inc.: Cambridge, MA, USA, 2019; Volume 114, pp. 1–60, ISBN 9780128155578.
185. Hendrick, J.P.; Hartl, F.U. Molecular chaperone functions of heat-shock proteins. *Annu. Rev. Biochem.* **1993**, *62*, 349–384. [CrossRef]
186. Becker, J.; Craig, E.A. Heat-shock proteins as molecular chaperones. *Eur. J. Biochem.* **1994**, *219*, 11–23. [CrossRef] [PubMed]
187. Nollen, E.A.; Morimoto, R.I. Chaperoning signaling pathways: Molecular chaperones as stress-sensing "heat shock" proteins. *J. Cell Sci.* **2002**, *115*, 2809–2816. [PubMed]
188. Jacob, P.; Hirt, H.; Bendahmane, A. The heat-shock protein/chaperone network and multiple stress resistance. *Plant. Biotechnol. J.* **2017**, *15*, 405–414. [CrossRef] [PubMed]
189. Ellis, J. Proteins as molecular chaperones. *Nature* **1988**, *328*, 378–379. [CrossRef] [PubMed]
190. Ellis, R.J. The molecular chaperone concept. *Semin. Cell Biol.* **1990**, *1*, 1–9.
191. Ellis, R.J. Molecular chaperones: Assisting assembly in addition to folding. *Trends Biochem. Sci.* **2006**, *31*, 395–401. [CrossRef]
192. Kim, Y.E.; Hipp, M.S.; Bracher, A.; Hayer-Hartl, M.; Ulrich Hartl, F. Molecular chaperone functions in protein folding and proteostasis. *Annu. Rev. Biochem.* **2013**, *82*, 323–355. [CrossRef]
193. Finka, A.; Sharma, S.K.; Goloubinoff, P. Multi-layered molecular mechanisms of polypeptide holding, unfolding and disaggregation by HSP70/HSP110 chaperones. *Front. Mol. Biosci.* **2015**, *2*, 29. [CrossRef]
194. Mogk, A.; Bukau, B.; Kampinga, H.H. Cellular handling of protein aggregates by disaggregation machines. *Mol. Cell* **2018**, *69*, 214–226. [CrossRef]
195. Kaushik, S.; Cuervo, A.M. The coming of age of chaperone-mediated autophagy. *Nat. Rev. Mol. Cell Biol.* **2018**, *19*, 365–381. [CrossRef]
196. Willison, K.R. The structure and evolution of eukaryotic chaperonin-containing TCP-1 and its mechanism that folds actin into a protein spring. *Biochem. J.* **2018**, *475*, 3009–3034. [CrossRef] [PubMed]
197. Adams, B.M.; Oster, M.E.; Hebert, D.N. Protein quality control in the endoplasmic reticulum. *Protein J.* **2019**, *38*, 317–329. [CrossRef] [PubMed]
198. Juste, Y.R.; Cuervo, A.M. Analysis of chaperone-mediated autophagy. In *Methods in Molecular Biology*; Humana Press Inc.: Totowa, NJ, USA, 2019; Volume 1880, pp. 703–727.
199. Pockley, A.G.; Henderson, B. Extracellular cell stress (Heat shock) proteins—immune responses and disease: An overview. *Philos. Trans. R. Soc. B Biol. Sci.* **2018**, *373*, 20160522. [CrossRef] [PubMed]
200. Henderson, B.; Calderwood, S.K.; Coates, A.R.M.; Cohen, I.; van Eden, W.; Lehner, T.; Pockley, A.G. Caught with their PAMPs down? The extracellular signalling actions of molecular chaperones are not due to microbial contaminants. *Cell Stress Chaperones* **2010**, *15*, 123–141. [CrossRef]
201. Henderson, B.; Pockley, A.G. Molecular chaperones and protein-folding catalysts as intercellular signaling regulators in immunity and inflammation. *J. Leukoc. Biol.* **2010**, *88*, 445–462. [CrossRef]

202. Macario, A.J.L.; Conway de Macario, E. Sick chaperones, cellular stress, and disease. *New Engl. J. Med.* **2005**, *353*, 1489–1501. [CrossRef]
203. Macario, A.J.L.; Conway de Macario, E.; Cappello, F. *The Chaperonopathies. Diseases with Defective Molecular Chaperones*; Springer: Berlin/Heidelberg, Germany, 2013.
204. Hermisson, M.; Strik, H.; Rieger, J.; Dichgans, J.; Meyermann, R.; Weller, M. Expression and functional activity of heat shock proteins in human glioblastoma multiforme. *Neurology* **2000**, *54*, 1357–1364. [CrossRef]
205. Yang, I.; Fang, S.; Parsa, A.T. Heat shock proteins in glioblastomas. *Neurosurg. Clin.* **2010**, *21*, 111–123. [CrossRef]
206. Rajesh, Y.; Biswas, A.; Mandal, M. Glioma progression through the prism of heat shock protein mediated extracellular matrix remodeling and epithelial to mesenchymal transition. *Exp. Cell Res.* **2017**, *359*, 299–311. [CrossRef]
207. Iglesia, R.P.; de Lima Fernandes, C.F.; Coelho, B.P.; Prado, M.B.; Escobar, M.I.M.; Almeida, G.H.D.R.; Lopes, M.H. Heat shock proteins in glioblastoma biology: Where do we stand? *Int. J. Mol. Sci.* **2019**, *20*, 5794. [CrossRef]
208. Alexiou, G.A. Role of heat shock proteins in brain tumors. In *Heat Shock Proteins in Neuroscience*; Springer International Publishing: Cham, Switzerland, 2019; pp. 23–28.
209. Khalid, H.; Tsutsumi, K.; Yamashita, H.; Kishikawa, M.; Yasunaga, A.; Shibata, S. Expression of the small heat shock protein (hsp) 27 in human astrocytomas correlates with histologic grades and tumor growth fractions. *Cell. Mol. Neurobiol.* **1995**, *15*, 257–268. [CrossRef] [PubMed]
210. Assimakopoulou, M.; Sotiropoulou-Bonikou, G.; Maraziotis, T.; Varakis, I. Prognostic significance of Hsp-27 in astrocytic brain tumors: An immunohistochemical study. *Anticancer Res.* **1997**, *17*, 2677–2682. [PubMed]
211. Nomura, N.; Nomura, M.; Sugiyama, K.; Hamada, J.I. Phorbol 12-myristate 13-acetate (PMA)-induced migration of glioblastoma cells is mediated via p38MAPK/Hsp27 pathway. *Biochem. Pharm.* **2007**, *74*, 690–701. [CrossRef] [PubMed]
212. Sang, D.P.; Li, R.J.; Lan, Q. Quercetin sensitizes human glioblastoma cells to temozolomide in vitro via inhibition of Hsp27. *Acta Pharm. Sin.* **2014**, *35*, 832–838. [CrossRef] [PubMed]
213. Belkacemi, L.; Hebb, M.O. HSP27 knockdown produces synergistic induction of apoptosis by HSP90 and kinase inhibitors in glioblastoma multiforme. *Anticancer Res.* **2014**, *34*, 4915–4927. [PubMed]
214. Önay Uçar, E.; Şengelen, A. Resveratrol and siRNA in combination reduces Hsp27 expression and induces caspase-3 activity in human glioblastoma cells. *Cell Stress Chaperones* **2019**, *24*, 763–775. [CrossRef]
215. Wu, Z.B.; Cai, L.; Lin, S.J.; Leng, Z.G.; Guo, Y.H.; Yang, W.L.; Chu, Y.W.; Yang, S.-H.; Zhao, W.G. Heat shock protein 47 promotes glioma angiogenesis. *Brain Pathol.* **2016**, *26*, 31–42. [CrossRef]
216. Zhao, D.; Jiang, X.; Yao, C.; Zhang, L.; Liu, H.; Xia, H.; Wang, Y. Heat shock protein 47 regulated by miR-29a to enhance glioma tumor growth and invasion. *J. Neurooncol.* **2014**, *118*, 39–47. [CrossRef]
217. Jiang, X.; Zhou, T.; Wang, Z.; Qi, B.; Xia, H. HSP47 promotes glioblastoma stemlike cell survival by modulating tumor microenvironment extracellular matrix through TGF-β pathway. *Acs Chem. Neurosci.* **2017**, *8*, 128–134. [CrossRef]
218. Ghosh, J.C.; Siegelin, M.D.; Dohi, T.; Altieri, D.C. Heat shock protein 60 regulation of the mitochondrial permeability transition pore in tumor cells. *Cancer Res.* **2010**, *70*, 8988–8993. [CrossRef]
219. Tang, H.; Li, J.; Liu, X.; Wang, G.; Luo, M.; Deng, H. Down-regulation of HSP60 suppresses the proliferation of glioblastoma cells via the ROS/AMPK/mTOR pathway. *Sci. Rep.* **2016**, *6*, 28388. [CrossRef] [PubMed]
220. Rappa, F.; Unti, E.; Baiamonte, P.; Cappello, F.; Scibetta, N. Different immunohistochemical levels of Hsp60 and Hsp70 in a subset of brain tumors and putative role of Hsp60 in neuroepithelial tumorigenesis. *Eur. J. Histochem.* **2013**, *57*, e20. [CrossRef] [PubMed]
221. Caruso Bavisotto, C.; Marino Gammazza, A.; Cascio, F.L.; Mocciaro, E.; Vitale, A.M.; Vergilio, G.; Pace, A.; Cappello, F.; Campanella, C.; Palumbo Piccionello, A. Curcumin affects HSP60 folding activity and levels in neuroblastoma cells. *Int. J. Mol. Sci.* **2020**, *21*, 661. [CrossRef]
222. Marino Gammazza, A.; Caruso Bavisotto, C.; David, S.; Barone, R.; Rappa, F.; Campanella, C.; Conway de Macario, E.; Cappello, F.; Macario, A.J.L. HSP60 is a ubiquitous player in the physiological and pathogenic interactions between the chaperoning and the immune systems. *Curr. Immunol. Rev.* **2017**, *13*, 44–55.
223. Caruso Bavisotto, C.; Alberti, G.; Vitale, A.M.; Paladino, L.; Campanella, C.; Rappa, F.; Gorska, M.; Conway de Macario, E.; Cappello, F.; Macario, A.J.L.; et al. Hsp60 Post-translational Modifications: Functional and Pathological Consequences. *Front. Mol. Biosci.* **2020**, *7*, 95. [CrossRef] [PubMed]

224. Vilasi, S.; Bulone, D.; Caruso Bavisotto, C.; Campanella, C.; Marino Gammazza, A.; San Biagio, P.L.; Cappello, F.; Conway de Macario, E.; Macario, A.J.L. Chaperonin of group I: Oligomeric spectrum and biochemical and biological implications. *Front. Mol. Biosci.* **2018**, *4*, 99. [CrossRef] [PubMed]
225. Albakova, Z.; Armeev, G.A.; Kanevskiy, L.M.; Kovalenko, E.I.; Sapozhnikov, A.M. HSP70 multi-functionality in cancer. *Cells* **2020**, *9*, 587. [CrossRef]
226. Thorsteinsdottir, J.; Stangl, S.; Fu, P.; Guo, K.; Albrecht, V.; Eigenbrod, S.; Erl, J.; Gehrmann, M.; Tonn, J.C.; Multhoff, G.; et al. Overexpression of cytosolic, plasma membrane bound and extracellular heat shock protein 70 (Hsp70) in primary glioblastomas. *J. Neurooncol.* **2017**, *135*, 443–452. [CrossRef]
227. Da Rocha, A.B.; Regner, A.; Grivicich, I.; Pretto Schunemann, D.; Diel, C.; Kovaleski, G.; De Farias, C.B.; Mondadori, E.; Almeida, L.; Braga Filho, A.; et al. Radioresistance is associated to increased Hsp70 content in human glioblastoma cell lines. *Int. J. Oncol.* **2004**, *25*, 777–785. [CrossRef]
228. Sun, G.; Cao, Y.; Xu, Y.; Huai, D.; Chen, P.; Guo, J.; Li, M.; Dai, Y. Overexpression of Hsc70 promotes proliferation, migration, and invasion of human glioma cells. *J. Cell. Biochem.* **2019**, *120*, 10707–10714. [CrossRef]
229. Li, G.; Xu, Y.; Guan, D.; Liu, Z.; Liu, D.X. HSP70 protein promotes survival of C6 and U87 glioma cells by inhibition of ATF5 degradation. *J. Biol. Chem.* **2011**, *286*, 20251–20259. [CrossRef] [PubMed]
230. Lazarev, V.F.; Nikotina, A.D.; Mikhaylova, E.R.; Nudler, E.; Polonik, S.G.; Guzhova, I.V.; Margulis, B.A. Hsp70 chaperone rescues C6 rat glioblastoma cells from oxidative stress by sequestration of aggregating GAPDH. *Biochem. Biophys. Res. Commun.* **2016**, *470*, 766–771. [CrossRef]
231. Gopal, U.; Bohonowych, J.E.; Lema-Tome, C.; Liu, A.; Garrett-Mayer, E.; Wang, B.; Isaacs, J.S. A novel extracellular Hsp90 mediated co-receptor function for LRP1 regulates EphA2 dependent glioblastoma cell invasion. *PloS ONE* **2011**, *6*, e17649. [CrossRef] [PubMed]
232. Thuringer, D.; Hammann, A.; Benikhlef, N.; Fourmaux, E.; Bouchot, A.; Wettstein, G.; Solary, E.; Garrido, C. Transactivation of the epidermal growth factor receptor by heat shock protein 90 via toll-like receptor 4 contributes to the migration of glioblastoma cells. *J. Biol. Chem.* **2011**, *286*, 3418–3428. [CrossRef] [PubMed]
233. Memmel, S.; Sisario, D.; Zöller, C.; Fiedler, V.; Katzer, A.; Heiden, R.; Becker, N.; Eing, L.; Ferreira, F.L.R.; Zimmermann, H.; et al. Migration pattern, actin cytoskeleton organization and response to PI3K-, mTOR-, and Hsp90-inhibition of glioblastoma cells with different invasive capacities. *Oncotarget* **2017**, *8*, 45298–45310. [CrossRef] [PubMed]
234. Snigireva, A.V.; Vrublevskaya, V.V.; Skarga, Y.Y.; Morenkov, O.S. Cell surface heparan sulfate proteoglycans are involved in the extracellular Hsp90-stimulated migration and invasion of cancer cells. *Cell Stress Chaperones* **2019**, *24*, 309–322. [CrossRef] [PubMed]
235. Wachsberger, P.R.; Lawrence, Y.R.; Liu, Y.; Rice, B.; Feo, N.; Leiby, B.; Dicker, A.P. Hsp90 inhibition enhances PI-3 kinase inhibition and radiosensitivity in glioblastoma. *J. Cancer Res. Clin. Oncol.* **2014**, *140*, 573–582. [CrossRef]
236. Sauvageot, C.M.-E.; Weatherbee, J.L.; Kesari, S.; Winters, S.E.; Barnes, J.; Dellagatta, J.; Ramakrishna, N.R.; Stiles, C.D.; Kung, A.L.-J.; Kieran, M.W.; et al. Efficacy of the HSP90 inhibitor 17-AAG in human glioma cell lines and tumorigenic glioma stem cells. *Neuro Oncol.* **2009**, *11*, 109–121. [CrossRef]
237. Zhu, H.; Woolfenden, S.; Bronson, R.T.; Jaffer, Z.M.; Barluenga, S.; Winssinger, N.; Rubenstein, A.E.; Chen, R.; Charest, A. The novel Hsp90 inhibitor NXD30001 induces tumor regression in a genetically engineered mouse model of glioblastoma multiforme. *Mol. Cancer Ther.* **2010**, *9*, 2618–2626. [CrossRef]
238. Canella, A.; Welker, A.M.; Yoo, J.Y.; Xu, J.; Abas, F.S.; Kesanakurti, D.; Nagarajan, P.; Beattie, C.E.; Sulman, E.P.; Liu, J.; et al. Efficacy of onalespib, a long-acting second-generation HSP90 inhibitor, as a single agent and in combination with temozolomide against malignant gliomas. *Clin. Cancer Res.* **2017**, *23*, 6215–6226. [CrossRef]
239. Hauser, P.; Hanzély, Z.; Jakab, Z.; Oláh, L.; Szabó, E.; Jeney, A.; Schuler, D.; Fekete, G.; Bognár, L.; Garami, M. Expression and prognostic examination of heat shock proteins (HSP 27, HSP 70, and HSP 90) in medulloblastoma. *J. Pediatr. Hematol. Oncol.* **2006**, *28*, 461–466. [CrossRef] [PubMed]
240. Alexiou, G.A.; Vartholomatos, G.; Stefanaki, K.; Patereli, A.; Dova, L.; Karamoutsios, A.; Lallas, G.; Sfakianos, G.; Moschovi, M.; Prodromou, N. Expression of heat shock proteins in medulloblastoma: Laboratory investigation. *J. Neurosurg. Pediatr.* **2013**, *12*, 452–457. [CrossRef] [PubMed]

241. Hallal, S.; Russell, B.P.; Wei, H.; Lee, M.Y.T.; Toon, C.W.; Sy, J.; Shivalingam, B.; Buckland, M.E.; Kaufman, K.L. Extracellular vesicles from neurosurgical aspirates identifies chaperonin containing TCP1 subunit 6A as a potential glioblastoma biomarker with prognostic significance. *Proteomics* **2019**, *19*, e1800157. [CrossRef] [PubMed]
242. Graner, M.W.; Cumming, R.I.; Bigner, D.D. The heat shock response and chaperones/heat shock proteins in brain tumors: Surface expression, release, and possible immune consequences. *J. Neurosci.* **2007**, *27*, 11214–11227. [CrossRef]
243. Graner, M.W.; Alzate, O.; Dechkovskaia, A.M.; Keene, J.D.; Sampson, J.H.; Mitchell, D.A.; Bigner, D.D. Proteomic and immunologic analyses of brain tumor exosomes. *FASEB J.* **2009**, *23*, 1541–1557. [CrossRef] [PubMed]
244. Epple, L.M.; Griffiths, S.G.; Dechkovskaia, A.M.; Dusto, N.L.; White, J.; Ouellette, R.J.; Anchordoquy, T.J.; Bemis, L.T.; Graner, M.W. Medulloblastoma exosome proteomics yield functional roles for extracellular vesicles. *PLoS ONE* **2012**, *7*, e42064. [CrossRef] [PubMed]
245. Graner, M.; Redzic, J.; Ung, T. Glioblastoma extracellular vesicles: Reservoirs of potential biomarkers. *Pharmgenomics Pers. Med.* **2014**, *7*, 65. [CrossRef]
246. Lane, R.; Simon, T.; Vintu, M.; Solkin, B.; Koch, B.; Stewart, N.; Benstead-Hume, G.; Pearl, F.M.G.; Critchley, G.; Stebbing, J.; et al. Cell-derived extracellular vesicles can be used as a biomarker reservoir for glioblastoma tumor subtyping. *Commun. Biol.* **2019**, *2*, 315. [CrossRef]
247. Wei, Z.; Batagov, A.O.; Schinelli, S.; Wang, J.; Wang, Y.; El Fatimy, R.; Rabinovsky, R.; Balaj, L.; Chen, C.C.; Hochberg, F.; et al. Coding and noncoding landscape of extracellular RNA released by human glioma stem cells. *Nat. Commun.* **2017**, *8*, 1145. [CrossRef]
248. Bronisz, A.; Godlewski, J.; Chiocca, E.A. Extracellular vesicles and microRNAs: Their role in Tumorigenicity and therapy for brain tumors. *Cell. Mol. Neurobiol.* **2016**, *36*, 361–376. [CrossRef]
249. Gourlay, J.; Morokoff, A.P.; Luwor, R.B.; Zhu, H.J.; Kaye, A.H.; Stylli, S.S. The emergent role of exosomes in glioma. *J. Clin. Neurosci.* **2017**, *35*, 13–23. [CrossRef]
250. Skog, J.; Würdinger, T.; van Rijn, S.; Meijer, D.H.; Gainche, L.; Sena-Esteves, M.; Curry, W.T.; Carter, B.S.; Krichevsky, A.M.; Breakefield, X.O. Glioblastoma microvesicles transport RNA and proteins that promote tumour growth and provide diagnostic biomarkers. *Nat. Cell Biol.* **2008**, *10*, 1470–1476. [CrossRef] [PubMed]
251. Gao, X.; Zhang, Z.; Mashimo, T.; Shen, B.; Nyagilo, J.; Wang, H.; Wang, Y.; Liu, Z.; Mulgaonkar, A.; Hu, X.L.; et al. Gliomas interact with non-glioma brain cells via extracellular vesicles. *Cell Rep.* **2020**, *30*, 2489–2500.e5. [CrossRef] [PubMed]
252. Zeng, A.L.; Yan, W.; Liu, Y.W.; Wang, Z.; Hu, Q.; Nie, E.; Zhou, X.; Li, R.; Wang, X.F.; Jiang, T.; et al. Tumour exosomes from cells harbouring PTPRZ1-MET fusion contribute to a malignant phenotype and temozolomide chemoresistance in glioblastoma. *Oncogene* **2017**, *36*, 5369–5381. [CrossRef] [PubMed]
253. Zhou, W.; Fong, M.Y.; Min, Y.; Somlo, G.; Liu, L.; Palomares, M.R.; Yu, Y.; Chow, A.; O'Connor, S.T.F.; Chin, A.R.; et al. Cancer-secreted miR-105 destroys vascular endothelial barriers to promote metastasis. *Cancer Cell* **2014**, *25*, 501–515. [CrossRef]
254. Tominaga, N.; Kosaka, N.; Ono, M.; Katsuda, T.; Yoshioka, Y.; Tamura, K.; Lötvall, J.; Nakagama, H.; Ochiya, T. Brain metastatic cancer cells release microRNA-181c-containing extracellular vesicles capable of destructing blood-brain barrier. *Nat. Commun.* **2015**, *6*, 6716. [CrossRef] [PubMed]
255. Xu, G.; Zhu, H.; Zhang, M.; Xu, J. Histone deacetylase 3 is associated with gastric cancer cell growth via the miR-454-mediated targeting of CHD5. *Int. J. Mol. Med.* **2018**, *41*, 155–163. [CrossRef]
256. Tytell, M. Release of heat shock proteins (Hsps) and the effects of extracellular Hsps on neural cells and tissues. *Int. J. Hyperth.* **2005**, *21*, 445–455. [CrossRef]
257. Van Den Boorn, J.G.; Schlee, M.; Coch, C.; Hartmann, G. SiRNA delivery with exosome nanoparticles. *Nat. Biotechnol.* **2011**, *29*, 325–326. [CrossRef]
258. Pullan, J.E.; Confeld, M.I.; Osborn, J.K.; Kim, J.; Sarkar, K.; Mallik, S. Exosomes as drug carriers for cancer therapy. *Mol. Pharm.* **2019**, *16*, 1789–1798. [CrossRef]
259. Kim, G.; Kim, M.; Lee, Y.; Byun, J.W.; Hwang, D.W.; Lee, M. Systemic delivery of microRNA-21 antisense oligonucleotides to the brain using T7-peptide decorated exosomes. *J. Control. Release* **2020**, *317*, 273–281. [CrossRef] [PubMed]

260. Bu, N.; Wu, H.; Sun, B.; Zhang, G.; Zhan, S.; Zhang, R.; Zhou, L. Exosome-loaded dendritic cells elicit tumor-specific CD8 + cytotoxic T cells in patients with glioma. *J. Neurooncol.* **2011**, *104*, 659–667. [CrossRef] [PubMed]
261. Bu, N.; Wu, H.; Zhang, G.; Zhan, S.; Zhang, R.; Sun, H.; Du, Y.; Yao, L.; Wang, H. Exosomes from dendritic cells loaded with chaperone-rich cell lysates elicit a potent T cell immune response Against intracranial glioma in mice. *J. Mol. Neurosci.* **2015**, *56*, 631–643. [CrossRef] [PubMed]
262. O'Brien, K.; Lowry, M.C.; Corcoran, C.; Martinez, V.G.; Daly, M.; Rani, S.; Gallagher, W.M.; Radomski, M.W.; MacLeod, R.A.F.; O'Driscoll, L. MiR-134 in extracellular vesicles reduces triple-negative breast cancer aggression and increases drug sensitivity. *Oncotarget* **2015**, *6*, 32774–32789. [CrossRef] [PubMed]
263. Su, S.F.; Chang, Y.W.; Andreu-Vieyra, C.; Fang, J.Y.; Yang, Z.; Han, B.; Lee, A.S.; Liang, G. MiR-30d, miR-181a and miR-199a-5p cooperatively suppress the endoplasmic reticulum chaperone and signaling regulator GRP78 in cancer. *Oncogene* **2013**, *32*, 4694–4701. [CrossRef]
264. Cai, H.; Xue, Y.; Wang, P.; Wang, Z.; Li, Z.; Hu, Y.; Li, Z.; Shang, X.; Liu, Y. The long noncoding RNA TUG1 regulates blood-tumor barrier permeability by targeting miR-144. *Oncotarget* **2015**, *6*, 19759–19779. [CrossRef]

applied sciences

MDPI

Review

Extracellular Chaperones as Novel Biomarkers of Overall Cancer Progression and Efficacy of Anticancer Therapy

Malgorzata Anna Krawczyk [1,*], Agata Pospieszynska [2], Małgorzata Styczewska [3], Ewa Bien [1], Sambor Sawicki [2], Antonella Marino Gammazza [4], Alberto Fucarino [4] and Magdalena Gorska-Ponikowska [5,*]

1 Department of Pediatrics, Hematology and Oncology, Medical University of Gdansk, 7 Debinki Street, 80-211 Gdansk, Poland; ewa.bien@gumed.edu.pl
2 Department of Gynecology, Oncologic Gynecology and Gynecologic Endocrinology, Medical University of Gdansk, 80-211 Gdansk, Poland; agata.pakulniewicz@gmail.com (A.P.); sambor.sawicki@gumed.edu.pl (S.S.)
3 The English Division Pediatric Oncology Scientific Circle, Medical University of Gdansk, 80-211 Gdansk, Poland; mstyczewska@gumed.edu.pl
4 Department of Biomedicine, Neurosciences and Advanced Diagnostics (BiND), University of Palermo, 90127 Palermo, Italy; antonella.marino@hotmail.it (A.M.G.); fucaro1984@gmail.com (A.F.)
5 Department of Medical Chemistry, Medical University of Gdansk, 17 Debinki Street, 80-211 Gdansk, Poland
* Correspondence: mkrawczyk@gumed.edu.pl (M.A.K.); m.gorska@gumed.edu.pl (M.G.-P.); Tel.: +48-58-349-28-80 (M.A.K.); +48-58-349-14-50 (M.G.-P.); Fax: +48-58-349-29-50 (M.A.K.); +48-58-349-14-56 (M.G.-P.)

Received: 29 June 2020; Accepted: 27 August 2020; Published: 30 August 2020

check for updates

Abstract: Exosomal heat shock proteins (Hsps) are involved in intercellular communication both in physiological and pathological conditions. They play a role in key processes of carcinogenesis including immune system regulation, cell differentiation, vascular homeostasis and metastasis formation. Thus, exosomal Hsps are emerging biomarkers of malignancies and possible therapeutic targets. Adolescents and young adults (AYAs) are patients aged 15–39 years. This age group, placed between pediatric and adult oncology, pose a particular challenge for cancer management. New biomarkers of cancer growth and progression as well as prognostic factors are desperately needed in AYAs. In this review, we attempted to summarize the current knowledge on the role of exosomal Hsps in selected solid tumors characteristic for the AYA population and/or associated with poor prognosis in this age group. These included malignant melanoma, brain tumors, and breast, colorectal, thyroid, hepatocellular, lung and gynecological tract carcinomas. The studies on exosomal Hsps in these tumors are limited; however; some have provided promising results. Although further research is needed, there is potential for future clinical applications of exosomal Hsps in AYA cancers, both as novel biomarkers of disease presence, progression or relapse, or as therapeutic targets or tools for drug delivery.

Keywords: exosomes; heat shock proteins; cancer; adolescents and young adults; AYA

1. Introduction

1.1. Pathophysiological Role of Heat Shock Proteins

Heat shock proteins (Hsps) are ubiquitously expressed housekeeping chaperones responsible for maintaining homeostasis. Hsps associate in protein folding and protection of the proteome against misfolding and aggregation dangers under physiological conditions. Hsps also promote survival of cells exposed to hypoxia, infection, increased temperature, and chemical and physical factors [1–3]. Consequently, Hsps have been found to be upregulated in cells subjected to various proteotoxic stressors as an adaptive response in order to promote cell survival.

Hsps overexpression seems to be well correlated with cancer development and resistance to chemotherapy, making Hsps promising tumor biomarkers [4,5]. Indeed, Hsps molecular chaperone levels have been found to be elevated in many cancers, and their overexpression has been associated with poor overall survival and poor response to therapy in specific cancer types [6–10].

The three major families of Hsps whose expressions and activities are abnormally enhanced in cancer are the Hsp60, the Hsp70, and the Hsp90 families [11]. Small Hsps and Hsp40 are also upregulated in malignancies. The small Hsps family includes molecules with a molecular weight of 12–43 kDa, e.g., Hsp20, Hsp22, and Hsp27 [12].

The most abundant eukaryotic Hsp protein is Hsp90, which is a ubiquitously expressed protein with a molecular weight of 90 kDa. Its expression level is estimated to be 1–2% of total proteins under physiological conditions [13–15]. Two cytoplasmic isoforms, Hsp90 alpha (Hsp90AA1, Hsp90a, HspC1) and Hsp90 beta (Hsp90AB1, Hsp90b, HspC3), and the ER homolog, glucose-regulated protein 94 (GRP94, Hsp90B1), have been distinguished [14,15]. Hsp90 alpha and Hsp90 beta are both constitutively expressed isoforms. However, Hsp90 alpha is believed to be more inducible in response to great variety of stimuli [14,16]. An increased level of Hsp90 alpha seems to be principally associated with tumor progression and sustained cancer cell proliferation. Hsp90 beta is primarily responsible for development of drug resistance and long-term cellular adaptations [17]. Hsp90 is expressed at 2–10 fold higher levels in cancer cells compared to normal cells and is suggested to be one of the key factors implicated in promotion of cancer cell survival and metastases [18].

Hsp70 is a chaperone that, together with Hsp90, constitutes part of the major chaperone machinery. Several Hsp70 family members have been distinguished in humans, including stress-inducible Hsp70 (Hsp72, HspA1) and constitutively expressed HSC70 (Hsp73, HspA8) [19,20]. Each Hsp70, similar to Hsp90 proteins, is an ATP-dependent chaperone. Hsp70 and Hsp90 proteins are believed to be antiapoptotic heat shock proteins. However, the mechanisms of their action still remain controversial. Hsp70 family members may be involved in passing the newly synthesized unfolded proteins to Hsp60 proteins under physiological conditions. Hsp70s are engaged in protein translocation, facilitating degradation, and control of the activity of a great number of proteins, mainly transcription factors [20–22].

Hsp60 is a highly conservative chaperonin that is primarily found in mitochondria. Mitochondrial Hsp60 functions as an oligomer of fourteen 60 kDa subunits. It cooperates with the heptameric 10 kDa Hsp10 complex, which forms a lid, closing the Hsp60 tetradecamer opening [23,24]. This Hsp60/Hsp10 complex is responsible for mitochondrial protein homeostasis, including active folding of unfolded proteins and ATP-dependent proteolysis of denatured or misfolded proteins [25,26]. It has been shown that mitochondrial production of Hsp60 is significantly increased under stress conditions. However, small amounts of Hsp60 protein have also been detected in the cytosol [26]. Cytosolic Hsp60 can be detected as a monomer, but only oligomers can be active [24]. The role of cytosolic Hsp60 is not clear; however, many studies have indicated its involvement in either pro- or anti-apoptotic processes [27–30]. Extracellular release of Hsp60 either through secretion or as a result of cell necrosis has been reported as well [31,32].

One of the best known small Hsps is Hsp27 with a molecular weight of 27 kDa. It is a multi-functional and ATP-independent protein that can be found in the nucleus and cytoplasm in nearly all tissues. However, the function of Hsp27 as a protein chaperone has not been extensively examined compared to other Hsps [33]. HSP27 is composed of the WDPF domain, α-crystallin domain, PSRLFDQXFGEXLL sequence, and C-terminus. The α-crystallin domain with an active structure is associated with oligomerization [34,35]. Hsp27 is able to form multimeric aggregates (alone or with other small Hsps, e.g., Hsp20) and thus stabilize different proteins by facilitating the refolding of denatured proteins into active forms [36] and protect cells from death [37]. Hsp27 is also involved in regulation of apoptosis mediated by increasing antioxidant defense of cells and interactions with STAT3, cytochrome c-1 and pro-caspase 3 [38–40]. It was shown that Hsp27 is overexpressed in various physiological (cell development, differentiation, aging) and pathological conditions, including cancers. High expression of Hsp27 is associated with increased tumorigenic potential of neoplastic cells, leading to a poor prognosis in many cancer types [41–49]. Moreover, upregulated Hsp27 is associated with resistance to cytostatics in neoplastic cells [50–55] and inhibits the apoptosis caused by radiation [56–58]. Hsp27, as a significant molecule involved in apoptosis, seems to be an interesting target in cancer therapy.

The Hsp40 family is a large group of a co-chaperones. Hsp40 has been found in the cytosol, nucleus, mitochondria, ribosomes, endosome and endoplasmic reticulum. It is involved in protein translation, folding, unfolding, translocation, and degradation. There are three subclasses of Hsp40: A, B and C. Subclasses A and B work independently of ATP and reduce stress in cells by inhibition of protein aggregation. Moreover, subclass A of Hsp40 can bind proteins and prevent their aggregation without Hsp70. Subclass B binds protein independently of Hsp70; however, it must be connected with Hsp70 to prevent the aggregation of proteins [59]. Subclass C of Hsp40 is strictly dependent on Hsp70 [59,60]. The complex of Hsp40-Hsp70 together with Hsp90 takes part in upregulation of the Akt pathway, associated with upregulation of cellular survival. However, the role of the Hsp40 family in cancers has not been extensively investigated and current data are rather controversial. Some authors underlined that Hsp40 co-chaperones are involved in tumor growth [61]. Furthermore, studies on different cell lines suggest that high expression of Hsp40 is correlated with invasion and progression of cancer cells [62–64]. From another point of view, the inhibition of neoplastic cell proliferation induced by overexpressed Hsp40 has been reported [65]. Therefore, further studies are important to understand the role of Hsp40 in the biology of malignancies.

1.2. The Role of Heat-Shock Proteins in Tumor-Associated Hypoxia and in Metastasis Formation

Hypoxia is a characteristic feature of solid tumors. It is caused by insufficient oxygen supply, not meeting the demands of rapidly proliferating cells. Both the production and extracellular release of Hsps are strongly dependent on the hypoxic tumor microenvironment [66–68]. Various Hsps have been found to be overexpressed upon hypoxic conditions, including Hsp27 [69], Hsp70 [70], and Hsp90 [71,72]. The main hypoxia-modulated proteins responsible for downstream induction of Hsps are heat shock factor 1 (HSF1) and hypoxia-inducible factor 1 (HIF1) [72,73].

On the other hand, Hsps are also involved in HIF1 regulation. HIF1, acting via downstream factors like vascular endothelial growth factor (VEGF) and nitric oxide synthase (NOS), induces neo-angiogenesis [74]. Both Hsp70 and Hsp90 were found to stabilize HIF1 subunit α (HIF1α), VEGF and NOS [75,76]. One of the reported mechanisms of HIF1α stabilization is direct binding of Hsp70 to its oxygen-dependent degradation domain [77]. In glioma, the modulation of HIF1 by Hsp90 was found to be involved in maintaining the tumor stem cell phenotype. The HIF1 stabilization led to increased expression of stem cell markers, including CD133 and nestin. Consequently, inhibition of Hsp90 impaired the HIF1 expression and resulted in the loss of stemness by glioma cells [78]. Thus, the Hsps undergo complex interactions with other hypoxia-induced factors, playing a significant role in tumor progression via promoting neo-angiogenesis and maintaining the stemness of cancer cells.

The hypoxia-induced Hsps are also known to promote cell motility both in physiological conditions and in cancer. It was reported that HIF1 induced the extracellular secretion of Hsp90α, which stimulated the migration of fibroblasts [72,79]. The secreted Hsp90α binds to the specific receptors on the membranes of target dermal and epidermal cells [80]. Hsps may also interact with extracellular matrix (ECM) proteins. Hsp90 was found to alter the fibronectin expression in prostate cancer and mediate the matrix metalloproteinase 2 (MMP2) activity in fibrosarcoma [81,82]. Hsp72 expressed on exosomes released from tumor cells may also activate myeloid-derived suppressor cells (MDSCs), contributing to the immune evasion of the tumor [83]. These mechanisms are involved in tumor invasiveness and metastasis formation.

1.3. Exosomal Heat Shock Proteins as Biomarkers of Cancer Progression and Efficacy of Anticancer Therapy

Exosomes are nanometer-sized bioactive vesicles secreted by a diverse range of cells and present in body fluids, such as: blood, urine, cerebrospinal fluid, breast milk and saliva, as well as bronchoalveolar lavage, ascitic and amniotic fluids [84].

Exosomes under an electron microscope have a "cupped" or "saucer-like" morphology [85,86]. There are various validated methods for exosome extraction. In most studies, exosomes were isolated from various cell lines [87–98], autologous tumor cells [99], blood [95,100,101], urine [95], or ascites samples [92,102]. Supernatants were separated by serial centrifugation at low speed to eliminate cellular debris and filtration [87–90,95,99–101]. Then, exosomes were ultracentrifuged and washed in phosphate-buffered saline solution. The assessment of the exosomal fraction was performed by electron microscopy [87,89,93,97,99,100,102], Western blot analysis [87–89,96,98–103], Bradford assay [89,91], bicinchoninic acid (BCA) assay [92,94], NanoSight LM10 tracking system [95,97], or acetylcholinesterase assay [100].

The molecular content of exosomes reflects the specialized functions of the original cells. However, the fate of exosomes remains enigmatic. They are able to bind target cells and/or exchange molecules, thus modulating the activity of other cells. Many families of proteins are expressed on the surface of exosomes, such as targeting/adhesion molecules, membrane and cytoskeleton molecules, Hsps and signal transduction proteins [104–106].

It has been demonstrated that various cancerous cells release exosomes containing Hsps—either by passive release, e.g., from damaged, stressed, or dead cells, or active release, including secretion of Hsp-containing exosomes [91,107–109]. It has been hypothesized that pathological conditions, such as fever and acidosis, would increase the leakiness of exosomes and release of Hsps e.g., Hsp60 [110]. As reported by the Knowlton group, different stimuli will result in different proteins being released in exosomes from the same cell type. They showed that exosomes may form in the multivesicular body, which either traffics to the lysosome or fuses with the plasma membrane, emptying its contents of exosomes into the extracellular space [110]. However, the regulation of the release of exosomal Hsps, remains to be elucidated.

These exosomal Hsps play many important roles in cancer, including: immune system regulation [91,111,112], cell differentiation [113,114], vascular homeostasis [109], angiogenesis, epithelial–mesenchymal transition (EMT), cell migration, invasion and metastasis formation [76,88,115–119]. Moreover, exosomes also contain lipids, and various types of RNA and DNA, which may be used for malignant growth and dissemination [120–126].

Several studies have reported that elevated levels of Hsps can protect malignant cells against therapy-induced apoptosis [3,127]. It has been suggested that exosomes released by tumor cells are essential factors involved in the resistance-associated secretory phenotype (RASP) of cells. Molecular co-transfer of Hsps with oncogenic factors to recipient cells can promote cancer progression and resistance against stresses such as hypoxia, radiation, drugs, and immune systems. In addition, the RASP of tumor cells can eject anticancer drugs, targeted therapeutics, and immune checkpoint inhibitors with oncosomes [128].

Recent studies have also confirmed that Hsps localized on the surface of exosomes, secreted by normal and tumor cells, are the key players in the intercellular cross-talk during the course of cancer. It has been reported that microvesicles expressing Hsp70 on their surface activate immune system cells, including natural killer (NK) cells [91,129] and macrophages [130]. Increased levels of exosomal Hsps, including Hsp90, Hsp70, and Hsp60 have been associated with poor outcome in several malignancies [3].

Therefore, it seems that exosomal Hsps offer significant opportunities for clinical applications, as potential novel biomarkers of the diagnoses or prognoses of different diseases, or for therapeutic applications and drug delivery.

1.4. Adolescents and Young Adults (AYA)

In this review we focus particularly on the role of exosomal Hsps as biomarkers in cancers developing in adolescents and young adults (AYA). AYA comprises patients aged 15–39 years. The most common cancers in AYA include: carcinomas (of the thyroid, breast, cervix, ovary, colon and liver), melanoma, brain and other central nervous system (CNS) tumors, germ cell tumors, sarcomas and lymphomas [131–133]. The distribution of particular tumors differs in the age subgroups of 15–19, 20–29 and 30–39 year-old patients [133].

The AYA age group, placed between pediatric and adult oncology, poses a particular challenge for diagnostics and therapy [134]. Consequently, the results of oncological treatment in AYA have been, for decades, worse than in other age groups with the same cancers. In addition, the progress in management of cancers occurring in AYA, measured as the mortality rate reduction, has been much smaller than in other age groups for many years [135]. Therefore, new biomarkers of cancer growth and progression as well as prognostic factors to predict response to therapy and final outcome are desperately needed in AYA.

2. Materials and Methods

A MEDLINE/PubMed database search was performed to identify studies reporting the role of exosomal Hsps (small Hsps, Hsp40, Hsp60, Hsp70 and Hsp90 families) in selected AYA cancers. The following key words were searched together with their appropriate synonyms, abbreviations and MeSH terms: "heat shock proteins" and "chaperones" combined with "exosomes" and "extracellular vesicles". All terms were searched in selected AYA neoplasms: carcinomas of the thyroid, breast, cervix, endometrium, ovary, colon, liver, lung, melanoma and malignant CNS tumors. The search was limited to manuscripts published in English from I' 1990 to V' 2020. The reference lists in all publications were also searched for additional studies. Articles in languages other than English, studies analyzing exosomal Hsps in benign diseases and malignancies not included in this review, studies analyzing the exosome-detection methodology only, studies not defining the precise location of analyzed Hsps and reviews were excluded from analysis. Summary of the methodology of the exosomes' determination and the clinical values of reviewed studies is presented in Table 1 and in Figure 1.

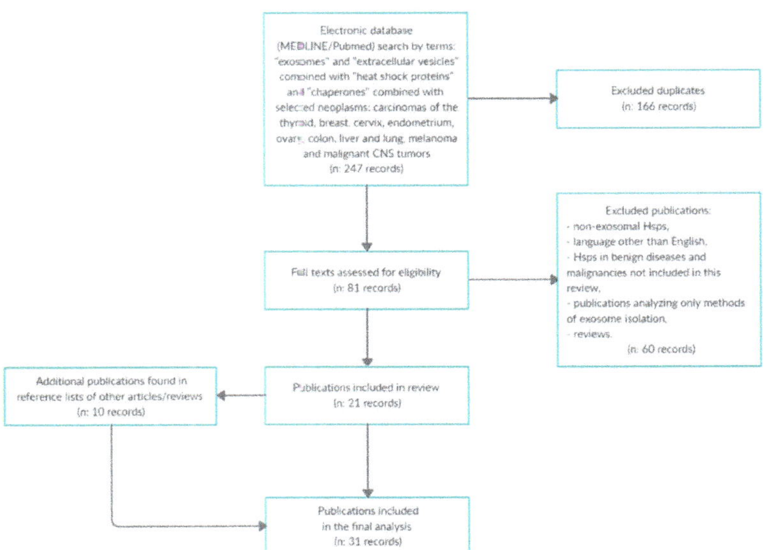

Figure 1. Summary of the methodology of the exosomes' determination and the clinical values of reviewed studies.

Table 1. Summary of the methodology of the exosomes determination and their clinical values.

Cancer Type	Analyzed Material	Exosomal Hsps	Clinical Values	Location	References
CNS					
glioma, glioblastoma multiforme, murine anaplastic astrocytoma	cell lines	Hsp27, Hsp60, Hsp70, Hsp90, others	-	not mentioned	[87,136]
glioblastoma multiforme	cell line	Hsp90α	increase in tumor cells motility by activating plasmin; tumor progression and metastasis formation	not mentioned	[88]
glioblastoma multiforme	human tumor tissue	Hsp70	antitumor immunostimulation (induction of glioma-specific cytotoxic CD8+ T-cells response)	not mentioned	[99]
Melanoma					
-	cell lines, human plasma	Hsp70, Hsp90	possible use as a diagnostic marker	not mentioned	[89,137,138]
-	human ascites aspirate	Hsp70, Hsp80	-	not mentioned	[102]
-	mice tumor tissue	Hsp70	inhibition of melanoma progression and dissemination via activation of NK cells	exosomal surface	[90]
-	cell line	Hsp70	inhibition of antitumor immune response via activation of the myeloid-derived suppressor cells	exosomal surface	[95]
Colorectal carcinoma					
-	human blood	Hsp60	possible biomarker of recurrence	exosomal surface	[100]
-	cell lines, human ascites aspirate	Hsp70	induction of anti-tumor immune response via activation of NK cells and Hsp70-dependent IL-6 secretion from dendritic cells	exosomal surface [91,95]/not mentioned [92]	[91,92,95]
-	cell lines	Hsp70	release of immunogenic exosomes containing Hsp70 by heat-stressed carcinoembryonic antigen (CEA)-positive tumor cells; possible use of exosomes as an anti-cancer vaccine	exosomal surface	[139]
Hepatocellular carcinoma					
-	cell line	Hsp70	-	not mentioned	[140]
-	cell line	Hsp70	promotion of angiogenesis, immune response	exosomal surface	[93]
-	cell line	Hsp60, Hsp70, Hsp90	enhancement of exosomal Hsps release under chemotherapy treatment; activation of NK cells resulting in stronger antitumor immune response	exosomal surface	[129]
-	cell line	Hsp70	upregulation of exosomal Hsp70 secretion and enhancement of NK-mediated immune response as a result of treatment with the epigenetic drug MS-275	exosomal surface	[94]

Table 1. *Cont.*

Cancer Type	Analyzed Material	Exosomal Hsps	Clinical Values	Location	References
-	cell lines	Hsp70	increase in the number of exosomes produced by HCC cells and the amount of exosomal Hsp70 after treatment with inhibitor of DNA methyltransferase, 5-Aza-2′-deoxycytidine	not mentioned	[141]
Thyroid cancer					
papillary thyroid carcinoma	human tumor tissue, human plasma	Hsp27, Hsp60, Hsp90	possible use as plasma-derived biomarkers of tumor presence/recurrence	not mentioned	[101]
Gynecological cancer					
ovarian papillary adenocarcinoma	human ascites aspirate	Hsp70, Hsp80	-	not mentioned	[102]
ovarian cancer	human urine	Hsp70	possible biomarker of tumor presence; correlation with tumor progression	exosomal surface	[95]
ovarian cancer, endometrial cancer	human serum	Hsp20, Hsp22	positive correlation with markers of cytotoxic immune response (Perforin and Granzyme B); increased exosomal Hsps expression in ovarian cancer patients compared to endometrial cancer patients	not mentioned	[142]
ovarian cancer	cell lines	Hsp27	dependence of exosomal Hsp27 release on intracellular Hsp27 concentration; potential use as a tumor biomarker	not mentioned	[96]
ovarian cancer	human blood, human urine	Hsp70	possible use as a biomarker of tumor presence/recurrence/dissemination	exosomal surface	[103,143,144]
cervical cancer	cell line	Hsp70, Hsp90	interaction of exosomal Hsp70 and Hsp90 with survivin	exosomal surface, inside exosomes	[145]
Breast cancer					
ductal adenocarcinoma	human ascites aspirate	Hsp70, Hsp80	-	not mentioned	[102]
-	cell line	Hsp90α	increase in tumor cells motility by activating plasmin; tumor progression and metastasis formation	not mentioned	[88]
-	human blood, human urine	Hsp70	possible use as a biomarker of tumor presence/recurrence/dissemination	exosomal surface	[103,143,144]
-	human urine	Hsp70	possible biomarker of tumor presence; correlation with tumor progression	exosomal surface	[95]
adenocarcinoma	cell line	Hsp72	release of exosomal Hsp72 triggered by IFN-γ cell stimulation; ability of Hsp72 to stimulate the release of IL-12 by naive dendritic cells.	exosomal surface	[146]

Table 1. *Cont.*

Cancer Type	Analyzed Material	Exosomal Hsps	Clinical Values	Location	References
Lung cancer					
adenocarcinoma	human ascites aspirate	Hsp70, Hsp80	-	not mentioned	[102]
-	human urine	Hsp70	possible biomarker of tumor presence; correlation with tumor progression	exosomal surface	[95]
-	cell line	Hsp60	nitration and exosomal release of Hsp60 after treatment of the tumor cells with suberoylanilide hydroxamic acid; possible role of exosomal Hsp60 in immune stimulation	inside exosomes	[147]
non-small cell lung cancer	human serum	Hsp70	correlation with response to radio- and chemotherapy; correlation with osteopontin plasma levels	lipid-bound plasma Hsp70	[148]
non-small cell lung cancer	cell line	Hsp70, Hsp90	increased concentrations of Hsp70 and Hsp90 in the exosomes derived from Rab27a-overexpressing tumor cells; role of these exosomes in immune antitumor stimulation via activation of dendritic cells	exosomal surface	[149]
non-small cell lung cancer	cell line	Hsp70	induction of pro-inflammatory phenotype by triggering TLR2 signaling and NF-κB pathway in mesenchymal stem cells; facilitating tumor growth and invasion	exosomal surface	[98]
non-small cell lung cancer	human blood, human urine	Hsp70	possible use as a biomarker of tumor presence/recurrence/dissemination	exosomal surface	[103,143,144]

Abbreviations: HCC = Hepatocellular carcinoma; Hsp = Heat shock protein; IL = interleukin; NK = natural killer; TLR = Toll-like receptor.

3. Exosomal Heat Shock Proteins as Biomarkers of Cancer Progression and Efficacy of Anticancer Therapy in AYA

3.1. Central Nervous System Tumors

Malignant CNS tumors are one of the most prevalent cancer types and the third most common cause of cancer-related death in AYA [131,150]. This heterogenous group consists of a variety of tumors; among them the major categories are gliomas, intracranial germ cell tumors and, in younger patients, embryonal tumors typical for pediatric age, such as medulloblastoma (MB). The prognosis in CNS malignancies in AYA is poorer than in children and the survival rates decrease with age [150].

The research on Hsps in malignant CNS tumors addressed mainly MB and high-grade gliomas. Various Hsps, including Hsp27, Hsp40, Hsp60, Hsp70, and Hsp90α were found to be expressed in MB specimens [151,152]. One study reported that expressions of particular Hsps significantly differed between MB histological subtypes [151]. In another, Hsp90β was one of the proteins detected in MB, but not in benign pilocytic astrocytoma [153].

The studies on glioblastoma multiforme (GBM) confirmed the high expressions of Hsps, including Hsp27, Hsp40, Hsp60, Hsp70, Hsp72, and Hsp90α in tumor specimens and GBM stem cells [154–159]. These proteins were found to be involved in pathways responsible for glioma cell invasiveness and motility and their overexpression may be associated with treatment resistance [160]. The spontaneous and radiation- or heat-shock induced release of Hsp27 and Hsp70 from glioma cells was also reported [87,161–163]. Some studies suggested the key role of extracellular Hsp90 in GBM metastasis formation [164–166]. These findings translated into in vitro and animal model studies on Hsps inhibitors and Hsp-peptide complex vaccines in GBM treatment [167–172]. Some of them recently reached the I and II phases of clinical trials [173–176].

Several studies also analyzed the role of exosomes in both pediatric and adult CNS tumors with a focus on the microRNA role [177,178] and the possible use of other exosomal proteins as biomarkers [179,180]. However, the field of exosomal Hsps in CNS malignancies was not extensively studied.

Graner et al. were the first to report the exosomal pathway of Hsp27, Hsp60, Hsp70, and Hsp90 secretion from malignant CNS tumor cells (adult glioma and murine anaplastic astrocytoma) [87]. Later, various Hsps were detected in the proteomic analysis of extracellular vesicles released by eight glioma cell lines [136]. However, the role of exosomal release of these proteins is not fully understood. McCready et al. confirmed that exosomal Hsp90α is released from various malignant cell lines, including glioma cells [88]. They also reported that, additionally to previously known mechanisms of activating matrix metalloproteinase 2 (MMP2) [81] and human epidermal growth factor receptor 2 (HER2) [181], extracellular Hsp90α increases motility of the tumor cells by also activating plasmin. Interestingly, Hsp90α-containing exosomes enhanced the movement of the glioma cells more than Hsp90α protein alone, which indicates the cooperation of Hsp90α with other exosomal proteins in promoting GBM progression and metastasis formation. Another study, however, showed the antitumor immune-stimulating function of exosomes carrying Hsp70. The glioma-derived exosomes were incubated with dendritic cells, leading to stimulation of the glioma-specific cytotoxic CD8+ T-cell response [99]. Further research is needed to better understand these mechanisms and, possibly, allow one to target exosomal Hsps in experimental therapies.

3.2. Melanoma

Malignant melanoma (MM) is one of the most prevalent cancers in AYA, constituting 14% of all malignancies in this age group. In general, it is associated with favorable prognosis, with 5-year overall survival rates exceeding 85% [131]. However, the biology and clinical course of MM in AYA have not been comprehensively studied. Young patients diagnosed at advanced stages of disease face specific obstacles, including lack of treatment standards and limited accessibility to clinical trials [182].

The aberrant expressions of Hsps, including Hsp27, Hsp40, Hsp60, Hsp70, Hsp90, and glycoprotein gp96 were reported in skin, conjunctival and uveal melanoma cells [183–191]. The expressions of Hsps were associated with clinical features of MM: worse prognosis (Hsp40 and Hsp90) [183,190,192], more advanced stage (Hsp40, Hsp60 and Hsp90) [186,190,191], and better response to treatment with bevacizumab (Hsp27) [193]. Based on these observations, some experimental studies evaluating the efficacy of Hsps inhibitors in MM treatment in vitro and on animal models were performed [183,194–197]. Some of them have been translated into clinical studies [198,199]. Immunotherapies based on tumor-derived Hsp peptide complexes (HspPCs) also reached clinical trials [200–204]. However, the role of extracellular Hsps released from MM cells in exosomes has not been extensively investigated.

Several studies detected various Hsps, including Hsp70 and Hsp90, in MM-derived exosomes [89,137]. The chaperones were more abundant in exosomes in comparison to microvesicles and apoptotic vesicles obtained from the same cell lines [138]. Exosomes containing Hsp70 and Hsp80 were also detected in malignant ascites in patients with solid tumors, including two with advanced MM [102]. The studies evaluating the role of MM-derived exosomes carrying Hsp70 yielded conflicting results. Elsner et al. found that Hsp70-overexpressing MM cells release Hsp70-positive exosomes, which in vitro activate mouse natural killer (NK) cells to kill mouse lymphoma and human melanoma cells expressing NK-activating receptors. Therefore, they suggested that Hsp70-carrying exosomes play a role in the inhibition of MM progression and dissemination [90]. Conversely, Gobbo et al. suggested that exosomal Hsp70 played a role in the inhibition of antitumor immune responses. They reported that MM-derived exosomes presenting Hsp70 on their surfaces might bind to toll-like receptor 2 (TLR2) on the myeloid-derived suppressor cells (MDSC), leading to their activation. In this study, inhibition of the extracellular domain of exosomal Hsp70 with specific peptide A8 led to the blockage of Hsp70/TLR2 association, decreasing the ability of exosomes to activate MDSCs. Consequently, treatment of melanoma-bearing mice with A8 peptide reduced the number of MDSCs in the mice spleen, inhibited MM progression and potentiated the antitumor effect of cisplatin [95].

3.3. Colorectal Carcinoma

Colorectal carcinoma (CRC) in AYA is rare. However, in contrary to the older patients benefiting from screening programs, the CRC incidence in this age group is constantly increasing. Moreover, due to experienced diagnostic delays, AYA are more likely to be diagnosed in the advanced stage of disease. The diagnosis of CRC in the young may be a sign of an underlying genetic disorder, such as familial adenomatous polyposis (FAP) or Lynch syndrome; however, most cases occur sporadically [205].

Many studies reported altered expression of various Hsps, including Hsp 20 [206], Hsp 40 [207], Hsp60 [208–210], Hsp70 [207,211], and Hsp90 [212] in the CRC specimens. The tissue Hsp status was found to be associated with tumor grade, lymph node involvement, distant metastases, response to CHT, recurrence and overall survival [206,209,212,213]. The serum concentration of Hsp60, Hsp70 and Hsp90α were assessed as possible CRC biomarkers [214–217]. Consequently, several in vitro trials with Hsps inhibitors have been conducted, yielding promising results [44,119,211,218,219]. However, the data regarding the presence and function of Hsps in the circulating exosomes released from CRC cells is scant.

Merendino et al. for the first time reported that Hsp60 is actively released from cancer cells via the exosomal pathway [107,108]. These studies were based on previous research reporting increased expression of Hsp60 in preneoplastic colon lesions and CRC, suggesting its role in tumorigenesis [208–210]. Subsequently, the same research group reported the presence of Hsp60 in the exosomes obtained from the peripheral blood of patients with CRC. Hsp60 was also detected in the inflammatory natural killer (NK) cells and macrophages within the tumor microenvironment. The levels of Hsp60 in the exosomes were significantly increased at diagnosis but comparable to healthy controls after the tumor removal. Moreover, Hsp60 was detected in the membrane of exosomes before the tumor excision, but neither in the same patients after the surgery nor in healthy controls [100].

The authors therefore suggest that the exosomal Hsp60 blood levels could serve as a promising biomarker of CRC recurrence.

Another chaperone that was found in exosomes derived from human CRC cell lines is Hsp70 [91,95,139]. Being presented at the exosomal plasma membrane, it was found to induce natural killer (NK) lymphocyte anti-tumor activity [91]. Later, Guo et al. reported that Hsp70-rich exosomes from a mice CRC cell line also elicit an immune anti-tumor response via inducing Hsp70-dependent interleukin (IL)-6 secretion from dendritic cells. The exosomes secreted after heat stress had higher concentration of Hsp70 and were more effective immune stimulators than exosomes released spontaneously [92].

3.4. Hepatocellular Carcinoma

Primary hepatic malignancies are very rare in AYA. Among them, HCC predominates. In developed countries, the incidence of HCC in AYA has been constantly increasing during the last decades [132,133].

HCC in AYA has distinct pathological and clinical features compared to older patients. In countries with low rates of hepatitis B virus (HBV) infections, it is usually of a fibrolamellar pathological subtype and not related to cirrhosis or underlying liver disease [220,221]. However, HCC has been observed in patients with Fanconi's syndrome and metabolic diseases (hereditary tyrosinemia, Wilson disease and glycogen storage disease type IA) [222–224]. As most patients are asymptomatic for a long time, the final diagnosis is delayed and the prognosis is poor [131,220].

There are several reports about exosomes in HCC [93,129,140,225–229]. They suggest that exosomes may be a new biomarker in diagnosis of early-stage HCC and potential target for the treatment of this type of the tumor. However, only single studies concern exosomal Hsps and their role in the biology and aggressiveness of HCC.

Yukawa et al. detected Hsp70 expression on the surface of hepatoma G2 (HepG2) cell-derived exosomes. They confirmed that exosomal Hsp70 can be considered as significant molecules connected with angiogenesis and immune response in HCC [93]. Another study found increased release of Hsp70-containing exosomes from HCC cell lines after the treatment with an inhibitor of DNA methyltransferase, 5-Aza-2′-deoxycytidine [141]. Lv et al. reported that release of Hsp60, Hsp70, and Hsp90 was enhanced in HCC cells under antineoplastic treatment. Additionally, secretion of exosomal Hsps caused by HCC cell-resistant anticancer drugs, such as carboplatin and irinotecan hydrochloride, was remarkably higher than sensitive drugs. The authors showed that Hsps released from exosomes can trigger activity of natural killer (NK) cells resulting in augmented cytolytic activity against targeted cells and stronger immune response [129].

Another group investigated the effect of the epigenetic drug MS-275 on exosomes from HepG2 cells [94]. The study revealed that MS-275 was responsible for higher secretion of Hsp70 in exosomes, resulting in enhancement of the exosome-induced cytotoxic activity of NK cells. Findings reported by Lv et al. and Xiao et al. can contribute to the innovative and effective treatment of HCC. Notably, further efforts are required in this field, in particular, studies on the fibrolamellar subtype of HCC, characteristic for AYA patients, are lacking.

3.5. Thyroid Cancer

Thyroid cancer (TC) is one of the most common cancers in AYA, especially in 15–24-year-old patients [133]. In this age group, TC accounts for approximately 13% of all malignant neoplasms [230]. A significant increase in the incidence of TC has been observed in recent decades [231]. There are four histological types of TC: papillary, follicular, medullary, and anaplastic. They can develop de novo or in individuals previously treated for cancer. Medullary TC can also be present in hereditary syndromes like type 2 multiple endocrine neoplasia (MEN) syndromes MEN2A, MEN2B, and the related multigenerational familial MTC (FMTC) syndrome.

The overall survival rate of TC is one of the highest among AYA cancers [232,233], but in anaplastic TC it is still unsatisfactory [234]. For this reason, there is a continuing need to understand the biology of thyroid cancer, find clinically useful biomarkers and develop novel therapies.

One of the promising options for therapeutic targets in TC are Hsps, which are overexpressed in thyroid malignancies. Many studies reported altered expression of various Hsps, including Hsp27 [235,236], Hsp60 [237], Hsp70 [238–242], and Hsp90 [241,243–245]. There were studies that showed that KU711 and WGA-TA, inhibitors of Hsp90, decrease migration and invasion of anaplastic TC cells in vitro. Moreover, these molecules lead to downregulation of β-catenin, BRAF, Akt, and phospho-Akt, leading to induction of apoptosis [246–248]. Kim et al. showed that apoptosis of anaplastic TC cells can be induced not only by inhibition of Hsp90, but also by Hsp70 inhibitors [247]. However, data on the role of exosomal Hsps in TC cells is scarce.

To our knowledge there is only one study on exosomal Hsps in TC. Bavisotto et al. for the first time reported significantly higher levels of Hsp27, Hsp60, and Hsp90 in tissue of papillary TC than in peritumoral tissue and non-toxic goiter. Moreover, the same Hsps were increased in exosomes obtained from plasma in patients with papillary TC. Furthermore, it was noticed that exosomal levels of Hsp27, Hsp60, and Hsp90 were higher before ablative surgery than after this procedure. These results suggest that exosomal Hsps play a role in carcinogenesis and may have clinical implications for the prognostic analysis and treatment of patients with TC. However, in the same study, no differences in Hsp70 level were found in thyroid cancer tissue and benign goiter. This is why the evaluation of Hsp70 in exosomes was not done [101]. The results obtained by the authors are interesting, but further studies are needed.

3.6. Gynecological Cancer

Exosomes play a significant role in intercellular female genital tract cancer interactions. They have shown immense impact on the early diagnosis, drug resistance, prognostic evaluation, metastasis and target therapies [249–251]. Exosomal proteins seem to be effective cancer biomarkers [250,252]. In gynecologic oncology, there are reports regarding tumor expression of Hsp22, Hsp27, Hsp60, Hsp70, and Hsp90 in ovarian, cervical and endometrial cancers as potential biomarkers [253]. However, there are only a few reports of exosomal HSP levels in these malignancies, mainly in ovarian cancer.

Ovarian cancer, with the high resistance to chemotherapeutic agents, remains the foremost cause of death in women globally. More than half of ovarian cancer patients are in an advanced stage at the moment of diagnosis [254] because there are no symptoms of early stage disease nor effective screening procedures. For this reason, finding a novel diagnostic method would have an impact on the treatment results.

Wyciszkiewicz et al. showed that the Hsp22 expression in peritoneal fluid samples and serum-derived exosomes in patients with ovarian cancer in comparison with endometrial cancer was higher, but the difference was statistically significant only in peritoneal fluid. They also found a positive correlation between markers of cytotoxic immune response (perforin and granzyme B) and exosomal Hsp20 and Hsp22 levels [142]. Chaperones, including Hsp70 and Hsp80, were also detected in exosomes isolated from ascites effusions in patients with advanced ovarian papillary adenocarcinoma [102]. Moreover, Stope et al. reported release of exosomal Hsp27 by ovarian cancer cell lines (OVCAR-3 and SK-OV-3). They found concentration-dependent incorporation of Hsp27 into exosomes, which suggests that Hsp27, liberated by exosomes from ovarian cancer cell lines, could serve as a noninvasive biomarker of ovarian cancer [96].

One study reported release of exosomal Hsp70 and Hsp90 from cervical cancer cells. It was suggested that these Hsps may interact with survivin, allowing for its release and intercellular transport in exosomes [145].

3.7. Breast Cancer

Breast cancer is the most common malignancy in women worldwide, representing about 25% of all cancers [255,256]. It is also the most prevalent cancer in AYA women, especially in the older subset of patients. Breast cancer in AYA is more often associated with familial cancer predisposition syndromes compared to older patients. Young females are also more likely to present with large tumors, unfavorable histological and molecular characteristics, and disseminated disease at diagnosis. Consequently, the outcomes in this age group are relatively unfavorable [257,258].

The studies on Hsps in breast cancer revealed altered expressions of Hsp90, Hsp70 and Hsp27 in human breast cancer tissues [259–261]. Yano et al. showed that a Hsp90α homolog may be involved in cell proliferation, while Hsp90ß was correlated with poor differentiated breast carcinomas [259]. Hsp27 has been also considered as a predictor for hormone sensitivity in breast cancer and also has been associated with short disease-free survival (DFS), similar to Hsp70 [260]. Heat shock proteins may be induced in response to anoxia. It has also been proven that exposure to hypoxia increases release of exosomes from breast cancer cells to promote their invasion [4,97]. However, only a few studies analyzing the role of Hsps secreted via the exosomal pathway in breast cancer have been published to date.

McCready et al. reported exosomal Hsp90α presence and secretion from solid tumor cells, including breast cancer. They found that Hsp90α alone does not affect tumor cell motility and invasiveness, unlike the addition of exosomes. This study is thoroughly described in the CNS tumors section [88]. Exosomes containing Hsp70 and Hsp80 were also detected in ascites fluid in two patients with advanced-stage ductal adenocarcinoma [102]. Bausero et al. reported that exosomal Hsp72 is released from breast cancer cells after interferon-γ stimulation and may be involved in tumor surveillance mechanisms [146]. Recently, a report from a pilot study ExoDiag (NCT02662621) was published. This project aims to establish whether it is possible to identify and quantify Hsp70-exosomes in blood and urine samples in patients with solid cancers, including breast, ovarian, and non-small cell lung cancer. This could help demonstrate that exosomal Hsp70 is a valuable biomarker of tumor presence, progression or recurrence [103,143,144].

3.8. Lung Cancer

Despite being the leading cause of cancer-related deaths in the entire population, lung cancer is exceedingly rare in AYA. Patients below 40 years of age make up only less than 1% of all lung cancer cases. In this age group, non-small cell lung cancer (NSCLC) predominates, with adenocarcinoma histology significantly more prevalent than in older adults. Probably due to better overall performance status, the prognosis in AYA with lung cancer is better than in older patients [262,263]. There are no reports of exosomal Hsps detection being performed on AYA patients' tumors. However, some more general studies on the role of exosomal Hsps in lung cancer have been published.

Campanella et al. reported that treatment of the lung carcinoma-derived cells with cytotoxic suberoylanilide hydroxamic acid (SAHA) resulted in posttranslational modification and increased exosomal release of Hsp60. They also suggested a possible role of this modified exosomal Hsp60 in immune antitumor stimulation [147]. Gobbo et al. described a higher level of exosomal Hsp70 in lung, breast and ovarian cancer patient urine samples than in healthy voluntary donors and reported exosomal Hsp70 correlation with tumor progression [95]. Exosomal Hsps were also found in malignant ascites in a patient with lung adenocarcinoma [102]. Ostheimer et al. showed a significant correlation between Hsp70 serum levels and treatment response and elevated level of osteopontin—a biomarker related to hypoxia in patients with NSCLC. Post-therapeutic plasma exosomal Hsp70 promoted a positive clinical result, defined as a better response to radio and chemotherapy. This study did not investigate exosomal Hsp70 directly; however, plasma Hsp70 was detected in two fractions: free Hsp70 and lipid-bound Hsp70, the latter corresponding to the exosomal Hsp70 [148]. Li et al. reported a relationship between lung cancer cell (A549)-derived exosomal Hsp70 presence and induction of the pro-inflammatory phenotype in mesenchymal stem cells (MSCs) by triggering TLR2 signaling and activation of the NF-κB pathway. This may facilitate lung cancer invasion and tumor growth [98]. Moreover, the elevated levels of Hsp70 and Hsp90 were detected in exosomes derived from Rab27a-overexpressing non-small cell lung carcinoma cells. These exosomes, via upregulating major histocompatibility complex class

II (MHCII) and co-stimulatory molecules on dendritic cells, were found to elicit potent antitumor immune responses [149].

Engineered exosomes have been also considered as a potential cargo for specific delivery of anticancer therapeutic drugs. Exosomes being loaded with Celastrol, which inhibits Hsp90 signaling pathways, showed higher antitumor potency in lung cancer cells. [264].

4. Conclusions and Future Directions

Exosomes and their molecular content, including Hsps, seem to be key players in intracellular communication. Notably, they can modulate antitumor immunity and may serve as potential tumor vaccines or immunotherapeutic vesicles. The future clinical application of exosomes also includes their use as plausible biomarkers for the diagnoses, prognoses and follow-up of different diseases, including cancer [106].

Author Contributions: Conceptualization, M.A.K. and M.G.-P.; database collections M.A.K., A.P.; analysis of literature data available in the selected topic—M.A.K. and E.B.; writing—original draft preparation, M.A.K., A.P., M.S., E.B., A.M.G., A.F., S.S. and M.G.-P; writing—review and editing according to reviewers' comments, M.A.K. and M.G.-P.; graphical part—M.A.K, supervision, M.A.K. and M.G.-P.; correspondence with the editor and reviewers—M.A.K. and M.G.-P.; funding acquisition M.G.-P. All authors have read and agreed to the published version of the manuscript.

Funding: The study was supported by the Iuventus Plus project of the Polish Ministry of Science and Higher Education No. IP2015 022074.

Conflicts of Interest: The authors declare no conflict of interest.

References

1. Khalil, A.A.; Kabapy, N.F.; Deraz, S.F.; Smith, C. Heat shock proteins in oncology: Diagnostic biomarkers or therapeutic targets? *Biochim. Biophys. Acta Rev. Cancer* **2011**, *1816*, 89–104. [CrossRef] [PubMed]
2. MacArio, A.J.L.; Cappello, F.; Zummo, G.; Conway De MacArio, E. Chaperonopathies of senescence and the scrambling of interactions between the chaperoning and the immune systems. *Ann. N. Y. Acad. Sci.* **2010**, *1197*, 85–93. [CrossRef] [PubMed]
3. Rappa, F.; Farina, F.; Zummo, G.; David, S.; Campanella, C.; Carini, F.; Tomasello, G.; Damiani, P.; Cappello, F.; De Macario, E.C.; et al. HSP-Molecular chaperones in cancer biogenesis and tumor therapy: An overview. *Anticancer Res.* **2012**, *32*, 5139–5150. [PubMed]
4. Vargas-Roig, L.M.; Gago, F.E.; Tello, O.; Aznar, J.C.; Ciocca, D.R. Heat shock protein expression and drug resistance in breast cancer patients treated with induction chemotherapy. *Int. J. Cancer* **1998**, *79*, 468–475. [CrossRef]
5. Uozaki, H.; Ishida, T.; Kakiuchi, C.; Horiuchi, H.; Gotoh, T.; Iijima, T.; Imamura, T.; Machinami, R. Expression of heat shock proteins in osteosarcoma and its relationship to prognosis. *Pathol. Res. Pract.* **2000**, *196*, 665–673. [CrossRef]
6. Ciocca, D.R.; Clark, G.M.; Tandon, A.K.; Fuqua, S.A.W.; Welch, W.J.; McGuire, W.L. Heat shock protein hsp70 in patients with axillary lymph node-negative breast cancer: Prognostic implications. *JNCI J. Natl. Cancer Inst.* **1993**, *85*, 570–574. [CrossRef] [PubMed]
7. Blagosklonny, M.V. Re: Role of the heat shock response and molecular chaperones in oncogenesis and cell death. *JNCI J. Natl. Cancer Inst.* **2001**, *93*, 239–240. [CrossRef] [PubMed]
8. Van de Vijver, M.J.; He, Y.D.; van't Veer, L.J.; Dai, H.; Hart, A.A.M.; Voskuil, D.W.; Schreiber, G.J.; Peterse, J.L.; Roberts, C.; Marton, M.J.; et al. A Gene-Expression Signature as a Predictor of Survival in Breast Cancer. *N. Engl. J. Med.* **2002**, *347*, 1999–2009. [CrossRef]
9. Van't Veer, L.J.; Dai, H.; Van de Vijver, M.J.; He, Y.D.; Hart, A.A.M.; Mao, M.; Peterse, H.L.; Van Der Kooy, K.; Marton, M.J.; Witteveen, A.T.; et al. Gene expression profiling predicts clinical outcome of breast cancer. *Nature* **2002**, *415*, 530–536. [CrossRef]
10. Cornford, P.A.; Dodson, A.R.; Parsons, K.F.; Desmond, A.D.; Woolfenden, A.; Fordham, M.; Neoptolemos, J.P.; Ke, Y.; Foster, C.S. Heat shock protein expression independently predicts clinical outcome in prostate cancer. *Cancer Res.* **2000**, *60*, 7099–7105.

11. Almeida, M.B.; Do Nascimento, J.L.M.; Herculano, A.M.; Crespo-López, M.E. Molecular chaperones: Toward new therapeutic tools. *Biomed. Pharmacother.* **2011**, *65*, 239–243. [CrossRef] [PubMed]
12. Huang, L.H.; Wang, H.S.; Kang, L. Different evolutionary lineages of large and small heat shock proteins in eukaryotes. *Cell Res.* **2008**, *13*, 1074–1075. [CrossRef] [PubMed]
13. Gorska, M.; Popowska, U.; Sielicka-Dudzin, A.; Kuban-Jankowska, A.; Sawczuk, W.; Knap, N.; Cicero, G.; Bucchieri, F.; Wozniak, M. Geldanamycin and its derivatives as Hsp90 inhibitors. *Front. Biosci.* **2011**, *17*, 2269–2277. [CrossRef] [PubMed]
14. Chang, Y.S.; Lo, C.W.; Sun, F.C.; Chang, M.D.T.; Lai, Y.K. Differential expression of Hsp90 isoforms in geldanamycin-treated 9 L cells. *Biochem. Biophys. Res. Commun.* **2006**, *344*, 37–44. [CrossRef]
15. Csermely, P.; Schnaider, T.; Soti, C.; Prohászka, Z.; Nardai, G. The 90-kDa molecular chaperone family: Structure, function, and clinical applications. A comprehensive review. *Pharmacol. Ther.* **1998**, *79*, 129–168. [CrossRef]
16. Sreedhar, A.S.; Kalmár, É.; Csermely, P.; Shen, Y.F. Hsp90 isoforms: Functions, expression and clinical importance. *FEBS Lett.* **2004**, *562*, 11–15. [CrossRef]
17. Bertram, J.; Palfner, K.; Hiddemann, W.; Kneba, M. Increase of P-glycoprotein-mediated drug resistance by hsp 90β. *Anticancer Drugs* **1996**, *7*, 838–845. [CrossRef]
18. Hanahan, D.; Weinberg, R.A. The hallmarks of cancer. *Cell* **2000**, *100*, 57–70. [CrossRef]
19. Mayer, M.P.; Bukau, B. Hsp70 chaperone systems: Diversity of cellular functions and mechanism of action. *Biol. Chem.* **1998**, *379*, 261–268.
20. Bukau, B.; Horwich, A.L. The Hsp70 and Hsp60 chaperone machines. *Cell* **1998**, *92*, 351–366. [CrossRef]
21. Kiang, J.G.; Tsokos, G.C. Heat shock protein 70 kDa: Molecular biology, biochemistry, and physiology. *Pharmacol. Ther.* **1998**, *80*, 183–201. [CrossRef]
22. Hartl, F.U. Molecular chaperones in cellular protein folding. *Nature* **1996**, *381*, 571–580. [CrossRef] [PubMed]
23. Vilasi, S.; Carrotta, R.; Mangione, M.R.; Campanella, C.; Librizzi, F.; Randazzo, L.; Martorana, V.; Gammazza, A.M.; Ortore, M.G.; Vilasi, A.; et al. Human Hsp60 with its mitochondrial import signal occurs in solution as heptamers and tetradecamers remarkably stable over a wide range of concentrations. *PLoS ONE* **2014**, *9*, e97657. [CrossRef] [PubMed]
24. Vilasi, S.; Bulone, D.; Bavisotto, C.C.; Campanella, C.; Gammazza, A.M.; San Biagio, P.L.; Cappello, F.; de Macario, E.C.; Macario, A.J.L. Chaperonin of Group I: Oligomeric spectrum and biochemical and biological implications. *Front. Mol. Biosci.* **2018**, *4*, 1–14. [CrossRef]
25. Knowlton, A.A.; Srivatsa, U. Heat-shock protein 60 and cardiovascular disease: A paradoxical role. *Future Cardiol.* **2008**, *4*, 151–161. [CrossRef] [PubMed]
26. Henderson, B.; Fares, M.A.; Lund, P.A. Chaperonin 60: A paradoxical, evolutionarily conserved protein family with multiple moonlighting functions. *Biol. Rev.* **2013**, *88*, 955–987. [CrossRef]
27. Deniset, J.F.; Hedley, T.E.; Hlaváčková, M.; Chahine, M.N.; Dibrov, E.; O'Hara, K.; Maddaford, G.G.; Nelson, D.; Maddaford, T.G.; Fandrich, R.; et al. Heat shock protein 60 involvement in vascular smooth muscle cell proliferation. *Cell. Signal.* **2018**, *47*, 44–51. [CrossRef]
28. Kirchhoff, S.R.; Gupta, S.; Knowlton, A.A. Cytosolic heat shock protein 60, apoptosis, and myocardial injury. *Circulation* **2002**, *105*, 2899–2904. [CrossRef]
29. Chandra, D.; Choy, G.; Tang, D.G. Cytosolic accumulation of HSP60 during apoptosis with or without apparent mitochondrial release: Evidence that its pro-apoptotic or pro-survival functions involve differential interactions with caspase-3. *J. Biol. Chem.* **2007**, *282*, 31289–31301. [CrossRef] [PubMed]
30. Shan, Y.X.; Liu, T.J.; Su, H.F.; Samsamshariat, A.; Mestril, R.; Wang, P.H. Hsp10 and Hsp60 modulate Bcl-2 family and mitochondria apoptosis signaling induced by doxorubicin in cardiac muscle cells. *J. Mol. Cell. Cardiol.* **2003**, *35*, 1135–1143. [CrossRef]
31. Quintana, F.J.; Cohen, I.R. The HSP60 immune system network. *Trends Immunol.* **2011**, *32*, 89–95. [CrossRef] [PubMed]
32. Calderwood, S.K.; Mambula, S.S.; Gray, P.J. Extracellular heat shock proteins in cell signaling and immunity. *Ann. N. Y. Acad. Sci.* **2007**, *1113*, 28–39. [CrossRef] [PubMed]
33. Seul-Ki, C.; Kam, H.; Kye-Young, K.; In Park, S.; Yun-Sil, L. Targeting heat shock protein 27 in cancer: A druggable target for cancer treatment? *Cancers Basel* **2019**, *11*, 1195. [CrossRef]

34. Kokolakis, G.; Tatari, M.; Zacharopoulou, A.; Mintzas, A.C. The *hsp27* gene of the Mediterranean fruit fly, *Ceratitis capitata*: Structural characterization, regulation and developmental expression. *Insect. Mol. Biol.* **2008**, *17*, 699–710. [CrossRef] [PubMed]
35. Delbecq, S.P.; Klevit, R.E. One size does not fit all: The oligomeric states of αb crystallin. *FEBS Lett.* **2013**, *587*, 1073–1080. [CrossRef] [PubMed]
36. Liu, Z.; Xi, D.; Kang, M.; Guo, X.; Xu, B. Molecular cloning and characterization of Hsp27.6: The first reported small heat shock protein from Apis cerana cerana. *Cell Stress Chaperones* **2012**, *17*, 539–551. [CrossRef] [PubMed]
37. Concannon, C.G.; Gorman, A.M.; Samali, A. On the role of Hsp27 in regulating apoptosis. *Apoptosis* **2003**, *8*, 61–70. [CrossRef] [PubMed]
38. Rocchi, P.; Beraldi, E.; Ettinger, S.; Fazli, L.; Vessella, R.L.; Nelson, C.; Gleave, M. Increased Hsp27 after androgen ablation facilitates androgen-independent progression in prostate cancer via signal transducers and activators of transcription 3-mediated suppression of apoptosis. *Cancer Res.* **2005**, *65*, 11083–11093. [CrossRef] [PubMed]
39. Rocchi, P.; Jugpal, P.; So, A.; Sinneman, S.; Ettinger, S.; Fazli, L.; Nelson, C.; Gleave, M. Small interference RNA targeting heat-shock protein 27 inhibits the growth of prostatic cell lines and induces apoptosis via caspase-3 activation in vitro. *BJU Int.* **2006**, *98*, 1082–1089. [CrossRef] [PubMed]
40. Pandey, P.; Farber, R.; Nakazawa, A.; Kumar, S.; Bharti, A.; Nalin, C.; Weichselbaum, R.; Kufe, D.; Kharbanda, S. Hsp27 functions as a negative regulator of cytochrome c-dependent activation of procaspase-3. *Oncogene* **2000**, *19*, 1975–1981. [CrossRef]
41. Shiota, M.; Bishop, J.L.; Nip, K.M.; Zardan, A.; Takeuchi, A.; Cordonnier, T.; Beraldi, E.; Bazov, J.; Fazli, L.; Chi, K.; et al. Hsp27 regulates epithelial mesenchymal transition, metastasis, and circulating tumor cells in prostate cancer. *Cancer Res.* **2013**, *73*, 3109–3119. [CrossRef] [PubMed]
42. Voll, E.A.; Ogden, I.M.; Pavese, J.M.; Huang, X.K.; Xu, L.; Jovanovic, B.D.; Bergan, R.C. Heat shock protein 27 regulates human prostate cancer cell motility and metastatic progression. *Oncotarget* **2014**, *5*, 2648–2663. [CrossRef] [PubMed]
43. Ye, H.; Huang, H.; Cao, F.; Chen, M.; Zheng, X.; Zhan, R. HSPB1 Enhances SIRT2-Mediated G6PD Activation and Promotes Glioma Cell Proliferation. *PLoS ONE* **2016**, *11*, e0164285. [CrossRef] [PubMed]
44. Huang, C.-Y.; Wei, P.-L.; Chen, W.-Y.; Chang, W.-C.; Chang, Y.-J. Silencing Heat Shock Protein 27 Inhibits the Progression and Metastasis of Colorectal Cancer (CRC) by Maintaining the Stability of Stromal Interaction Molecule 1 (STIM1) Proteins. *Cells* **2018**, *7*, 262. [CrossRef]
45. Sheng, B.; Qi, C.; Liu, B.; Lin, Y.; Fu, T.; Zeng, Q. Increased HSP27 correlates with malignant biological behavior of non-small cell lung cancer and predicts patient's survival. *Sci. Rep.* **2017**, *7*, 13807. [CrossRef]
46. Ge, H.; He, X.; Guo, L.; Yang, X. Clinicopathological significance of HSP27 in gastric cancer: A meta-analysis. *Onco. Targets. Ther.* **2017**, *10*, 4543–4551. [CrossRef]
47. Tweedle, E.M.; Khattak, I.; Ang, C.W.; Nedjadi, T.; Jenkins, R.; Park, B.K.; Kalirai, H.; Dodson, A.; Azadeh, B.; Terlizzo, M.; et al. Low molecular weight heat shock protein HSP27 is a prognostic indicator in rectal cancer but not colon cancer. *Gut* **2010**, *59*, 1501–1510. [CrossRef]
48. Wu, P.-S.; Chang, Y.-H.; Pan, C.-C. High expression of heat shock proteins and heat shock factor-1 distinguishes an aggressive subset of clear cell renal cell carcinoma. *Histopathology* **2017**, *71*, 711–718. [CrossRef]
49. Feng, J.T.; Liu, Y.K.; Song, H.Y.; Dai, Z.; Qin, L.X.; Almofti, M.R.; Fang, C.Y.; Lu, H.J.; Yang, P.Y.; Tang, Z.Y. Heat-shock protein 27: A potential biomarker for hepatocellular carcinoma identified by serum proteome analysis. *Proteomics* **2005**, *5*, 4581–4588. [CrossRef]
50. Stope, M.B.; Weiss, M.; Preuss, M.; Streitbörger, A.; Ritter, C.A.; Zimmermann, U.; Walther, R.; Burchardt, M. Immediate and transient phosphorylation of the heat shock protein 27 initiates chemoresistance in prostate cancer cells. *Oncol. Rep.* **2014**, *32*, 2380–2386. [CrossRef]
51. Chauhan, D.; Li, G.; Auclair, D.; Hideshima, T.; Podar, K.; Mitsiades, N.; Mitsiades, C.; Chen, L.B.; Munshi, N.; Saxena, S.; et al. 2-Methoxyestardiol and bortezomib/proteasome-inhibitor overcome dexamethasone-resistance in multiple myeloma cells by modulating Heat Shock Protein-27. *Apoptosis* **2004**, *9*, 149–155. [CrossRef] [PubMed]
52. Kamada, M.; So, A.; Muramaki, M.; Rocchi, P.; Beraldi, E.; Gleave, M. Hsp27 knockdown using nucleotide-based therapies inhibit tumor growth and enhance chemotherapy in human bladder cancer cells. *Mol. Cancer Ther.* **2007**, *6*, 299–308. [CrossRef] [PubMed]

53. Zhang, D.; Putti, T.C. Over-expression of ERp29 attenuates doxorubicin-induced cell apoptosis through up-regulation of Hsp27 in breast cancer cells. *Exp. Cell Res.* **2010**, *316*, 3522–3531. [CrossRef] [PubMed]
54. Xia, Y.; Rocchi, P.; Iovanna, J.L.; Peng, L. Targeting heat shock response pathways to treat pancreatic cancer. *Drug Discov. Today* **2012**, *17*, 35–43. [CrossRef]
55. Kuramitsu, Y.; Wang, Y.; Taba, K.; Suenaga, S.; Ryozawa, S.; Kaino, S.; Sakaida, I.; Nakamura, K. Heat-shock protein 27 plays the key role in gemcitabine-resistance of pancreatic cancer cells. *Anticancer Res.* **2012**, *32*, 2295–2299.
56. Aloy, M.-T.; Hadchity, E.; Bionda, C.; Diaz-Latoud, C.; Claude, L.; Rousson, R.; Arrigo, A.-P.; Rodriguez-Lafrasse, C. Protective Role of Hsp27 Protein Against Gamma Radiation–Induced Apoptosis and Radiosensitization Effects of Hsp27 Gene Silencing in Different Human Tumor Cells. *Int. J. Radiat. Oncol.* **2008**, *70*, 543–553. [CrossRef]
57. Hadchity, E.; Aloy, M.T.; Paulin, C.; Armandy, E.; Watkin, E.; Rousson, R.; Gleave, M.; Chapet, O.; Rodriguez-Lafrasse, C. Heat shock protein 27 as a new therapeutic target for radiation sensitization of head and neck squamous cell carcinoma. *Mol. Ther.* **2009**, *17*, 1387–1394. [CrossRef]
58. Teimourian, S.; Jalal, R.; Sohrabpour, M.; Goliaei, B. Down-regulation of Hsp27 radiosensitizes human prostate cancer cells. *Int. J. Urol.* **2006**, *13*, 1221–1225. [CrossRef]
59. Fan, C.-Y.; Lee, S.; Ren, H.-Y.; Cyr, D.M. Exchangeable Chaperone Modules Contribute to Specification of Type I and Type II Hsp40 Cellular Function. *Mol. Biol. Cell* **2004**, *15*, 761–773. [CrossRef]
60. Fan, C.Y.; Lee, S.; Cyr, D.M. Mechanisms for regulation of Hsp70 function by Hsp40. *Cell Stress Chaperones* **2003**, *8*, 309–316. [CrossRef]
61. Qiu, X.B.; Shao, Y.M.; Miao, S.; Wang, L. The diversity of the DnaJ/Hsp40 family, the crucial partners for Hsp70 chaperones. *Cell. Mol Life Sci.* **2006**, *63*, 2560–2570. [CrossRef] [PubMed]
62. Wang, C.C.; Tsai, M.F.; Hong, T.M.; Chang, G.C.; Chen, C.Y.; Yang, W.M.; Chen, J.J.W.; Yang, P.C. The transcriptional factor YY1 upregulates the novel invasion suppressor HLJ1 expression and inhibits cancer cell invasion. *Oncogene* **2005**, *24*, 4081–4093. [CrossRef] [PubMed]
63. Kurzik-Dumke, U.; Hörner, M.; Czaja, J.; Nicotra, M.R.; Simiantonaki, N.; Koslowski, M.; Natali, P.G. Progression of colorectal cancers correlates with overexpression and loss of polarization of expression of the htid-1 tumor suppressor. *Int. J. Mol. Med.* **2008**, *21*, 19–31. [CrossRef] [PubMed]
64. Lianos, G.D.; Alexiou, G.A.; Mangano, A.; Mangano, A.; Rausei, S.; Boni, L.; Dionigi, G.; Roukos, D.H. The role of heat shock proteins in cancer. *Cancer Lett.* **2015**, *360*, 114–118. [CrossRef] [PubMed]
65. Cheng, H.; Cenciarelli, C.; Nelkin, G.; Tsan, R.; Fan, D.; Cheng-Mayer, C.; Fidler, I.J. Molecular Mechanism of hTid-1, the Human Homolog of Drosophila Tumor Suppressor l(2)Tid, in the Regulation of NF-κB Activity and Suppression of Tumor Growth. *Mol. Cell. Biol.* **2005**, *25*, 44–59. [CrossRef]
66. Semenza, G.L. The hypoxic tumor microenvironment: A driving force for breast cancer progression. *Biochim. Biophys. Acta Mol. Cell Res.* **2016**, *1863*, 382–391. [CrossRef]
67. Koong, A.C.; Denko, N.C.; Hudson, K.M.; Schindler, C.; Swiersz, L.; Koch, C.; Evans, S.; Ibrahim, H.; Le, Q.T.; Terris, D.J.; et al. Candidate genes for the hypoxic tumor phenotype. *Cancer Res.* **2000**, *60*, 883–887. [CrossRef]
68. Keith, B.; Simon, M.C. Hypoxia inducible factors, stem cells and cancer. *Cell* **2007**, *129*, 465–472. [CrossRef]
69. Son, T.W.; Yun, S.P.; Yong, M.S.; Seo, B.N.; Ryu, J.M.; Youn, H.Y.; Oh, Y.M.; Han, H.J. Netrin-1 protects hypoxia-induced mitochondrial apoptosis through HSP27 expression via DCC- and integrin α6β4-dependent AKt, GSK-3β, and HSF-1 in mesenchymal stem cells. *Cell Death Dis.* **2013**, *4*, 1–12. [CrossRef]
70. Patel, B.; Khaliq, A.; Jarvis-Evans, J.; Boulton, M.; Arrol, S.; Mackness, M.; McLeod, D. Hypoxia induces HSP 70 gene expression in human hepatoma (HEP G2) cells. *Biochem. Mol. Biol. Int.* **1995**, *36*, 907–912.
71. Liu, X.; Chen, S.; Tu, J.; Cai, W.; Xu, Q. HSP90 inhibits apoptosis and promotes growth by regulating HIF-1 abundance in hepatocellular carcinoma. *Int. J. Mol. Med.* **2016**, *37*, 825–835. [CrossRef] [PubMed]
72. Li, W.; Li, Y.; Guan, S.; Fan, J.; Cheng, C.F.; Bright, A.M.; Chinn, C.; Chen, M.; Woodley, D.T. Extracellular heat shock protein-90α: Linking hypoxia to skin cell motility and wound healing. *EMBO J.* **2007**, *26*, 1221–1233. [CrossRef] [PubMed]
73. Murshid, A.; Eguchi, T.; Calderwood, S.K. Stress proteins in aging and life span. *Int. J. Hyperth.* **2013**, *29*, 442–447. [CrossRef] [PubMed]

74. Krawczyk, M.A.; Styczewska, M.; Sokolewicz, E.M.; Kunc, M.; Gabrych, A.; Fatyga, A.; Izycka-Swieszewska, E.; Kazanowska, B.; Adamkiewicz-Drozynska, E.; Bien, E. Tumour expressions of hypoxic markers predict the response to neo-adjuvant chemotherapy in children with inoperable rhabdomyosarcoma. *Biomarkers* **2019**, *24*, 538–548. [CrossRef] [PubMed]
75. Guo, W.; Yang, Z.; Xia, Q.; Liu, J.; Yu, Y.; Li, J.; Zuo, Z.; Zhang, D.; Li, X.; Shi, X.; et al. Arsenite stabilizes HIF-1α protein through p85α-mediated up-regulation of inducible Hsp70 protein expression. *Cell. Mol. Life Sci.* **2011**, *68*, 475–488. [CrossRef] [PubMed]
76. Nolan, K.D.; Franco, O.E.; Hance, M.W.; Hayward, S.W.; Isaacs, J.S. Tumor-secreted Hsp90 subverts polycomb function to drive prostate tumor growth and invasion. *J. Biol. Chem.* **2015**, *290*, 8271–8282. [CrossRef]
77. Zhou, J.; Schmid, T.; Frank, R.; Brüne, B. PI3K/Akt Is Required for Heat Shock Proteins to Protect Hypoxia-inducible Factor 1α from pVHL-independent Degradation. *J. Biol. Chem.* **2004**, *279*, 13506–13513. [CrossRef]
78. Filatova, A.; Seidel, S.; Böğürcü, N.; Gräf, S.; Garvalov, B.K.; Acker, T. Acidosis acts through HSP90 in a PHD/VHL-independent manner to promote HIF function and stem cell maintenance in glioma. *Cancer Res.* **2016**, *76*, 5845–5856. [CrossRef]
79. Jayaprakash, P.; Dong, H.; Zou, M.; Bhatia, A.; O'Brien, K.; Chen, M.; Woodley, D.T.; Li, W. Hsp90α and Hsp90β Co-Operate a Stress-Response Mechanism to Cope With Hypoxia and Nutrient Paucity during Wound Healing. *J. Cell Sci.* **2015**, *128*, 1475–1480. [CrossRef]
80. Cheng, C.-F.; Fan, J.; Fedesco, M.; Guan, S.; Li, Y.; Bandyopadhyay, B.; Bright, A.M.; Yerushalmi, D.; Liang, M.; Chen, M.; et al. Transforming Growth Factor α (TGFα)-Stimulated Secretion of HSP90α: Using the Receptor LRP-1/CD91 To Promote Human Skin Cell Migration against a TGFβ-Rich Environment during Wound Healing. *Mol. Cell. Biol.* **2008**, *28*, 3344–3358. [CrossRef]
81. Eustace, B.K.; Sakurai, T.; Stewart, J.K.; Yimlamai, D.; Unger, C.; Zehetmeier, C.; Lain, B.; Torella, C.; Henning, S.W.; Beste, G.; et al. Functional proteomic screens reveal an essential extracellular role for hsp90α in cancer cell invasiveness. *Nat. Cell Biol.* **2004**, *6*, 507–514. [CrossRef] [PubMed]
82. Armstrong, H.K.; Gillis, J.L.; Johnson, I.R.D.; Nassar, Z.D.; Moldovan, M.; Levrier, C.; Sadowski, M.C.; Chin, M.Y.; Tomlinson Guns, E.S.; Tarulli, G.; et al. Dysregulated fibronectin trafficking by Hsp90 inhibition restricts prostate cancer cell invasion. *Sci. Rep.* **2018**, *8*, 1–14. [CrossRef] [PubMed]
83. Chalmin, F.; Ladoire, S.; Mignot, G.; Vincent, J.; Bruchard, M.; Remy-Martin, J.-P.; Boireau, W.; Rouleau, A.; Simon, B.; Lanneau, D.; et al. Membrane-associated Hsp72 from tumor-derived exosomes mediates STAT3-dependent immunosuppressive function of mouse and human myeloid-derived suppressor cells. *J. Clin. Investig.* **2010**, *120*, 457–471. [CrossRef] [PubMed]
84. Vlassov, A.V.; Magdaleno, S.; Setterquist, R.; Conrad, R. Exosomes: Current knowledge of their composition, biological functions, and diagnostic and therapeutic potentials. *Biochim. Biophys. Acta Gen. Subj.* **2012**, *1820*, 940–948. [CrossRef] [PubMed]
85. Tauro, B.J.; Greening, D.W.; Mathias, R.A.; Ji, H.; Mathivanan, S.; Scott, A.M.; Simpson, R.J. Comparison of ultracentrifugation, density gradient separation, and immunoaffinity capture methods for isolating human colon cancer cell line LIM1863-derived exosomes. *Methods* **2012**, *56*, 293–304. [CrossRef] [PubMed]
86. Pan, B.T.; Teng, K.; Wu, C.; Adam, M.; Johnstone, R.M. Electron microscopic evidence for externalization of the transferrin receptor in vesicular form in sheep reticulocytes. *J. Cell Biol.* **1985**, *101*, 942–948. [CrossRef]
87. Graner, M.W.; Cumming, R.I.; Bigner, D.D. The heat shock response and chaperones/heat shock proteins in brain tumors: Surface expression, release, and possible immune consequences. *J. Neurosci.* **2007**, *27*, 11214–11227. [CrossRef] [PubMed]
88. McCready, J.; Sims, J.D.; Chan, D.; Jay, D.G. Secretion of extracellular hsp90α via exosomes increases cancer cell motility: A role for plasminogen activation. *BMC Cancer* **2010**, *10*, 294. [CrossRef]
89. Mears, R.; Craven, R.A.; Hanrahan, S.; Totty, N.; Upton, C.; Young, S.L.; Patel, P.; Selby, P.J.; Banks, R.E. Proteomic analysis of melanoma-derived exosomes by two-dimensional polyacrylamide gel electrophoresis and mass spectrometry. *Proteomics* **2004**, *4*, 4019–4031. [CrossRef]
90. Elsner, L.; Muppala, V.; Gehrmann, M.; Lozano, J.; Malzahn, D.; Bickeböller, H.; Brunner, E.; Zientkowska, M.; Herrmann, T.; Walter, L.; et al. The Heat Shock Protein HSP70 Promotes Mouse NK Cell Activity against Tumors That Express Inducible NKG2D Ligands. *J. Immunol.* **2007**, *179*, 5523–5533. [CrossRef]

91. Gastpar, R.; Gehrmann, M.; Bausero, M.A.; Asea, A.; Gross, C.; Schroeder, J.A.; Multhoff, G. Heat shock protein 70 surface-positive tumor exosomes stimulate migratory and cytolytic activity of natural killer cells. *October* **2005**, *65*, 5238–5247. [CrossRef] [PubMed]
92. Guo, D.; Chen, Y.; Wang, S.; Yu, L.; Shen, Y.; Zhong, H.; Yang, Y. Exosomes from heat-stressed tumour cells inhibit tumour growth by converting regulatory T cells to Th17 cells via IL-6. *Immunology* **2018**, *154*, 132–143. [CrossRef] [PubMed]
93. Yukawa, H.; Suzuki, K.; Aoki, K.; Arimoto, T.; Yasui, T.; Kaji, N.; Ishikawa, T.; Ochiya, T.; Baba, Y. Imaging of angiogenesis of human umbilical vein endothelial cells by uptake of exosomes secreted from hepatocellular carcinoma cells. *Sci. Rep.* **2018**, *8*, 1–12. [CrossRef] [PubMed]
94. Xiao, W.; Dong, W.; Zhang, C.; Saren, G.; Geng, P.; Zhao, H.; Li, Q.; Zhu, J.; Li, G.; Zhang, S.; et al. Effects of the epigenetic drug MS-275 on the release and function of exosome-related immune molecules in hepatocellular carcinoma cells. *Eur. J. Med. Res.* **2013**, *18*, 1–7. [CrossRef]
95. Gobbo, J.; Marcion, G.; Cordonnier, M.; Dias, A.M.M.; Pernet, N.; Hammann, A.; Richaud, S.; Mjahed, H.; Isambert, N.; Clausse, V.; et al. Restoring Anticancer Immune Response by Targeting Tumor-Derived Exosomes with a HSP70 Peptide Aptamer. *J. Natl. Cancer Inst.* **2016**, *108*, 1–11. [CrossRef]
96. Stope, M.B.; Klinkmann, G.; Diesing, K.; Koensgen, D.; Burchardt, M.; Mustea, A. Heat shock protein HSP27 secretion by ovarian cancer cells is linked to intracellular expression levels, occurs independently of the endoplasmic reticulum pathway and HSP27's phosphorylation status, and is mediated by exosome liberation. *Dis. Markers* **2017**, 1575374. [CrossRef]
97. King, H.W.; Michael, M.Z.; Gleadle, J.M. Hypoxic enhancement of exosome release by breast cancer cells. *BMC Cancer* **2012**, *12*, 1–10. [CrossRef]
98. Li, X.; Wang, S.; Zhu, R.; Li, H.; Han, Q.; Zhao, R.C. Lung tumor exosomes induce a pro-inflammatory phenotype in mesenchymal stem cells via NFκB-TLR signaling pathway. *J. Hematol. Oncol.* **2016**, *9*, 42. [CrossRef]
99. Bu, N.; Wu, H.; Sun, B.; Zhang, G.; Zhan, S.; Zhang, R.; Zhou, L. Exosome-loaded dendritic cells elicit tumor-specific CD8 + cytotoxic T cells in patients with glioma. *J. Neurooncol.* **2011**, *104*, 659–667. [CrossRef]
100. Campanella, C.; Rappa, F.; Sciumè, C.; Marino Gammazza, A.; Barone, R.; Bucchieri, F.; David, S.; Curcurù, G.; Caruso Bavisotto, C.; Pitruzzella, A.; et al. Heat shock protein 60 levels in tissue and circulating exosomes in human large bowel cancer before and after ablative surgery. *Cancer* **2015**, *121*, 3230–3239. [CrossRef]
101. Bavisotto, C.C.; Cipolla, C.; Graceffa, G.; Barone, R.; Bucchieri, F.; Bulone, D.; Cabibi, D.; Campanella, C.; Gammazza, A.M.; Pitruzzella, A.; et al. Immunomorphological pattern of molecular chaperones in normal and pathological thyroid tissues and circulating exosomes: Potential use in clinics. *Int. J. Mol. Sci.* **2019**, *20*, 4496. [CrossRef] [PubMed]
102. Andre, F.; Schartz, N.E.C.; Movassagh, M.; Flament, C.; Pautier, P.; Morice, P.; Pomel, C.; Lhomme, C.; Escudier, B.; Le Chevalier, T.; et al. Malignant effusions and immunogenic tumour-derived exosomes. *Lancet* **2002**, *360*, 295–305. [CrossRef]
103. Chanteloup, G.; Cordonnier, M.; Isambert, N.; Bertaut, A.; Marcion, G.; Garrido, C.; Gobbo, J. Membrane-bound exosomal HSP70 as a biomarker for detection and monitoring of malignant solid tumours: A pilot study. *Pilot Feasibility Stud.* **2020**, *6*, 1–7. [CrossRef] [PubMed]
104. Théry, C.; Zitvogel, L.; Amigorena, S. Exosomes: Composition, biogenesis and function. *Nat. Rev. Immunol.* **2002**, *2*, 569–579. [CrossRef] [PubMed]
105. Théry, C.; Ostrowski, M.; Segura, E. Membrane vesicles as conveyors of immune responses. *Nat. Rev. Immunol.* **2009**, *9*, 581–593. [CrossRef] [PubMed]
106. Matsuo, H.; Chevallier, J.; Mayran, N.; Le Blanc, I.; Ferguson, C.; Fauré, J.; Blanc, N.S.; Matile, S.; Dubochet, J.; Sadoul, R.; et al. Role of LBPA and alix in multivesicular liposome formation and endosome organization. *Science* **2004**, *303*, 531–534. [CrossRef] [PubMed]
107. Merendino, A.M.; Bucchieri, F.; Campanella, C.; Marcianò, V.; Ribbene, A.; David, S.; Zummo, G.; Burgio, G.; Corona, D.F.V.; De Macario, E.C.; et al. Hsp60 is actively secreted by human tumor cells. *PLoS ONE* **2010**, *5*, e9247. [CrossRef]
108. Campanella, C.; Bucchieri, F.; Merendino, A.M.; Fucarino, A.; Burgio, G.; Corona, D.F.V.; Barbieri, G.; David, S.; Farina, F.; Zummo, G.; et al. The odyssey of Hsp60 from tumor cells to other destinations includes plasma membrane-associated stages and Golgi and exosomal protein-trafficking modalities. *PLoS ONE* **2012**, *7*, e42008. [CrossRef]

109. Lancaster, G.I.; Febbraio, M.A. Exosome-dependent trafficking of HSP70: A novel secretory pathway for cellular stress proteins. *J. Biol. Chem.* **2005**, *280*, 23349–23355. [CrossRef]
110. Malik, Z.A.; Kott, K.S.; Poe, A.J.; Kuo, T.; Chen, L.; Ferrara, K.W.; Knowlton, A.A. Cardiac myocyte exosomes: Stability, HSP60, and proteomics. *Am. J. Physiol. Hear. Circ. Physiol.* **2013**, *304*, H954–H965. [CrossRef]
111. Tamura, Y.; Torigoe, T.; Kutomi, G.; Hirata, K.; Sato, N. New Paradigm for Intrinsic Function of Heat Shock Proteins as Endogenous Ligands in Inflammation and Innate Immunity. *Curr. Mol. Med.* **2012**, *12*, 1198–1206. [CrossRef] [PubMed]
112. Van Eden, W.; Wick, G.; Albani, S.; Cohen, I. Stress, heat shock proteins, and autoimmunity: How immune responses to heat shock proteins are to be used for the control of chronic inflammatory diseases. *Ann. N. Y. Acad. Sci.* **2007**, *1113*, 217–237. [CrossRef] [PubMed]
113. Fagone, P.; Di Rosa, M.; Palumbo, M.; De Gregorio, C.; Nicoletti, F.; Malaguarnera, L. Modulation of heat shock proteins during macrophage differentiation. *Inflamm. Res.* **2012**, *61*, 1131–1139. [CrossRef] [PubMed]
114. Nolan, K.D.; Kaur, J.; Isaacs, J.S. Secreted heat shock protein 90 promotes prostate cancer stem cell heterogeneity. *Oncotarget* **2017**, *8*, 19323–19341. [CrossRef]
115. Campanella, C.; Bucchieri, F.; Ardizzone, N.M.; Marino Gammazza, A.; Montalbano, A.; Ribbene, A.; Di Felice, V.; Bellafiore, M.; David, S.; Rappa, F.; et al. Upon oxidative stress, the antiapoptotic Hsp60/procaspase-3 complex persists in mucoepidermoid carcinoma cells. *Eur. J. Histochem.* **2008**, *52*, 221–228. [CrossRef]
116. Czarnecka, A.M.; Campanella, C.; Zummo, G.; Cappello, F. Mitochondrial chaperones in cancer: From molecular biology to clinical diagnostics. *Cancer Biol. Ther.* **2006**, *5*, 714–720. [CrossRef] [PubMed]
117. Cappello, F.; De Macario, E.C.; Marasà, L.; Zummo, G.; Macario, A.J.L. Hsp60 expression, new locations, functions and perspectives for cancer diagnosis and therapy. *Cancer Biol. Ther.* **2008**, *7*, 801–809. [CrossRef]
118. Hance, M.W.; Dole, K.; Gopal, U.; Bohonowych, J.E.; Jezierska-Drutel, A.; Neumann, C.A.; Liu, H.; Garraway, I.P.; Isaacs, J.S. Secreted Hsp90 is a novel regulator of the epithelial to mesenchymal transition (EMT) in prostate cancer. *J. Biol. Chem.* **2012**, *287*, 37732–37744. [CrossRef]
119. Nagaraju, G.P.; Long, T.E.; Park, W.; Landry, J.C.; Taliaferro-Smith, L.; Farris, A.B.; Diaz, R.; El-Rayes, B.F. Heat shock protein 90 promotes epithelial to mesenchymal transition, invasion, and migration in colorectal cancer. *Mol. Carcinog.* **2015**, *54*, 1147–1158. [CrossRef]
120. Ogorevc, E.; Kralj-Iglic, V.; Veranic, P. The role of extracellular vesicles in phenotypic cancer transformation. *Radiol. Oncol.* **2013**, *47*, 197–205. [CrossRef]
121. Keerthikumar, S.; Chisanga, D.; Ariyaratne, D.; Al Saffar, H.; Anand, S.; Zhao, K.; Samuel, M.; Pathan, M.; Jois, M.; Chilamkurti, N.; et al. ExoCarta: A web-based compendium of exosomal cargo. *J. Mol. Biol.* **2016**, *428*, 688–692. [CrossRef] [PubMed]
122. Skotland, T.; Sandvig, K.; Llorente, A. Lipids in exosomes: Current knowledge and the way forward. *Prog. Lipid Res.* **2017**, *66*, 30–41. [CrossRef] [PubMed]
123. Sato-Kuwabara, Y.; Melo, S.A.; Soares, F.A.; Calin, G.A. The fusion of two worlds: Non-coding RNAs and extracellular vesicles diagnostic and therapeutic implications (review). *Int. J. Oncol.* **2015**, *46*, 17–27. [CrossRef] [PubMed]
124. Kahlert, C.; Melo, S.A.; Protopopov, A.; Tang, J.; Seth, S.; Koch, O.; Zhang, J.; Weitz, J.; Chin, L.; Futreal, A.; et al. Identification of double-stranded genomic DNA spanning all chromosomes with mutated KRAS and p53 DNA in the serum exosomes of patients with pancreatic cancer. *J. Biol. Chem.* **2014**, *289*, 3869–3875. [CrossRef]
125. Mathivanan, S.; Fahner, C.J.; Reid, G.E.; Simpson, R.J. ExoCarta 2012: Database of exosomal proteins, RNA and lipids. *Nucleic Acids Res.* **2012**, *40*, D1241–D1244. [CrossRef]
126. Thakur, B.K.; Zhang, H.; Becker, A.; Matei, I.; Huang, Y.; Costa-Silva, B.; Zheng, Y.; Hoshino, A.; Brazier, H.; Xiang, J.; et al. Double-stranded DNA in exosomes: A novel biomarker in cancer detection. *Cell Res.* **2014**, *24*, 766–769. [CrossRef]
127. Jego, G.; Chiron, D.; Berthenet, K.; Pellat-Decceunynck, C. Modulation of normal and malignant plasma cells function by toll-like receptors. *Front. Biosci.* **2012**, *4*, 2289–2301. [CrossRef]
128. Taha, E.A.; Ono, K.; Eguchi, T. Roles of extracellular HSPs as biomarkers in immune surveillance and immune evasion. *Int. J. Mol. Sci.* **2019**, *20*, 4588. [CrossRef]

129. Lv, L.H.; Wan, Y.L.; Lin, Y.; Zhang, W.; Yang, M.; Li, G.N.; Lin, H.M.; Shang, C.Z.; Chen, Y.J.; Min, J. Anticancer drugs cause release of exosomes with heat shock proteins from human hepatocellular carcinoma cells that elicit effective natural killer cell antitumor responses in vitro. *J. Biol. Chem.* **2012**, *287*, 15874–15885. [CrossRef]
130. Vega, V.L.; Rodríguez-Silva, M.; Frey, T.; Gehrmann, M.; Diaz, J.C.; Steinem, C.; Multhoff, G.; Arispe, N.; De Maio, A. Hsp70 Translocates into the Plasma Membrane after Stress and Is Released into the Extracellular Environment in a Membrane-Associated Form that Activates Macrophages. *J. Immunol.* **2008**, *180*, 4299–4307. [CrossRef]
131. Desandes, E.; Stark, D.P. Epidemiology of Adolescents and Young Adults with Cancer in Europe. *Prog. Tumor Res.* **2016**, *43*, 1–15. [CrossRef] [PubMed]
132. Bleyer, A.; Ferrari, A.; Whelan, J.; Barr, R.D. Global assessment of cancer incidence and survival in adolescents and young adults. *Pediatr. Blood Cancer* **2017**, *64*, 1–9. [CrossRef] [PubMed]
133. Trama, A.; Botta, L.; Steliarova-Foucher, E. Cancer Burden in Adolescents and Young Adults: A Review of Epidemiological Evidence. *Cancer J. USA* **2018**, *24*, 256–266. [CrossRef]
134. Sender, L.; Zabokrtsky, K.B. Adolescent and young adult patients with cancer: A milieu of unique features. *Nat. Rev. Clin. Oncol.* **2015**, *12*, 465–480. [CrossRef] [PubMed]
135. Barr, R.D.; Ferrari, A.; Ries, L.; Whelan, J.; Bleyer, W.A. Cancer in adolescents and young adults: A narrative review of the current status and a view of the future. *JAMA Pediatr.* **2016**, *170*, 495–501. [CrossRef]
136. Mallawaaratchy, D.M.; Hallal, S.; Russell, B.; Ly, L.; Ebrahimkhani, S.; Wei, H.; Christopherson, R.I. Comprehensive proteome profiling of glioblastoma-derived extracellular vesicles identifies markers for more aggressive disease. *J. Neuro-Oncol.* **2017**, *131*, 233–244. [CrossRef]
137. Peinado, H.; Aleckovic, M.; Lavotshkin, S.; Matei, I.; Costa-Silva, B.; Moreno-Bueno, G.; Hergueta-Redondo, M.; Williams, C.; Garcia-Santos, G.; Nitadori-Hoshino, A.; et al. Melanoma exosomes educate bone marrow progenitor cells. *Nat. Med.* **2012**, *18*, 883–891. [CrossRef]
138. Muhsin-Sharafaldine, M.R.; Saunderson, S.C.; Dunn, A.C.; Faed, J.M.; Kleffmann, T.; McLellan, A.D. Procoagulant and immunogenic properties of melanoma exosomes, microvesicles and apoptotic vesicles. *Oncotarget* **2016**, *7*, 56279–56294. [CrossRef]
139. Dai, S.; Wan, T.; Wang, B.; Zhou, X.; Xiu, F.; Chen, T.; Wu, Y.; Cao, X. More efficient induction of HLA-A*0201-restricted and carcinoembryonic antigen (CEA)—Specific CTL response by immunization with exosomes prepared from heat-stressed CEA-positive tumor cells. *Clin. Cancer Res.* **2005**, *11*, 7554–7563. [CrossRef]
140. Rao, Q.; Zuo, B.; Lu, Z.; Gao, X.; You, A.; Wu, C.; Du, Z.; Yin, H.F. Tumor-derived exosomes elicit tumor suppression in murine hepatocellular carcinoma models and humans in vitro. *Hepatology* **2016**, *64*, 456–472. [CrossRef]
141. Xiao, W.H.; Sanren, G.W.; Zhu, J.H.; Li, Q.W.; Kang, H.R.; Wang, R.L.; Song, L.P.; Ye, M. Effect of 5-Aza-2′-deoxycytidine on immune-associated proteins in exosomes from hepatoma. *World J. Gastroenterol.* **2010**, *16*, 2371–2377. [CrossRef] [PubMed]
142. Wyciszkiewicz, A.; Kalinowska-Łyszczarz, A.; Nowakowski, B.; Kaźmierczak, K.; Osztynowicz, K.; Michalak, S. Expression of small heat shock proteins in exosomes from patients with gynecologic cancers. *Sci. Rep.* **2019**, *9*, 1–9. [CrossRef] [PubMed]
143. Cordonnier, M.; Chanteloup, G.; Isambert, N.; Seigneuric, R.; Fumoleau, P.; Garrido, C.; Gobbo, J. Exosomes in cancer theranostic: Diamonds in the rough. *Cell Adhes. Migr.* **2017**, *11*, 151–163. [CrossRef] [PubMed]
144. Chanteloup, G.; Cordonnier, M.; Isambert, N.; Bertaut, A.; Hervieu, A.; Hennequin, A.; Luu, M.; Zanetta, S.; Coudert, B.; Bengrine, L.; et al. Monitoring HSP70 exosomes in cancer patients' follow up: A clinical prospective pilot study. *J. Extracell. Vesicles* **2020**, *9*. [CrossRef]
145. Khan, S.; Jutzy, J.M.S.; Aspe, J.R.; McGregor, D.W.; Neidigh, J.W.; Wall, N.R. Survivin is released from cancer cells via exosomes. *Apoptosis* **2011**, *16*, 1–12. [CrossRef]
146. Bausero, M.A.; Gastpar, R.; Multhoff, G.; Asea, A. Alternative mechanism by which IFN-γ enhances tumor recognition: Active release of heat shock protein 72. *J. Immunol.* **2005**, *175*, 2900–2912. [CrossRef]
147. Campanella, C.; D'Anneo, A.; Gammazza, A.M.; Bavisotto, C.C.; Barone, R.; Emanuele, S.; Lo Cascio, F.; Mocciaro, E.; Fais, S.; De Macario, E.C.; et al. The histone deacetylase inhibitor SAHA induces HSP60 nitration and its extracellular release by exosomal vesicles in human lung-derived carcinoma cells. *Oncotarget* **2016**, *7*, 28849–28867. [CrossRef]

148. Ostheimer, C.; Gunther, S.; Bache, M.; Vordermark, D.; Multhoff, G. Dynamics of heat shock protein 70 serum levels as a predictor of clinical response in non-small-cell lung cancer and correlation with the hypoxia-related marker osteopontin. *Front. Immunol.* **2017**, *8*. [CrossRef]
149. Li, W.; Mu, D.; Tian, F.; Hu, Y.; Jiang, T.; Han, Y.; Chen, J.; Han, G.; Li, X. Exosomes derived from Rab27a-overexpressing tumor cells elicit efficient induction of antitumor immunity. *Mol. Med. Rep.* **2013**, *8*, 1876–1882. [CrossRef]
150. Ostrom, Q.T.; Gittleman, H.; De Blank, P.M.; Finlay, J.L.; Gurney, J.G.; McKean-Cowdin, R.; Stearns, D.S.; Wolff, J.E.; Liu, M.; Wolinsky, Y.; et al. American Brain Tumor Association Adolescent and Young Adult Primary Brain and Central Nervous System Tumors Diagnosed in the United States in 2008-2012. *Neuro. Oncol.* **2015**, *18*, i1–i50. [CrossRef]
151. Alexiou, G.A.; Vartholomatos, G.; Stefanaki, K.; Patereli, A.; Dova, L.; Karamoutsios, A.; Lallas, G.; Sfakianos, G.; Moschovi, M.; Prodromou, N. Expression of heat shock proteins in medulloblastoma. *J. Neurosurg. Pediatr.* **2013**, *12*, 452–457. [CrossRef] [PubMed]
152. Hauser, P.; Hanzély, Z.; Jakab, Z.; Oláh, L.; Szabó, E.; Jeney, A.; Schuler, D.; Fekete, G.; Bognár, L.; Garami, M. Expression and prognostic examination of heat shock proteins (HSP 27, HSP 70, and HSP 90) in medulloblastoma. *J. Pediatr. Hematol. Oncol.* **2006**, *28*, 461–466. [CrossRef]
153. Martelli, C.; Iavarone, F.; D'Angelo, L.; Arba, M.; Vincenzoni, F.; Inserra, I.; Delfino, D.; Rossetti, D.V.; Caretto, M.; Massimi, L.; et al. Integrated proteomic platforms for the comparative characterization of medulloblastoma and pilocytic astrocytoma pediatric brain tumors: A preliminary study. *Mol. Biosyst.* **2015**, *11*, 1668–1683. [CrossRef] [PubMed]
154. Alexiou, G.A.; Karamoutsios, A.; Lallas, G.; Ragos, V.; Goussia, A.; Kyritsis, A.P.; Voulgaris, S.; Vartholomatos, G. Expression of heat shock proteins in brain tumors. *Turk. Neurosurg.* **2014**, *24*, 745–749. [CrossRef] [PubMed]
155. Jiang, X.; Zhou, T.; Wang, Z.; Qi, B.; Xia, H. HSP47 Promotes Glioblastoma Stemlike Cell Survival by Modulating Tumor Microenvironment Extracellular Matrix through TGF-β Pathway. *ACS Chem. Neurosci.* **2017**, *8*, 128–134. [CrossRef] [PubMed]
156. Kato, M.; Herz, F.; Kato, S.; Hirano, A. Expression of stress-response (heat-shock) protein 27 in human brain tumors: An immunohistochemical study. *Acta Neuropathol.* **1992**, *83*, 420–422. [CrossRef]
157. Kato, S.; Hirano, A.; Kato, M.; Herz, F.; Ohama, E. Stress-response (heat-shock) protein 72 expression in tumors of the central nervous system: An immunohistochemical investigation. *Acta Neuropathol.* **1992**, *84*, 261–264. [CrossRef]
158. Kato, S.; Kato, M.; Hirano, A.; Takikawa, M.; Ohama, E. The immunohistochemical expression of stress-response protein (srp) 60 in human brain tumours: Relationship of srp 60 to the other five srps, proliferating cell nuclear antigen and p53 protein. *Histol. Histopathol.* **2001**, *16*, 809–820. [CrossRef]
159. Hermisson, M.; Strik, H.; Rieger, J.; Dichgans, J.; Meyermann, R.; Weller, M. Expression and functional activity of heat shock proteins in human glioblastoma multiforme. *Neurology* **2000**, *54*, 1357–1364. [CrossRef]
160. Bavisotto, C.C.; Graziano, F.; Rappa, F.; Gammazza, A.M.; Logozzi, M.; Fais, S.; Maugeri, R.; Bucchieri, F.; de Macario, E.C.; Macario, A.J.L.; et al. Exosomal chaperones and miRNAs in gliomagenesis: State-of-art and theranostics perspectives. *Int. J. Mol. Sci.* **2018**, *19*, 2626. [CrossRef]
161. Thorsteinsdottir, J.; Stangl, S.; Fu, P.; Guo, K.; Albrecht, V.; Eigenbrod, S.; Erl, J.; Gehrmann, M.; Tonn, J.C.; Multhoff, G.; et al. Overexpression of cytosolic, plasma membrane bound and extracellular heat shock protein 70 (Hsp70) in primary glioblastomas. *J. Neurooncol.* **2017**, *135*, 443–452. [CrossRef] [PubMed]
162. Guzhova, I.; Kislyakova, K.; Moskaliova, O.; Fridlanskaya, I.; Tytell, M.; Cheetham, M.; Margulis, B. In vitro studies show that Hsp70 can be released by glia and that exogenous Hsp70 can enhance neuronal stress tolerance. *Brain Res.* **2001**, *914*, 66–73. [CrossRef]
163. Pasi, F.; Paolini, A.; Nano, R.; Di Liberto, R.; Capelli, E. Effects of single or combined treatments with radiation and chemotherapy on survival and danger signals expression in glioblastoma cell lines. *Biomed Res. Int.* **2014**. [CrossRef] [PubMed]
164. Gopal, U.; Bohonowych, J.E.; Lema-Tome, C.; Liu, A.; Garrett-Mayer, E.; Wang, B.; Isaacs, J.S. A novel extracellular Hsp90 mediated co-receptor function for LRP1 regulates EphA2 dependent glioblastoma cell invasion. *PLoS ONE* **2011**, *6*, e17649. [CrossRef] [PubMed]

165. Thuringer, D.; Hammann, A.; Benikhlef, N.; Fourmaux, E.; Bouchot, A.; Wettstein, G.; Solary, E.; Garrido, C. Transactivation of the epidermal growth factor receptor by heat shock protein 90 via toll-like receptor 4 contributes to the migration of glioblastoma cells. *J. Biol. Chem.* **2011**, *286*, 3418–3428. [CrossRef] [PubMed]
166. Snigireva, A.V.; Vrublevskaya, V.V.; Skarga, Y.Y.; Morenkov, O.S. Cell surface heparan sulfate proteoglycans are involved in the extracellular Hsp90-stimulated migration and invasion of cancer cells. *Cell Stress Chaperones* **2019**, *24*, 309–322. [CrossRef] [PubMed]
167. Ampie, L.; Choy, W.; Lamano, J.B.; Fakurnejad, S.; Bloch, O.; Parsa, A.T. Heatshock protein vaccines against Glioblastoma: From bench to bedside. *J. Neuro-Oncol.* **2016**, *123*, 441–448. [CrossRef]
168. Canella, A.; Welker, A.M.; Yoo, J.Y.; Xu, J.; Abas, F.S.; Kesanakurti, D.; Nagarajan, P.; Beattie, C.E.; Sulman, E.P.; Liu, J.; et al. Efficacy of Onalespib a Long-acting Second Generation HSP90 Inhibitor as a Single Agent and in Combination with Temozolomide against Malignant Gliomas. *Clin. Cancer Res.* **2017**, *23*, 6215–6226. [CrossRef]
169. Gaspar, N.; Sharp, S.Y.; Eccles, S.A.; Gowan, S.; Popov, S.; Jones, C.; Pearson, A.; Vassal, G.; Workman, P. Mechanistic evaluation of the novel HSP90 inhibitor NVP-AUY922 in adult and pediatric glioblastoma. *Mol. Cancer Ther.* **2010**, *9*, 1219–1233. [CrossRef]
170. Sauvageot, C.M.-E.; Weatherbee, J.L.; Kesari, S.; Winters, S.E.; Barnes, J.; Dellagatta, J.; Ramakrishna, N.R.; Stiles, C.D.; Kung, A.L.-J.; Kieran, M.W.; et al. Efficacy of the HSP90 inhibitor 17-AAG in human glioma cell lines and tumorigenic glioma stem cells. *Neuro. Oncol.* **2009**, *11*, 109–121. [CrossRef]
171. Wu, J.; Wang, W.; Shao, Q.; Xiao, G.; Cheng, J.; Yuan, Y.; Zhang, M. Irradiation facilitates the inhibitory effect of the heat shock protein 90 inhibitor NVP-BEP800 on the proliferation of malignant glioblastoma cells through attenuation of the upregulation of heat shock protein 70. *Exp. Ther. Med.* **2014**, *8*, 893–898. [CrossRef] [PubMed]
172. Ohba, S.; Hirose, Y.; Yoshida, K.; Yazaki, T.; Kawase, T. Inhibition of 90-kD heat shock protein potentiates the cytotoxicity of chemotherapeutic agents in human glioma cells. Laboratory investigation. *J. Neurosurg.* **2010**, *112*, 33–42. [CrossRef] [PubMed]
173. Ji, N.; Zhang, Y.; Liu, Y.; Xie, J.; Wang, Y.; Hao, S.; Gao, Z. Heat shock protein peptide complex-96 vaccination for newly diagnosed glioblastoma: A phase I, single-arm trial. *JCI Insight* **2018**, *3*, 1–9. [CrossRef] [PubMed]
174. Bloch, O.; Crane, C.A.; Fuks, Y.; Kaur, R.; Aghi, M.K.; Berger, M.S.; Butowski, N.A.; Chang, S.M.; Clarke, J.L.; McDermott, M.W.; et al. Heat-shock protein peptide complex-96 vaccination for recurrent glioblastoma: A phase II, single-arm trial. *Neuro. Oncol.* **2014**, *16*, 274–279. [CrossRef] [PubMed]
175. Bloch, O.; Lim, M.; Sughrue, M.E.; Komotar, R.J.; Abrahams, J.M.; Rourke, D.M.O.; Ambrosio, A.D.; Bruce, J.N.; Parsa, A.T. Autologous Heat Shock Protein Peptide Vaccination for Newly Diagnosed Glioblastoma: Impact of Peripheral PD-L1 Expression on Response to Therapy. *Clin. Cancer Res.* **2017**, *23*, 3575–3584. [CrossRef] [PubMed]
176. Crane, C.A.; Han, S.J.; Ahn, B.; Oehlke, J.; Kivett, V.; Fedoroff, A.; Butowski, N.; Chang, S.M.; Clarke, J.; Berger, M.S.; et al. Individual patient-specific immunity against high-grade glioma after vaccination with autologous tumor derived peptides bound to the 96 KD chaperone protein. *Clin. Cancer Res.* **2013**, *19*, 205–214. [CrossRef] [PubMed]
177. Tuzesi, A.; Kling, T.; Wenger, A.; Lunavat, T.R.; Jang, S.C.; Rydenhag, B.; Lötvall, J.; Pollard, S.M.; Danielsson, A.; Carén, H. Pediatric brain tumor cells release exosomes with a miRNA repertoire that differs from exosomes secreted by normal cells. *Oncotarget* **2017**, *8*, 90164–90175. [CrossRef]
178. Skog, J.; Wurdinger, T.; van Rijn, S.; Meijer, D.; Gainche, L.; Sena-Esteves, M.; Curry, W.T., Jr.; Carter, R.S.; Krichevsky, A.M.; Breakefield, X.O. Glioblastoma microvesicles transport RNA and protein that promote tumor growth and provide diagnostic biomarkers. *Nat. Cell Biol.* **2008**, *10*, 1470–1476. [CrossRef]
179. Yang, J.; Song, J.; Huo, H.; Zhao, Y.; Zhang, G.; Zhao, Z.; Sun, G.; Jiao, B. DNM3, p65 and p53 from exosomes represent potential clinical diagnosis markers for glioblastoma multiforme Jian-kai. *Ther. Adv. Med. Oncol.* **2017**, *9*, 741–754. [CrossRef]
180. Saadatpour, L.; Fadaee, E.; Fadaei, S.; Nassiri Mansour, R.; Mohammadi, M.; Mousavi, S.M.; Goodarzi, M.; Verdi, J.; Mirzaei, H. Glioblastoma: Exosome and microRNA as novel diagnosis biomarkers. *Cancer Gene Ther.* **2016**, *23*, 415–418. [CrossRef]
181. Sidera, K.; Gaitanou, M.; Stellas, D.; Matsas, R.; Patsavoudi, E. A critical role for HSP90 in cancer cell invasion involves interaction with the extracellular domain of HER-2. *J. Biol. Chem.* **2008**, *283*, 2031–2041. [CrossRef] [PubMed]

182. Indini, A.; Brecht, I.; Del Vecchio, M.; Sultan, I.; Signoroni, S.; Ferrari, A. Cutaneous melanoma in adolescents and young adults. *Pediatr. Blood Cancer* **2018**, *65*, 1–6. [CrossRef] [PubMed]
183. Faingold, D.; Marshall, J.C.; Antecka, E.; Di Cesare, S.; Odashiro, A.N.; Bakalian, S.; Fernandes, B.F.; Burnier, M.N. Immune expression and inhibition of heat shock protein 90 in uveal melanoma. *Clin. Cancer Res.* **2008**, *14*, 847–855. [CrossRef]
184. Becker, B.; Multhoff, G.; Farkas, B.; Wild, P.J.; Landthaler, M.; Stolz, W.; Vogt, T. Induction of Hsp90 protein expression in malignant melanomas and melanoma metastases. *Exp. Dermatol.* **2004**, *13*, 27–32. [CrossRef] [PubMed]
185. Missotten, G.S.; Journée-de Korver, J.G.; De Wolff-Rouendaal, D.; Keunen, J.E.; Schlingemann, R.O.; Jager, M.J. Heat shock protein expression in the eye and in uveal melanoma. *Investig. Ophthalmol. Vis. Sci.* **2003**, *44*, 3059–3065. [CrossRef] [PubMed]
186. Nájera, L.; Alonso-Juarranz, M.; Garrido, M.; Ballestín, C.; Moya, L.; Martínez-Díaz, M.; Carrillo, R.; Juarranz, A.; Rojo, F.; Cuezva, J.M.; et al. Prognostic implications of markers of the metabolic phenotype in human cutaneous melanoma. *Br. J. Dermatol.* **2019**, *181*, 114–127. [CrossRef]
187. Westekemper, H.; Karimi, S.; Süsskind, D.; Anastassiou, G.; Freistühler, M.; Steuhl, K.P.; Bornfeld, N.; Schmid, K.W.; Grabellus, F. Expression of HSP 90, PTEN and Bcl-2 in conjunctival melanoma. *Br. J. Ophthalmol.* **2011**, *95*, 853–858. [CrossRef]
188. Park, H.S.; Park, C.H.; Choi, B.R.; Lim, M.S.; Heo, S.H.; Kim, C.H.; Kang, S.G.; Whang, K.U.; Cho, M.K. Expression of heat shock protein 105 and 70 in malignant melanoma and benign melanocytic nevi. *J. Cutan. Pathol.* **2009**, *36*, 511–516. [CrossRef]
189. Strickler, A.G.; Vasquez, J.G.; Yates, N.; Ho, J. Potential diagnostic significance of HSP90, ACS/TMS1, and L-plastin in the identification of melanoma. *Melanoma Res.* **2014**, *24*, 535–544. [CrossRef]
190. Shipp, C.; Weide, B.; Derhovanessian, E.; Pawelec, G. Hsps are up-regulated in melanoma tissue and correlate with patient clinical parameters. *Cell Stress Chaperones* **2013**, *18*, 145–154. [CrossRef]
191. McCarthy, M.M.; Pick, E.; Kluger, Y.; Gould-Rothberg, B.; Lazova, R.; Camp, R.L.; Rimm, D.L.; Kluger, H.M. HSP90 as a marker of progression in melanoma. *Ann. Oncol.* **2008**, *19*, 590–594. [CrossRef] [PubMed]
192. Košec, A.; Novak, R.; Konjevoda, P.; Trkulja, V.; Bedeković, V.; Grgurević, L. Tumor tissue hnRNP M and HSP 90α as potential predictors of disease-specific mortality in patients with early-stage cutaneous head and neck melanoma: A proteomics-based study. *Oncotarget* **2019**, *10*, 6713–6722. [CrossRef] [PubMed]
193. Schuster, C.; Akslen, L.A.; Straume, O. Expression of Heat Shock Protein 27 in melanoma metastases is associated with overall response to bevacizumab monotherapy: Analyses of predictive markers in a clinical phase II study. *PLoS ONE* **2016**, *11*, e0155242. [CrossRef]
194. Wu, X.; Marmarelis, M.E.; Hodi, F.S. Activity of the Heat Shock Protein 90 Inhibitor Ganetespib in Melanoma. *PLoS ONE* **2013**, *8*, e56134. [CrossRef] [PubMed]
195. Shin, M.K.; Jeong, K.H.; Choi, H.; Ahn, H.J.; Lee, M.H. Heat shock protein 90 inhibitor enhances apoptosis by inhibiting the AKT pathway in thermal-stimulated SK-MEL-2 human melanoma cell line. *J. Dermatol. Sci.* **2018**, *90*, 357–360. [CrossRef]
196. Liu, K.S.; Ding, W.C.; Wang, S.X.; Liu, Z.; Xing, G.W.; Wang, Y.; Wang, Y.F. The heat shock protein 90 inhibitor SNX-2112 inhibits B16 melanoma cell growth in vitro and in vivo. *Oncol. Rep.* **2012**, *27*, 1904–1910. [CrossRef]
197. Schmitt, E.; Maingret, L.; Puig, P.E.; Rerole, A.L.; Ghiringhelli, F.; Hammann, A.; Solary, E.; Kroemer, G.; Garrido, C. Heat shock protein 70 neutralization exerts potent antitumor effects in animal models of colon cancer and melanoma. *Cancer Res.* **2006**, *66*, 4191–4197. [CrossRef]
198. Eroglu, Z.; Chen, Y.A.; Gibney, G.T.; Weber, J.S.; Kudchadkar, R.R.; Khushalani, N.I.; Markowitz, J.; Brohl, A.S.; Tetteh, L.F.; Ramadan, H.; et al. Combined BRAF and HSP90 inhibition in patients with unresectable BRAF V600E mutant melanoma. *Clin. Cancer Res.* **2018**, *24*, 5516–5524. [CrossRef]
199. Shah, S.; Luke, J.J.; Jacene, H.A.; Chen, T.; Giobbie-Hurder, A.; Ibrahim, N.; Buchbinder, E.L.; McDermott, D.F.; Flaherty, K.T.; Sullivan, R.J.; et al. Results from phase II trial of HSP90 inhibitor, STA-9090 (ganetespib), in metastatic uveal melanoma. *Melanoma Res.* **2018**, *28*, 605–610. [CrossRef]
200. Tosti, G.; Cocorocchio, E.; Pennacchioli, E.; Ferrucci, P.F.; Testori, A.; Martinoli, C. Heat-shock proteins-based immunotherapy for advanced melanoma in the era of target therapies and immunomodulating agents. *Expert Opin. Biol. Ther.* **2014**, *14*, 955–967. [CrossRef]

201. Eton, O.; Ross, M.I.; East, M.J.; Mansfield, P.F.; Papadopoulos, N.; Ellerhorst, J.A.; Bedikian, A.Y.; Lee, J.E. Autologous tumor-derived heat-shock protein peptide complex-96 (HSPPC-96) in patients with metastatic melanoma. *J. Transl. Med.* **2010**, *8*, 1–13. [CrossRef] [PubMed]
202. Belli, F.; Testori, A.; Rivoltini, L.; Maio, M.; Andreola, G.; Sertoli, M.R.; Gallino, G.; Piris, A.; Cattelan, A.; Lazzari, I.; et al. Vaccination of metastatic melanoma patients with autologous tumor-derived heat shock protein gp96-peptide complexes: Clinical and immunologic findings. *J. Clin. Oncol.* **2002**, *20*, 4169–4180. [CrossRef] [PubMed]
203. Testori, A.; Richards, J.; Whitman, E.; Mann, G.B.; Lutzky, J.; Camacho, L.; Parmiani, G.; Tosti, G.; Kirkwood, J.M.; Hoos, A.; et al. Phase III comparison of vitespen, an autologous tumor-derived heat shock protein gp96 peptide complex vaccine, with physician's choice of treatment for stage IV melanoma: The C-100-21 study group. *J Clin. Oncol.* **2008**, *26*, 955–962. [CrossRef] [PubMed]
204. Pilla, L.; Patuzzo, R.; Rivoltini, L.; Maio, M.; Pennacchioli, E.; Lamaj, E.; Maurichi, A.; Massarut, S.; Marchianò, A.; Santantonio, C.; et al. A phase II trial of vaccination with autologous, tumor-derived heat-shock protein peptide complexes Gp96, in combination with GM-CSF and interferon-α in metastatic melanoma patients. *Cancer Immunol. Immunother.* **2006**, *55*, 958–968. [CrossRef]
205. Levine, O.; Zbuk, K. Colorectal cancer in adolescents and young adults: Defining a growing threat. *Pediatr. Blood Cancer* **2019**, *66*, e27941. [CrossRef]
206. Ju, Y.T.; Kwag, S.J.; Park, H.J.; Jung, E.J.; Jeong, C.Y.; Jeong, S.H.; Lee, Y.J.; Choi, S.K.; Kang, K.R.; Hah, Y.S.; et al. Decreased expression of heat shock protein 20 in colorectal cancer and its implication in Tumorigenesis. *J. Cell. Biochem.* **2015**, *116*, 277–286. [CrossRef] [PubMed]
207. Kanazawa, Y.; Isomoto, H.; Oka, M.; Yano, Y.; Soda, H.; Shikuwa, S.; Takeshima, F.; Omagari, K.; Mizuta, Y.; Murase, K.; et al. Expression of heat shock protein (Hsp) 70 and Hsp 40 in colorectal cancer. *Med. Oncol.* **2003**, *20*, 157–164. [CrossRef]
208. Cappello, F.; Bellafiore, M.; Palma, A.; David, S.; Marcianò, V.; Bartolotta, T.; Sciumè, C.; Modica, G.; Farina, F.; Zummo, G.; et al. 60KDa chaperonin (HSP60) is over-expressed during colorectal carcinogenesis. *Eur. J. Histochem.* **2003**, *47*, 105–109. [CrossRef]
209. Cappello, F.; David, S.; Rappa, F.; Bucchieri, F.; Marasà, L.; Bartolotta, T.E.; Farina, F.; Zummo, G. The expression of Hsp60 and Hsp10 in large bowel carcinomas with lymph node metastases. *BMC Cancer* **2005**, *5*, 1–10. [CrossRef]
210. He, Y.; Wu, Y.; Mou, Z.; Li, W.; Zou, L.; Fu, T.; Zhang, A.; Xiang, D.; Xiao, H.; Wang, X. Proteomics-based identification of HSP60 as a tumor-associated antigen in colorectal cancer. *Proteom.- Clin. Appl.* **2007**, *1*, 336–342. [CrossRef]
211. Jagadish, N.; Parashar, D.; Gupta, N.; Agarwal, S.; Suri, V.; Kumar, R.; Suri, V.; Sadasukhi, T.C.; Gupta, A.; Ansari, A.S.; et al. Heat shock protein 70-2 (HSP70-2) is a novel therapeutic target for colorectal cancer and is associated with tumor growth. *BMC Cancer* **2016**, *16*, 1–13. [CrossRef] [PubMed]
212. Drecoll, E.; Nitsche, U.; Bauer, K.; Berezowska, S.; Slotta-Huspenina, J.; Rosenberg, R.; Langer, R. Expression analysis of heat shock protein 90 (HSP90) and Her2 in colon carcinoma. *Int. J. Colorectal Dis.* **2014**, *29*, 663–671. [CrossRef] [PubMed]
213. Choi, D.H.; Ha, J.S.; Lee, W.H.; Song, J.K.; Kim, G.Y.; Park, J.H.; Cha, H.J.; Lee, B.J.; Park, J.W. Heat shock protein 27 is associated with irinotecan resistance in human colorectal cancer cells. *FEBS Lett.* **2007**, *581*, 1649–1656. [CrossRef] [PubMed]
214. Gunaldi, M.; Kocoglu, H.; Okuturlar, Y.; Gedikbasi, A.; Karabulut, M.; Alis, H.; Hursitoglu, M. Heat shock protein 70 is a useful marker for predicting colorectal cancer. *J. BUON* **2015**, *20*, 1464–1470. [PubMed]
215. Hamelin, C.; Cornut, E.; Poirier, F.; Pons, S.; Beaulieu, C.; Charrier, J.P.; Haïdous, H.; Cotte, E.; Lambert, C.; Piard, F.; et al. Identification and verification of heat shock protein 60 as a potential serum marker for colorectal cancer. *FEBS J.* **2011**, *278*, 4845–4859. [CrossRef]
216. Kasanga, M.; Liu, L.; Xue, L.; Song, X. Serum heat shock protein 90α have an advantage in diagnosis of colorectal cancer at early stage. *Biomark. Med.* **2018**, *12*, 881–890. [CrossRef] [PubMed]
217. Kocsis, J.; Madaras, B.; Tóth, É.K.; Füst, G.; Prohászka, Z. Serum level of soluble 70-kD heat shock protein is associated with high mortality in patients with colorectal cancer without distant metastasis. *Cell Stress Chaperones* **2010**, *15*, 143–151. [CrossRef]

218. Dinić, J.; Podolski-Renić, A.; Jovanović, M.; Musso, L.; Tsakovska, I.; Pajeva, I.; Dallavalle, S.; Pešić, M. Novel heat shock protein 90 inhibitors suppress P-glycoprotein activity and overcome multidrug resistance in cancer cells. *Int. J. Mol. Sci.* **2019**, *20*, 4575. [CrossRef]
219. Moradi-Marjaneh, R.; Paseban, M.; Moradi Marjaneh, M. Hsp70 inhibitors: Implications for the treatment of colorectal cancer. *IUBMB Life* **2019**, *71*, 1834–1845. [CrossRef]
220. Chaudhari, V.A.; Khobragade, K.; Bhandare, M.; Shrikhande, S.V. Management of fibrolamellar hepatocellular carcinoma. *Chinese Clin. Oncol.* **2018**, *7*. [CrossRef]
221. Riggle, K.M.; Turnham, R.; Scott, J.D.; Yeung, R.S.; Riehle, K.J. Fibrolamellar Hepatocellular Carcinoma: Mechanistic Distinction From Adult Hepatocellular Carcinoma. *Pediatr. Blood Cancer* **2016**, *63*, 1163–1167. [CrossRef] [PubMed]
222. Emre, S.; McKenna, G.J. Liver tumors in children. *Pediatr. Transplant.* **2004**, *8*, 632–638. [CrossRef] [PubMed]
223. Franco, L.M.; Krishnamurthy, V.; Bali, D.; Weinstein, D.A.; Arn, P.; Clary, B.; Boney, A.; Sullivan, J.; Frush, D.P.; Chen, Y.T.; et al. Hepatocellular carcinoma in glycogen storage disease type Ia: A case series. *J. Inherit. Metab. Dis.* **2005**, *28*, 153–162. [CrossRef] [PubMed]
224. Gunjan, D.; Shalimar; Nadda, N.; Kedia, S.; Nayak, B.; Paul, S.B.; Gamanagatti, S.R.; Acharya, S.K. Hepatocellular Carcinoma: An Unusual Complication of Longstanding Wilson Disease. *J. Clin. Exp. Hepatol.* **2017**, *7*, 152–154. [CrossRef] [PubMed]
225. He, M.; Qin, H.; Poon, T.C.W.; Sze, S.C.; Ding, X.; Co, N.N.; Ngai, S.M.; Chan, T.F.; Wong, N. Hepatocellular carcinoma-derived exosomes promote motility of immortalized hepatocyte through transfer of oncogenic proteins and RNAs. *Carcinogenesis* **2015**, *36*, 1008–1018. [CrossRef]
226. Wang, S.; Chen, G.; Lin, X.; Xing, X.; Cai, Z.; Liu, X.; Liu, J. Role of exosomes in hepatocellular carcinoma cell mobility alteration. *Oncol. Lett.* **2017**, *14*, 8122–8131. [CrossRef] [PubMed]
227. Zhang, J.; Lu, S.; Zhou, Y.; Meng, K.; Chen, Z.; Cui, Y.; Shi, Y.; Wang, T.; He, Q.Y. Motile hepatocellular carcinoma cells preferentially secret sugar metabolism regulatory proteins via exosomes. *Proteomics* **2017**, *17*, 13–14. [CrossRef]
228. Wang, S.; Xu, M.; Li, X.; Su, X.; Xiao, X.; Keating, A.; Zhao, R.C. Exosomes released by hepatocarcinoma cells endow adipocytes with tumor-promoting properties. *J. Hematol. Oncol.* **2018**, *11*, 1–14. [CrossRef]
229. Sasaki, R.; Kanda, T.; Yokosuka, O.; Kato, N.; Matsuoka, S.; Moriyama, M. Exosomes and hepatocellular carcinoma: From bench to bedside. *Int. J. Mol. Sci.* **2019**, *20*, 1406. [CrossRef] [PubMed]
230. Massimino, M.; Evans, D.B.; Podda, M.; Spinelli, C.; Collini, P.; Pizzi, N.; Bleyer, A. Thyroid cancer in adolescents and young adults. *Pediatr. Blood Cancer* **2018**, *65*, 1–9. [CrossRef]
231. Seib, C.D.; Sosa, J.A. Evolving Understanding of the Epidemiology of Thyroid Cancer. *Endocrinol. Metab. Clin. N. Am.* **2019**, *48*, 23–35. [CrossRef] [PubMed]
232. Gatta, G.; Capocaccia, R.; Stiller, C.; Kaatsch, P.; Berrino, F.; Terenziani, M. Childhood cancer survival trends in Europe: A EUROCARE working group study. *J. Clin. Oncol.* **2005**, *23*, 3742–3751. [CrossRef] [PubMed]
233. Collini, P.; Massimino, M.; Leite, S.F.; Mattavelli, F.; Seregni, E.; Zucchini, N.; Spreafico, F.; Ferrari, A.; Castellani, M.R.; Cantù, G.; et al. Papillary thyroid carcinoma of childhood and adolescence: A 30-year experience at the istituto nazionale tumori in Milan. *Pediatr. Blood Cancer* **2006**, *46*, 300–306. [CrossRef] [PubMed]
234. Patel, K.N.; Shaha, A.R. Poorly differentiated and anaplastic thyroid cancer. *Cancer Control* **2006**, *13*, 119–128. [CrossRef]
235. Mo, X.M.; Li, L.; Zhu, P.; Dai, Y.J.; Zhao, T.T.; Liao, L.Y.; Chen, G.G.; Liu, Z.M. Up-regulation of Hsp27 by ERα/Sp1 facilitates proliferation and confers resistance to apoptosis in human papillary thyroid cancer cells. *Mol. Cell. Endocrinol.* **2016**, *431*, 71–87. [CrossRef]
236. Wang, S.; Yang, S.; Vlantis, A.C.; Liu, S.Y.W.; Ng, E.K.W.; Chan, A.B.W.; Wu, J.; Du, J.; Wei, W.; Liu, X.; et al. Expression of Antioxidant Molecules and Heat Shock Protein 27 in Thyroid Tumors. *J. Cell. Biochem.* **2016**, *2481*, 2473–2481. [CrossRef]
237. Park, W.S.; Chung, K.-W.; Young, M.S.; Kim, S.-K.; Lee, Y.J.; Lee, E.K. Differential protein expression of lymph node metastases of papillary thyroid carcinoma harboring the BRAF mutation. *Anticancer Res.* **2013**, *33*, 4357–4364.
238. Abdullah, M.I.; Lee, C.C.; Mat Junit, S.; Ng, K.L.; Hashim, O.H. Tissue and serum samples of patients with papillary thyroid cancer with and without benign background demonstrate different altered expression of proteins. *PeerJ* **2016**, *4*, e2450. [CrossRef]

239. Fouad, E.M.; Harb, O.A.; Salem, R.A.; El Farargy, O.M.; Habib, F.M.; Gertallah, L.M. The expression of FOXE-1 and STIP-1 in papillary thyroid carcinoma and their relationship with patient prognosis. *Iran. J. Pathol.* **2018**, *13*, 256–271. [CrossRef]
240. Mo, J.-H.; Choi, I.J.; Jeong, W.-J.; Jeon, E.-H.; Ahn, S.-H. HIF-1α and HSP90: Target molecules selected from a tumorigenic papillary thyroid carcinoma cell line. *Cancer Sci.* **2012**, *103*, 464–471. [CrossRef]
241. Soudry, E.; Stern Shavit, S.; Hardy, B.; Morgenstern, S.; Hadar, T.; Feinmesser, R. Heat shock proteins HSP90, HSP70 and GRP78 expression in medullary thyroid carcinoma. *Ann. Diagn. Pathol.* **2017**, *26*, 52–56. [CrossRef] [PubMed]
242. Yuan, M.H.; Zhou, R.S.; She, B.; Xu, H.F.; Wang, J.Y.; Wei, L.X. Expression and clinical significance of STIP1 in papillary thyroid carcinoma. *Tumor Biol.* **2014**, *35*, 2391–2395. [CrossRef] [PubMed]
243. Song, H.Y.; Dunbar, J.D.; Zhang, Y.X.; Guo, D.; Donner, D.B. Identification of a protein with homology to hsp90 that binds the type 1 tumor necrosis factor receptor. *J. Biol. Chem.* **1995**, *270*, 3574–3581. [CrossRef] [PubMed]
244. Felts, S.J.; Owen, B.A.L.; Nguyen, P.M.; Trepel, J.; Donner, D.B.; Toft, D.O. The hsp90-related protein TRAP1 is a mitochondrial protein with distinct functional properties. *J. Biol. Chem.* **2000**, *275*, 3305–3312. [CrossRef] [PubMed]
245. Palladino, G.; Notarangelo, T.; Pannone, G.; Piscazzi, A.; Lamacchia, O.; Sisinni, L.; Spagnoletti, G.; Toti, P.; Santoro, A.; Storto, G.; et al. TRAP1 regulates cell cycle and apoptosis in thyroid carcinoma cells. *Endocr. Relat. Cancer* **2016**, *23*, 699–709. [CrossRef]
246. Liu, J.; Sun, W.; Dong, W.; Wang, Z.; Qin, Y.; Zhang, T.; Zhang, H. HSP90 inhibitor NVP-AUY922 induces cell apoptosis by disruption of the survivin in papillary thyroid carcinoma cells. *Biochem. Biophys. Res. Commun.* **2017**, *487*, 313–319. [CrossRef]
247. Kim, S.H.; Kang, J.G.; Kim, C.S.; Ihm, S.H.; Choi, M.G.; Yoo, H.J.; Lee, S.J. Novel heat shock protein 90 inhibitor NVP-AUY922 synergizes with the histone deacetylase inhibitor PXD101 in induction of death of anaplastic thyroid carcinoma cells. *J. Clin. Endocrinol. Metab.* **2015**, *100*, E253–E261. [CrossRef]
248. White, P.T.; Subramanian, C.; Zhu, Q.; Zhang, H.; Gallagher, R.; Timmermann, B.M.; Blagg, B.S.J.; Cohen, M.S. Novel HSP90 inhibitors effectively target thyroid cancer stem cell function preventing migration and invasion. *Surgery* **2016**, *159*, 142–151. [CrossRef]
249. Keller, S.; König, A.K.; Marmé, F.; Runz, S.; Wolterink, S.; Koensgen, D.; Mustea, A.; Sehouli, J.; Altevogt, P. Systemic presence and tumor-growth promoting effect of ovarian carcinoma released exosomes. *Cancer Lett.* **2009**, *278*, 73–81. [CrossRef]
250. Cheng, L.; Wu, S.; Zhang, K.; Qing, Y.; Xu, T. A comprehensive overview of exosomes in ovarian cancer: Emerging biomarkers and therapeutic strategies. *J. Ovarian Res.* **2017**, *10*, 1–9. [CrossRef]
251. Nieuwland, R.; Van Der Post, J.A.M.; Lok Gemma, C.A.R.; Kenter, G.; Sturk, A. Microparticles and exosomes in gynecologic neoplasias. *Semin. Thromb. Hemost.* **2010**, *36*, 925–929. [CrossRef]
252. Ciocca, D.R.; Calderwood, S.K. Heat shock proteins in cancer: Diagnostic, prognostic, predictive, and treatment implications. Targeting heat shock proteins in cancer. Heat shock proteins in obesity: Links to cardiovascular disease. *Cell Stress Chaperones* **2005**, *10*, 86–103. [CrossRef] [PubMed]
253. Hjerpe, E.; Egyhazi, S.; Carlson, J.; Stolt, M.F.; Schedvins, K.; Johansson, H.; Shoshan, M.; Åvall-Lundqvist, E. HSP60 predicts survival in advanced serous ovarian cancer. *Int. J. Gynecol. Cancer* **2013**, *23*, 448–455. [CrossRef] [PubMed]
254. Hoter, A.; Naim, H.Y. Heat shock proteins and ovarian cancer: Important roles and therapeutic opportunities. *Cancers* **2019**, *11*, 1389. [CrossRef] [PubMed]
255. Bray, F.; Ferlay, J.; Soerjomataram, I.; Siegel, R.L.; Torre, L.A.; Jemal, A. Global cancer statistics 2018: GLOBOCAN estimates of incidence and mortality worldwide for 36 cancers in 185 countries. *CA Cancer J. Clin.* **2018**, *68*, 394–424. [CrossRef] [PubMed]
256. Ghoncheh, M.; Pournamdar, Z.; Salehiniya, H. Incidence and mortality and epidemiology of breast cancer in the world. *Asian Pacific J. Cancer Prev.* **2016**, *17*, 43–46. [CrossRef]
257. Johnson, R.H.; Anders, C.K.; Litton, J.K.; Ruddy, K.J.; Bleyer, A. Breast cancer in adolescents and young adults. *Pediatr. Blood Cancer* **2018**, *65*. [CrossRef]
258. Murphy, B.L.; Day, C.N.; Hoskin, T.L.; Habermann, E.B.; Boughey, J.C. Adolescents and Young Adults with Breast Cancer have More Aggressive Disease and Treatment Than Patients in Their Forties. *Ann. Surg. Oncol.* **2019**, *26*, 3920–3930. [CrossRef]

259. Yano, M.; Naito, Z.; Tanaka, S.; Asano, G. Expression and roles of heat shock proteins in human breast cancer. *Jpn. J. Cancer Res.* **1996**, *87*, 908–915. [CrossRef]
260. Conroy, S.E.; Latchman, D.S. Do heat shock proteins have a role in breast cancer? *Br. J. Cancer* **1996**, *74*, 717–721. [CrossRef]
261. Banerjee, S.; Lin, C.F.L.; Skinner, K.A.; Schiffhauer, L.M.; Peacock, J.; Hicks, D.G.; Redmond, E.M.; Morrow, D.; Huston, A.; Shayne, M.; et al. Heat shock protein 27 differentiates tolerogenic macrophages that may support human breast cancer progression. *Cancer Res.* **2011**, *71*, 318–327. [CrossRef] [PubMed]
262. Rich, A.L.; Khakwani, A.; Free, C.M.; Tata, L.J.; Stanley, R.A.; Peake, R.B.; Hubbard, R.B.; Baldwin, D.R. Non-small cell lung cancer in young adults: Presentation and survival in the English National Lung Cancer Audit. *An Int. J. Med.* **2015**, *108*, 891–897. [CrossRef] [PubMed]
263. Subramanian, J.; Morgensztern, D.; Goodgame, B.; Baggstrom, M.Q.; Gao, F.; Piccirillo, J.; Govindan, R. Distinctive characteristics of non-small cell lung cancer (NSCLC) in the young: A Surveillance, Epidemiology, and End Results (SEER) analysis. *J. Thorac. Oncol.* **2010**, *5*, 23–28. [CrossRef] [PubMed]
264. Srivastava, A.; Amreddy, N.; Razaq, M.; Towner, R.; Zhao, Y.D.; Ahmed, R.A.; Munshi, A.; Ramesh, R. Exosomes as Theranostics for Lung Cancer. In *Advances in Cancer Research*; Academic Press Inc.: Cambridge, MA, USA, 2018; Volume 139, pp. 1–33, ISBN 9780128141694.

applied sciences

MDPI

Review

Biogenesis, Biologic Function and Clinical Potential of Exosomes in Different Diseases

Amany Magdy Beshbishy [1,*,†], Saad Alghamdi [2], ThankGod E. Onyiche [3,4], Muhammad Zahoor [5], Nallely Rivero-Perez [6], Adrian Zaragoza-Bastida [6], Mohamed A. Ghorab [7,8,9], Ahmed Kh. Meshaal [10], Mohamed A. El-Esawi [11], Helal F. Hetta [12,13] and Gaber El-Saber Batiha [14,†]

1 National Research Center for Protozoan Diseases, Obihiro University of Agriculture and Veterinary Medicine, Nishi 2-13, Inada-cho, Obihiro 080-8555, Hokkaido, Japan
2 Laboratory Medicine Department, Faculty of Applied Medical Sciences, Umm Al-Qura University, P. O. BOX 715, Makkah 21955, Saudi Arabia; ssalghamdi@uqu.edu.sa
3 Unit for Environmental Sciences and Management, North-West University, Potchefstroom Campus, Private Bag X6001, Potchefstroom 2520, South Africa; et.onyiche@unimaid.edu.ng
4 Department of Veterinary Parasitology and Entomology, University of Maiduguri, P. M. B. 1069, Maiduguri 600230, Nigeria
5 Department of Biochemistry, University of Malakand, Chakdara, Dir Lower, Khyber Pakhtunkhwa 18800, Pakistan; mohammadzahoorus@yahoo.com
6 Área Académica de Medicina Veterinaria y Zootecnia, Instituto de Ciencias Agropecuaria, Universidad Autónoma del Estado de Hidalgo, Av. Universidad Km 1, Ex-Hda. de Aquetzalpa, Tulancingo 43600, Hgo, Mexico; nallely_rivero@uaeh.edu.mx (N.R.-P.); adrian_zaragoza@uaeh.edu.mx (A.Z.-B.)
7 Pegasus c/o US Environmental Protection Agency (EPA), Office of Research and Development (ORD), Cincinnati, OH 45220, USA; ghorabmo@msu.edu
8 Wildlife Toxicology Laboratory, Department of Animal Science, Institute for Integrative Toxicology (IIT), Michigan State University, East Lansing, MI 48824, USA
9 Environmental Toxicology Lab, Marine Environment Department, National Institute of Oceanography, and Fisheries (NIOF), Alexandria 21556, Egypt
10 Department of Botany and Microbiology, Faculty of Science, Al-Azhar University, Assiut 71524, Egypt; ahmad_meshaal1650@yahoo.com
11 Botany Department, Faculty of Science, Tanta University, Tanta 31527, Egypt; mohamed.elesawi@science.tanta.edu.eg
12 Department of Internal Medicine, University of Cincinnati College of Medicine, Cincinnati, OH 45267, USA; helal.hetta@uc.edu
13 Department of Medical Microbiology and Immunology, Faculty of Medicine, Assiut University, Assiut 71515, Egypt
14 Department of Pharmacology and Therapeutics, Faculty of Veterinary Medicine, Damanhour University, Damanhour 22511, AlBeheira, Egypt; gaberbatiha@gmail.com
* Correspondence: amanimagdi2008@gmail.com; Tel.: +20-45-271-6024; Fax: +20-45-271-6024
† These authors contributed equally.

Received: 9 June 2020; Accepted: 24 June 2020; Published: 27 June 2020

check for updates

Abstract: Exosomes are extracellular vesicles (EVs) belonging to the nanovesicles family that function as signaling molecules between cells. After their first description in the late 1960s, interest in their potential as a research target has steadily increased. They are small secreted organelles with a single membrane that are well enriched in lipids, proteins, nucleic acids, and glycoconjugates. Exosomes take part in a larger communication network in which cells communicate between one another by DNA shuttling, proteins, RNA, and membrane-bound factors. The machinery of protein quality control occurs through the process termed "exosome biogenesis". Furthermore, the pathway involved in intercellular movement of vesicles is vital in various aspects of human health and diseases. Due to their inherent properties, exosomes are currently being developed as potential therapeutic agents in a wide range of diseases including infectious and non-infectious diseases. Exosomes and other EVs sourced from Mesenchymal stem cells (MSCs) have been shown in different studies to possess therapeutic effects in diverse disease models either in vivo or in vitro. Some mechanisms and/or pathways that MSC-derived exosomes use to illustrate their therapeutic effect against some diseases have also been summarized. This review aims to highlight the recent findings and potential therapeutic application of exosomes in different diseases such as autoimmune, cardiovascular, obesity, neural, soft tissues, bone, and cartilage.

Keywords: exosomes; adipose-derived stem cell released exosomes; tumor-derived exosomes; metabolic diseases; mesenchymal stem cells

1. Introduction

Before the late 1960s, little was known about exosomes until their first description by two investigators namely Bonucci and Anderson [1,2]. Right from these early studies, they were known to be the initiators of hydroxyapatite crystals which were formed as a result of direct budding of the plasma membrane [2]. Furthermore, in the 1980s, the term, exosome biogenesis was revealed following a study conducted by Trams et al. [3] who discovered that exosomes in the seminal fluid were derived from prostrate/epididymis.

Exosomes are extracellular vesicles (EVs) that function as signaling molecules between cells. These exosomes are secreted organelles with a single membrane usually small in size with a diameter of ~30 to 200 nm. They are well enriched in lipids, proteins, nucleic acids, and glycoconjugates. Structurally, exosomes are a series of membrane-associated, oligomeric complexes of protein with a marked molecular diversity formed by budding in endosome and plasma membranes [4]. The term EV includes exosomes, prostasomes, ectosomes, microvesicles, microparticles, tolerosomes, apoptotic bodies, and nanovesicles. Currently, available technology cannot separate exosomes from the other EVs mentioned earlier. Thus, exosomes isolated from cells and used for research related activities are best regarded as EVs, as stated in the guidelines formulated by the International Society of Extracellular Vesicles (ISEV) [5]. Cell-based membrane vesicles deliver microRNA (miRNA), messenger RNA (mRNA), and proteins through intracellular organelles. Microparticles, microvesicles (100–1000 nm), and exosomes (20–200 nm) are three examples of membrane vesicles [6]. Exosomes undergo a cycle called exosome biogenesis to ensure quality control. Once they are released, exosomes have diverse activities such as extracellular matrix remodeling and signal transmission. Intercellular vesicles play vital roles in various health and disease aspects including immunity, cancer, tissue homeostasis, and neurodegenerative diseases. Additionally, pathogens such as viruses utilize exosome biogenesis pathways viz-a-viz by establishing host permissiveness and by assembling infectious particles [7]. Exosome isolation from diverse body fluids such as plasma, cerebrospinal fluid, semen, blood, amniotic fluid, saliva, urine, epididymal and synovial fluid, breast milk, and bronchoalveolar lavage makes them an ideal model for the development of therapeutic agents against multiple diseases [8].

Exosomes are particularly regulated indifferent organisms during stresses [9,10], and they participate in a larger communication network in which cell connects to each other by DNA shuttling, proteins, RNA, and membrane-bound factors. Exosomes derived from tumors, commonly known as Tumor-derived exosomes (TEX), affect the surrounding tumor microenvironment (TME). As a result of the deep distribution of TEX in the blood and lymph, they have a big role in remote tissue sites and in the creation of their pre-metastatic niche referred to as the tumor macroenvironment (TMaE) [11]. TEX delivers tolerogenic signals to immune cells, prevents the proliferation of immune cells, interferes with the differentiation of monocyte, and induces apoptosis in activated CD^{8+} T lymphocytes leading to immune suppression via paracrine effect [12]. TEX exhibit some ligands (e.g., programmed death-ligand 1 (PD-L1)) to produce beneficial endocrine signals which provide pre-metastatic TMaE, expanding far from the primary tumor [11]. Furthermore, it was shown that the expression of PD-L1 is enhanced in melanoma cells exposed to interferon-γ (IFNγ), leading to the expression of PD-L1 in circulating melanoma exosomes derived from humans (HMEX) [13].

2. Biogenesis of Exosomes

According to the International Society for Extracellular Vesicles (ISEV), extracellular vesicle (EV) is the generic name used to refer to particles naturally released from a cell that is composed of a lipid bilayer and is unable to replicate (lack of functional nucleus). It is difficult to assign an EV to a particular pathway of biogenesis, except that the process was observed using live imaging techniques [5]. The nomenclature of exosomes is derived in 3 different ways: based on their biogenesis, physiological functions within cells, and lastly, empirical definition based on isolation [14]. In this review, exosomes will be defined by their biogenesis. Cells are professionals in producing and exporting molecular products, for instance, transporting insulin to the bloodstream [15]. Before transportation, these molecules are sent first for cell packaging, and they are the exosomes that are first released intracellularly. Multivesicular endosomes (MVEs) also encompass endosomes which are then formed. There are three ways in which exosome biogenesis is formed. First, vesicular budding into discrete endosomes with multivesicular body (MVB) maturation, which releases exosomes after plasma membrane fusion. Second, their immediate release from the plasma membrane by vesicular budding, and lastly, its release is delayed due to budding at the intracellular plasma membrane-connected compartments (IPMCs) after which there is a deconstruction of IPMC necks [16], as shown in Figure 1. Several molecules—such as endosomal sorting complexes required for transport (ESCRT) machinery, tetraspanins, and lipids—were involved in intraluminal vesicle (ILV) biogenesis, [17]. Whereas two different mechanisms, namely inducible release and trans-Golgi network, were assumed to participate in exosome secretion [18].

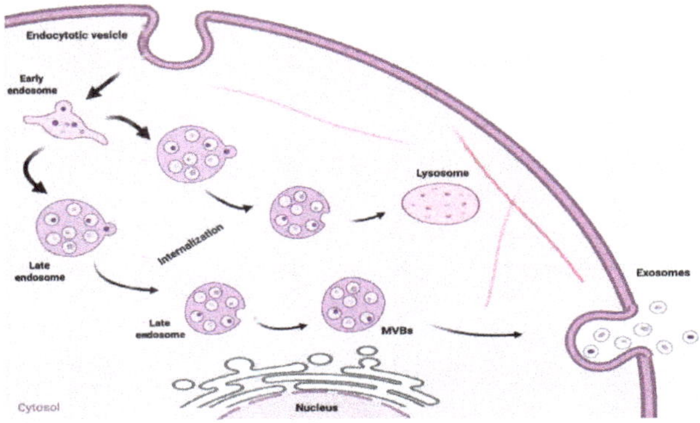

Figure 1. Biogenesis of exosomes.

Intracellular exosomes are fused into MVEs in the cytoplasm. The mechanisms for exosome generation and release have been well described [19]. Interestingly, exosomes are released into the extracellular matrix after the fusion of the MVEs with the cell membrane [20]. The peripheral membrane proteins (Rab GTPases) were found to facilitate the fusion of the MVEs with plasma membranes including RAB11 and RAB35 after which flotillin- and other cell-specific protein-rich exosomes are released [21], while the second mechanism facilitating the release of exosomes loaded with CD63, TSG1010, and ALIX involves RAB27A and RAB27B [19]. Microvesicles undergo a different mode of biogenesis in contrast to exosomes. They are formed by direct blebbing from the plasma membrane. In summary, diverse complex pathways are used to generate exosomes from endosomes. The formation of the exosomes also varies considerably depending on the type as well as the physiological state of the origin of the cell [22].

Limited MVB is subjected to internal budding leading to late endosome formation, which in turn generates exosomes in a constitutive way. The formation of intraluminal vesicles (ILVs) within MVBs from late endosomal membranes is caused by invagination [23]. During invagination, some proteins are incorporated into the membrane and the cytosolic components are engulfed and enclosed in ILVs. The release of ILVs is indicated in the extracellular space after fusion with the exosome's plasma membrane. Degradation of the components occurs in the lysosomes. Canonical exosomes are biconcave in shape produced by artificial drying, but they appeared spherical when viewed using transmission electron microscopy [4].

Recent scientific evidence indicates the role of the alternative pathway in exosomal cargo into MVBs. This process occurs in an ESCRT-independent manner dependent upon microdomains based on raft. These microdomains are enriched by sphingomyelinases that form ceramides [24]. Hydrolytic removal of phosphocholine moiety results in the formation of ceramides which is known to promote lateral phase separation and microdomains coalescence. Tetraspanins are proteins that play a key role in protein loading and exosome biogenesis. In the plasma membrane, tetraspanin-enriched microdomains (TEMs) are specialized platforms for protein signaling and receptor compartmentalization. Tetraspanin CD^{8+} and TEM together play a key role in sorting target receptors and intracellular components into the exosomes [25]. Bioactive molecules sorting into exosomes occurs either through ESCRT-dependent or ESCRT-independent mechanisms depending on the cell type origin. Plasma membrane-budded microvesicles (MVs) are a heterogeneous group of membrane vesicles that are produced with exosomes. They are generated by external budding of the plasma membrane and are of variable shapes, 100–1000 nm in size and are products of endothelial cells, red blood cells, and platelets. The small size and shape of exosomes make it easier for them to escape from the mononuclear phagocytic system cells, allowing them to stay longer in the general circulation [26].

3. Therapeutic Potential of Exosomes in Different Diseases

Attention is currently focused on the clinical application of exosomes for the treatment of disease. Exosomes may be included in useful applications for managing some diseases, including cardiac regeneration, obesity, soft tissues, bone/cartilage injury, brain/neural injury, etc. Figure 2 presents a pictorial presentation of certain conditions of disease in which exosomes were used. Exosomes and other EVs have been also used in stem cell therapy. Several reports have documented the application of Mesenchymal stem cells (MSCs) derived from secretome in various diseases such as skeletal, cardiac, nervous, ovarian, renal, hepatic, infectious, pulmonary, and soft tissue diseases. A list of some of the commonly used stem cells, their application, and mechanisms of actions are presented in Table 1.

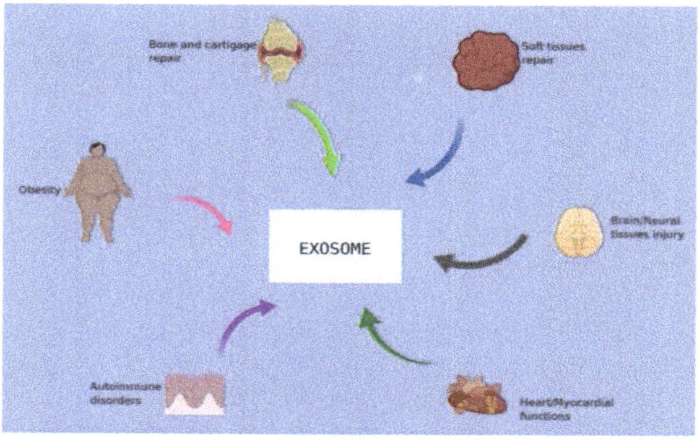

Figure 2. Applications and use of exosomes in different disease conditions.

Table 1. A summary of some therapeutic applications of Mesenchymal stem cells (MSC)-derived exosomes and other extracellular vesicles in different diseases.

Stem Cells	Study	Targeted Tissue/Cells	Mechanism	Reference
Tendon Stem Cells	in vitro and in vivo	Tendon healing	Through balancing synthesis and degradation of the tendon extracellular matrix	[27]
Neural Stem Cell	in vitro and in vivo	Spinal cord	Reduce neuronal apoptosis, inhibit neuroinflammation, and promote functional recovery in SCI model rats at an early stage by promoting autophagy	[28]
Bone marrow mesenchymal Stem Cells (BMSCs)	in vivo	Brain	miRNA129-5p's anti-inflammation and anti-apoptosis effects via quenching the activity of HMGB1-TLR4 pathway.	[29]
Bone Marrow Mesenchymal Stem Cells	in vivo	Spinal cord injury	Inhibiting complement mRNA synthesis and release and by inhibiting SCI-activated NF-κB by binding to microglia	[30]
Bone Marrow Mesenchymal Stem Cells	in vivo	Spinal cord	Exosome target M2-type macrophages at the site of SCI, support the idea that extracellular vesicles	[31]
Bone Marrow Mesenchymal Stem Cells (BMSCs)	in vivo	Chondrocytes/osteoarthritis	Restoration of cartilage and subchondral bone via the adenylate kinase and nucleoside-diphosphate kinase-dependent pathways	[32]
Bone Marrow Mesenchymal Stem Cells	in vitro and in vivo	Diabetes-induced cognition impairment	NA	[33]
Bone Marrow Mesenchymal Stem Cells (BMSCs)	in vitro and in vivo	Ovary	miR-644-5p carried by BMSC-derived exosomes inhibited the apoptosis of ovarian granulosa cell by targeting p53 of cells, suggesting that miR-644-5p had the potential to treat POF and restore ovarian function.	[34]
Adipose-derived mesenchymal Stem Cells	in vitro	Cardiomyocytes	Reduced apoptosis in myocardial cells subjected to oxidative stress	[33]
Adipose tissue, bone marrow, and umbilical cord blood mesenchymal Stem Cells	in vitro and in vivo	Myocardium	NA	[35]
Adipose-Derived Stem Cells (ADSCs)	in vitro and in vivo	Adipose tissue	Exosomes from ADSCs transferred into macrophages to induce anti-inflammatory M2 phenotypes through the transactivation of arginase-1 by exosome-carried active STAT3.	[36]
Umbilical cord mesenchymal Stem Cells	in vitro and in vivo	Brain	Decreased inflammation-related genes and pro-inflammatory cytokines. Exosomes interfered within the Toll-like receptor 4 signalings of BV-2 microglia, as they prevented the degradation of the NFκB inhibitor IκBα and the phosphorylation of molecules of the mitogen-activated protein kinase family in response to LPS stimulation	[37]
Umbilical cord mesenchymal Stem Cells (3D)	in vitro and in vivo	Cartilage	Stimulate chondrocyte proliferation, migration, and matrix synthesis, and inhibit apoptosis	[38]
miRNA-29b-modified mesenchymal Stem Cells	in vivo	Spinal cord	Improved SCI, including increased BBB score, NF200, and GAP-43 positive neurons, as well as decreased contractile nerve cell numbers and GFAP positive neurons	[39]

Table 1. *Cont.*

Stem Cells	Study	Targeted Tissue/Cells	Mechanism	Reference
Lipopolysaccharide-preconditioned human dental pulp Stem Cells		Schwann cells	Promote the proliferation, migration and odontogenic differentiation of SCs	[40]
Human Bone Marrow Mesenchymal Stromal Cells (MSCs)	in vitro and in vivo	Renal cells	Extracellular vesicles (EVs) derived from murine and/or human BM-MSCs reversed the radiation damage to marrow hematopoietic cells both in vitro and in vivo.	[41]
Human Umbilical Cord blood Mesenchymal Stem Cells (hUCB-MSCs)	in vitro and in vivo	Hepatic cells	hUCB-MSCs released exosomes that transferred miR-1246 to hepatocytes and subsequently produced a protective effect against hepatic IRI via regulating GSK3β-mediated Wnt/β-catenin signaling pathway	[42]
Human-induced Pluripotent Stem Cells	in vitro and in vivo	Cardiomyocytes	t iPSC-Pg-derived EV is effective in the treatment of CHF, possibly, in part, through their specific miRNA signature and the associated stimulation of cardioprotective pathways	[43]
Induced Pluripotent Stem Cells (iPSC)	in vitro and in vivo	Cardiac cells	The iPSC-derived EVs induced angiogenic, migratory, and antiapoptotic properties in cardiac endothelial cells in vitro; and induced superior infarct repair in vivo compared with iPSCs	[44]
Human-induced Pluripotent Stem Cells (iPSCs)	in vitro	Human Mesenchymal Stem Cells (MSCs)	Purified EVs secreted from both iPSCs and young MSCs alleviated senescence-associated cellular phenotypes of aged MSCs through reducing intracellular ROS level via the exosomal transfer of PRDXs	[33]
Human Mesenchymal Stem Cells (MSCs)	in vitro and in vivo	Autoimmune uveoretinitis	Inhibiting the migration of inflammatory cells and	[45]
Bone marrow stromal cells (BMSCs)	In vivo	Bronchopulmonary dysplasia	Paracrine manner via the release of immunomodulatory factors to ameliorate the parenchymal and vascular injury	[46]
Exosomes derived from *Mycobacterium bovis*-infected macrophages	In vivo	Macrophages	promote DC activation as well as generating an antibacterial T cell response	[47]
Human umbilical cord mesenchymal stem cells	In vitro and in vivo	Hepatocytes	HucMSC-Ex ameliorate CCl4-induced liver fibrosis by inhibiting epithelial-to-mesenchymal transition (EMT) and protecting hepatocytes.	[48]

NA: Not available restoration of homeostasis in bioenergetics and cell metabolism/restoration of cartilage and subchondral bone adenylate kinase and nucleoside-diphosphate kinase-dependent pathways.

Nucleic acids and proteins are naturally carried by exosomes making them potential vehicles for the intentional incorporation of proteins and nucleic acids for therapeutic purposes [49]. Synthetic nanoparticles, numerous small chemotherapeutic molecules, and natural compounds are other therapeutic molecules that are efficiently delivered via exosomes [49]. Altogether, these stacks of therapeutic cargos loaded into exosomes are delivered efficiently to host cells with diverse effects [50]. However, concerns are that exosomes may generate immune responses. Therefore, to overcome this challenge, certain factors must be taken into consideration such as the cell source and site of formation. Exosomes should be carefully selected from the donor cells. In recent times, clinical application and use of exosomes in therapeutics have grown rapidly. The targeted use of exosomes is one of the promising methods for treating cancer [51], and these applications have also been extended to other disease conditions such as soft tissues [52], cardiovascular disease [53], obesity [54], autoimmune diseases [55], and others. In summary, we can say that exosomes are potential carriers of minor interfering RNAs and have a great potential for clinical uses in treatments, and some of these useful applications will be discussed as follows.

3.1. Therapeutic Potential of Exosomes in Bone and Cartilage

Skeletal health is dependent on the fact that successful blood vessels and muscle generation are physiological processes that also include exosomal signals. Exosomes are composed of significant amounts of transforming growth factor-β1 (TGF-β1), promyogenic molecules (VEGF), proangiogenic, human T-cell factor 4 (TCF4), miR-494, interleukin-8 (IL-8), IL-6, miR-181, multiple miRNAs, and hepatocyte growth factor (HGF), most of which were thrown by Mesenchymal stem cells (MSCs) [56]. MSC-derived exosomes arising either from bone marrow, adipose tissue umbilical cord, or stimulating stem cells can enhance the bone differentiation of MSCs or primary bone cells. This is illustrated by the osteogenic genes and osteocalcin upregulation and elevated osteoblast multiplication and migration [57]. The changes in gene expression in MSCs, following MSC-derived exosomes absorption, indicates cell adhesion activation, PI3K-Akt signaling, and ECM-receptor interaction pathways, which are finally related to bone differentiation [53]. Exosomes originated from MSCs can also be absorbed immediately by osteoblasts, enhancing their multiplication and stimulating protein synthesis associated with the GLUT3 and MAPK pathways [58]. Adaptation to MSC-derived exosomes leads to tissue changes, such as increased matrix vascularization and mineralization along with bone restoration in rats with bone disorders [59]. Strikingly, MSC-derived exosomes are directly bound to fibronectin and type I collagen proteins [60]. The development of matrix mineralization shows a feature of late exosomes rather than the early stage of bone differentiation. Tendon stem cell (TSC) and exosome injections in rodents led to a decrease in the matrix metalloproteinases (MMP)-3 expression and an increase in the collagenase I (1a1) expression as well as the tissue metalloproteinase-3 (TIMP-3) inhibitor [20]. Furthermore, in vitro studies using 2D-Exo and 3D-Exo umbilical Mesenchymal stem cell (MSC) cultures indicated that both cultures stimulated the proliferation and migration of chondrocytes as well as matrix synthesis and apoptosis inhibition [25]. Finally, exosomes also appear to be part of cartilage growth. Exosomes have an altered miRNA cargo produced by MSCs derived from bone marrow with an activated chondrogenic phenotype (marked by a reduction in miR-6891-5p and miR-377-3p and an elevation in miR-1290, miR-92a, miR-1246, miR-320c, and miR-193a-5p levels), and hence promote chondrogenesis [60]. The most obvious chondrogenic event due to miR-320c exosomal is controlled by the up-regulation of SOX9 and the down-regulation of MMP13 metalloproteinase [61,62].

3.2. Therapeutic Potential of Exosomes in Myocardial Diseases

Exosomes play a crucial participatory role in the pathophysiological pathway of the cardiovascular system, and can become effective targets for clinical applications [63]. Vascular endothelial dysfunction is a critical primary mechanism at the onset of atherosclerosis (AS), and intercellular communication has been identified as a significant part of the pathological process [64,65]. First, endothelial cells (ECs) can express vesicles, as well as absorbing vesicles derived from EC under oxidative pressure, reducing the production of nitric oxide (NO) that has a protective role [66]. Furthermore, it was observed in a previous study that endothelial microbes of coronary artery disease (CAD) were diffused in patients containing miR-92a-3p; a fine vesicle in selected CAD patients, largely controlled by vesicular performance and internal integrity [67]. Similarly, atherosclerotic stimuli as IL-6 or ox-LDL (oxidized low-density lipoprotein) that specifically enhance miR-92a-3p encapsulation in EMV, and EMVs mediated with EMV-3p eventually strengthen the vascular responses in targeted ECs through a THBS1-based mechanism (thrombospondin 1) [68]. Additionally, the information exchange between cracker and plaque cells involves the modification of the endothelial balance. In a previous study, for example, it was observed that AS endothelial apoptosis was caused by growth mass specificity following exosomal lncRNA GAS5 transfer from macrophages to ECs. Furthermore, endothelial apoptosis is initiated by exosomal lncRNA GAS5-overexpressing THP-1 cells while exosomal lncRNA GAS5 knockout cells inhibit this pathway [69]. Meanwhile, the krüppel-like factor 5 (KLF5) up-regulates ox-LDL in vascular smooth muscle cells (SMCs), resulting in endothelial dysfunction. Vascular remodeling is regulated by zinger fingers and KLF5, a known transcription factor. Studies have found that SMCs expressing KLF5 affect the endothelial functions by controlling multiple microRNAs. Of these multiple microRNAs, the most important of them is miR-155 (a known proinflammatory factor). The resulting exosomes produced by KLF5-overexpressing SMCs are well enriched with miR-155. Poor barrier function and weak reproductive capacity are expressed by the ECs [70,71]. Nevertheless, these effects can be prevented by miR-155 suppression or KLF5 disability. In addition, exosomes resulting from SMC (SMC-exos) enriched with miR-221/222 have been identified to interrupt autophagy of endothelial cells through the PTEN/Akt signaling mechanism making ECs more likely to stimulate atherosclerosis. Exosomes derived from platelets help to maintain the function of endothelial vasculature, but whether the modulatory effect is harmful or beneficial is still contested [72]. Some results suggest that the absorption of external platelets can lead to endothelial adhesion and programmed cell death, while others draw inconsistent conclusions [73]. For instance, platelet-exos enriched with miR-320 or miR-223 inhibited inflammation and EC mobility by reducing the appearance of ICAM-1 in EC recipients [74,75]. The reason for the discrepancies is related to the complexity of the packed charges in the platelets, especially the ubiquitin/protease system found in the outer platelets that stimulate the degradation of the inflammatory substances [76,77]. Exosomes from adipose-derived stem cells have been shown to express CD9 and CD29 proteins that protect cardiomyocytes during oxidative stress [22]. Furthermore, exosomes obtained from adipose tissue-derived Mesenchymal stem cells (ADMSCs) promoted cardiac function, particularly myocardium, by inhibiting cardiomyocytes apoptosis and promoting angiogenesis [32]. Finally, induced pluripotent stem cells (iPSCs) obtained from EVs ensure the cardioprotection of the cardiac cells with reference to the left ventricular function in vivo [33]. Prolonged cholesterol precipitation in macrophages leads to atheroma. Exosomes are believed to be incorporated in lipid modulation. Studies on neurodegenerative diseases have shown that the reproductive mechanism of cholesterol-enriched exosomes serves as a balance regulating the homeostasis of intracellular cholesterol. Finally, macrophage foam mediated by exosomes has gained a lot of attention in recent times [78,79].

3.3. Therapeutic Potential of Exosomes in Neural Diseases

Peripheral nerve injury (PNI) is a well-known neurological disorder that adversely affects human health. It affects about 2.8%–5% of patients with polytrauma resulting in sensory-motor dysfunction [80]. Many factors such as immune cell deposition and conversion of Schwann cell (SC) phenotypes lead to axonal development following PNI. It is estimated that the daily growth of regenerating axons is only 1 mm, which could be difficult to manage clinically [81]. Additionally, proximal nerves extended denervation may prolong the possibility of nervous organ irreversible atrophy [82]. Furthermore, nerve tissue damage associated with PNI increases the complexity of nerve regeneration [83,84]. Although the gold standard in the management of PNI is the transplantation of autologous nerves, this technique is rarely performed because there is a shortage of donor sources and possible neurological damage to the donor site [85]. Many studies have incorporated new techniques for starting axon rejuvenation without sacrificing the functions of healthy nerves. Another approach is to use synthetic nerve ducts with numerous stem cells that can be beneficial, but the results are far from ideal [86]. Interestingly, another strategy for the management of peripheral nerve regeneration is the use of exosomes derived from MSC. Another study examined MSC use by modifying them to increase bioactive molecular levels which will promote PNI recovery. Furthermore, the in vitro internalization of MSC exosomes by SCs leads to a significant increase in SC proliferation and subsequently leading to axonal regeneration [87]. The nerve growth factor (NGF), fibroblast growth factor-1 (FGF-1), glial cell-derived neurotrophic factor (GDNF), insulin-like growth factor-1 (IGF-1), and brain-derived neurotrophic factor (BDNF) are some of the multiple neurotrophic causes of MSC exosomes. Nevertheless, MSC exosomes have beneficial effects which include promoting recovery from neural survival, denervation muscle atrophy, and axonal development by the use of bioactive molecules [88].

It has been demonstrated previously that axonal regeneration is controlled by macrophages and not by SCs [89]. Immune cells, primarily phagocytic neutrophils, and macrophages are the first to migrate to the degenerative site within hours or days after peripheral nerve damage, whether caused by a neurological disorder or infection. The critical role of neuroinflammation in PNI recovery cannot be overemphasized, and it starts with a process called Wallerian degeneration (WD). In this process, myelinating SCs undergo differentiation and are mainly seen in the distal portion of the nerve [90]. Thereafter, proinflammatory cytokines and chemokines are produced by differentiated SCs resulting in activation of the neuroinflammatory event. Neuroinflammatory response leads to the hiring of peripheral immune cells and circulating macrophages to the affected site [91]. The combination of resident and circulating macrophages activate the clearance of axonal debris and myelin. It is worthwhile to note here that damaged axons must be cleared of debris and myelin because they contain molecules that exert an inhibitory effect on the development of axons, most especially at the late stages of WD [92]. At the distal stump and neuronal cell body, macrophages and other phagocytic cells precipitate to promote lesion response which is the mechanism adopted by neurons to activate regeneration. Nonetheless, neuroinflammation is a double-edged sword in this scenario as it is a critical response for nerve regeneration, but the excess of it prevents nerve regeneration as well as the incorporation of pain in the neuropathic events. An important goal in PNI management is the creation of a microenvironment allowing a reasonable level of inflammation and regeneration [93]. During peripheral nerve repair, the rebuilding of the vascular network is a vital microenvironment that promotes regeneration and axonal development. Therefore, the integrity of the vascular network plays a key role in peripheral nerve regeneration and a therapeutic option in PNI management [94]. Insults to the neural tissue caused by spinal cord injury increase the number of complement factors. Bone Mesenchymal stem cells (BMSCs) inhibit the levels of C4–C6 as well as C4 binding proteins and complement factor H in rats [21]. Furthermore, the application and use of neural stem cell-derived extracellular vesicles significantly reduce the severity and extent of spinal cord injury by reducing neuronal apoptosis, activation of microglia, and neuroinflammation in rats [23].

Interestingly, MSC exosomes promote angiogenesis through paracrine activation leading to vascular remodeling. Through the P13K/AKT signaling pathway, exosomes are induced by the pluripotent Mesenchymal stem cells which activate local angiogenesis. [95]. Furthermore, vascular plasticity is through MSC exosomes and endothelial cell-transferred proangiogenic miRNAs. Additionally, other studies observed activation of angiogenesis by MSC-derived exosomes decreasing neurological deficits [96]. Intravenous injection of MSC-derived exosomes activates neurite remodeling, neurogenesis, and angiogenesis leading to functional recovery [52]. The same effects were explained in another study in which MSC exosomes decreased inflammation in rat models by activating the endogenous angiogenesis including neurogenesis [95]. In summary, reports from all studies revealed that MSC-derived exosomes are likely to be regulators by promoting communication with vascular endothelial cells thereby enhancing vascular plasticity after nerve injury [96]. Nanoparticles provide beneficial therapeutic effects such as those enhanced by MSC and may be a promising adjuvant for treatment due to the inherent advantages they possessed over the MSCs [97]. MSC-derived exosomes play an important role in the intercellular communication [80] by utilizing a number of neurotrophic factors, protein, and genetic materials to axons. The introduction of MSC-derived exosomes can inhibit the problems arising from stem cell transplantation [86,98].

3.4. Therapeutic Potential of Exosomes in Soft Tissue

Adipose-derived stem cell released exosomes (ADSC-Exos) have a wide range of potential therapeutic potentials and have applications in regenerative engineering, tumor disorders, and skin repairs. Previous studies have reported the applications of naturally derived exosomes in diverse disease models with modifications [54,99]. However, recent researches have brought to light the contents and characteristics of exosomes with adjustments achieved with the use of other substances. The ADSC-Exos have important functions in biological and pathological pathways [100]. However, their applications in clinics are yet to be identified and research is currently ongoing, including the therapeutic doses used. Nevertheless, it was found that the ideal dose was 50 μg/mL when using exosomes for wound healing, but in a separate experiment, ADSC-Exos were used at a dose of 40 μg/mL and effectively enabled adipogenic differentiation [101]. It was observed that several intra-articular injections of human MSC-derived exosome have been shown to promote cartilage repair and regeneration [32]. After 12 weeks of treatment, accelerated tissue filling and increased collagen synthesis were observed in osteoarthritis rat lesions leading to the restoration of cartilage and bone [32].

Regeneration and cutaneous healing are complex mechanisms and need a well-orchestrated integration of many molecular and biological processes, including differentiation, proliferation, cell migration, extracellular matrix deposition, and apoptosis [102]. ADSC-Exos have been found to solve problems mainly resulting from excessive scar formation and delayed cutaneous healing [103]. The internalization of ADSC-Exos by fibroblasts leads to proliferation and migration of collagen type I and III increasing via the PI3K/Akt signaling process. Furthermore, ADSC-Exos could induce migration, proliferation as well as suppressing HaCaT cell apoptosis through the signaling of Wnt/β-catenin [104]. All findings explained the promising potentials of ADSC-Exos in promoting cutaneous wound healing [105]. In ADSC-Exo treated mice, scar formation and clear zone of re-epithelialization were promoted, and the scar areas became narrower in the wounds. Furthermore, ADSC-Exosome could enhance the reconstruction of the extracellular matrix in the regeneration of cutaneous wound by modulating collagen type I: type III proportions, and also transforming growth factor-beta 3 (TGF-β3): TGF-β1 and MMP3: metalloproteinases 1 (TIMP1) tissue inhibitor, and inhibiting the fibroblasts differentiation into myofibroblasts decreasing scar formation [106]. Furthermore, it was also observed that the use of ADSC-Exos leads to the attenuation of the necrotic flap, neovascularization activation, and alleviation of inflammatory reaction after I/R disorder skin flap apoptosis [103]. In summary, it highlights the potentials of ADSC-Exo use in promoting cutaneous regeneration and complete healing. However, the molecular mechanism will need to be evaluated for clinical applications [36].

3.5. Therapeutic Potential of Exosomes in Obesity

In obese individuals, inflammation resulting from adipose tissue has been documented as the main factor in the progression of type 2 diabetes and insulin resistance [107]. Researchers have identified that ADSCs have key roles in metabolic diseases and inflammation resulting from obesity. ADSC-Exos were explained to affect diabetes and obesity. In white adipose tissue (WAT), ADSC-Exos were shown to dominate the anti-inflammatory (M2) macrophage phenotype polarization and regulation of immune homeostasis. In WAT obese mice, exosomes have been observed to promote polarization of the M2 macrophage and inflammation prevention [55,108]. Of note, M2 macrophages resulted from ADSC-Exos exerting high levels of tyrosine hydroxylase potentiating the production of catecholamine, which promotes brown adipose tissue-specific uncoupling protein 1 expression in WAT which activates fat burning leading to energy reserve dissipation. This observation highlights the indispensable exosome-mediated crosstalk existing between macrophages and ADSCs and attracts attention to the potential role of an exosome-based treatment approach to treat obesity [109].

3.6. Therapeutic Potential of Exosomes in Autoimmune Disease

The beneficial roles of exosomes in immune tolerance and stimulation cannot be overemphasized as they are involved in several pathways such as angiogenesis, inflammation, and immune signaling. Exosomes have also acted as good tools for drug delivery due to their inherent biological features such as biocompatibility and stability including the capacity for permeation [110]. Sjögren's syndrome (SS) is an important autoimmune disorder characterized by lymphocytic infiltration in the salivary and lacrimal glands as well as the existence of numerous autoantibodies. These autoantibodies include anti-La (SS-B) or anti-Ro (SS-A)) with clinical syndromes such as ocular and oral dryness [111]. SS pathogenesis involves both innate and adaptive immune mechanisms such as NF-kB signaling, interferon (IFN) signatures, and B cell-activating factor (BAFF)/BAFF receptor axis [112]. Furthermore, salivary gland epithelial cells (SGECs) play an important role in both autoimmune and inflammatory responses in SS by inducible expression of different immunoreactive factors such as autoantigenic ribonucleoproteins (RNPs), many Toll-like receptors (TLRs), and BAFF. Lymphocytic infiltration, mainly B and $CD4^+$ T cells, occurs and invades epithelial cells. This highlights the correlation between immune and epithelial cells [110]. A previous study documented the continuous release of autoantigenic exosomes La/SS-B, Sm RNPs, and Ro/SS-A by SGECs, documenting the transfer of intracellular autoantigens to autoreactive lymphocytes within RNP-containing exosomes [60]. It was also observed that EBV-miRBART13-3p in exosomes is transferrable from B cells to SGECs in EBV infected B cells. The stromal interacting molecule 1 (STIM1) and aquaporin 5 (AQP5) targeted by functional miRNA could significantly affect salivary secretion [12]. Recently, it was observed that subconjunctival administration of MSC-Exos to rabbits inhibited lacrimal gland inflammation in the affected group compared with the control group administered normal saline. It was concluded that MSC-Exo effects were partly due to the polarization of the lacrimal macrophage as well as the responses of Treg and Th2 by NF-kB signaling pathways. Thus, MSC-Exos may be useful in the treatment of dry eye [113]. Experimental autoimmune uveoretinitis (EAU) could be effectively resolved by the use of bone marrow-derived MSC-Exos [114]. It was observed that in experimentally induced autoimmune uveitis in rats, human-derived Mesenchymal stem cell-derived exosomes inhibited the migration of inflammatory cells, mainly T cell subsets, in the eyes. Additionally, the MSC-Exo inhibited the chemo attractive effects of CCL2 and CCL21 on inflammatory cells [45]. Another experimental study conducted on EAU found that human umbilical cord-derived MSC-Exos (hUC-MSC-Exos) did not inhibit the progression of conA-stimulated T cells, but suppressed inflammatory cell migration [115]. The suppressive effect of hUC-MSC-Exos on the interphotoreceptor retinoid-binding protein (IRBP)-specific Th17 responses via the regulation of DC-derived Th17-polarizing cytokines IL-1β, IL-6, and IL-23 led to the suppression of DC-driven Th17 responses [116]. Thus, it is clear that MSC-Exos can play a role in the treatment of auto-immune uveitis, however, further trials are essential to identify their immunomodulatory and anti-inflammatory effects [112].

3.7. Therapeutic Potential of Exosomes in Infectious (Parasitic and Bacterial) Conditions

Parasites secrete exosomes that interact effectively, as a medium of cell to cell communication, within their host. Logically, it is most probable that the host utilizes this pathway of cellular communication as a defense mechanism [117]. Using plasma cell-derived microvesicles, it was observed that *Plasmodium berghei* induces the production of CD40 in antigen-presenting cells, leading to generation of a potent inflammatory response and subsequent clearance of the parasite by macrophage activation [118]. Furthermore, immune cell-derived microvesicles from *Plasmodium vivax*, one of the species of *Plasmodium* responsible for human malaria infection, was associated with acute inflammation in the course of parasite eradication [119]. Additionally, a massive increase and release of antimicrobial peptide-containing exosomes was observed in intestinal epithelial cells in response to infection with *Cryptosporidium* [120].

Exosomes are also useful in vaccines and vaccination. In a vaccine trial using *Leishmania major*-pulsed DC exosomes, it was observed that the DC-derived exosomes mediate protective *Th*1 immunity against cutaneous leishmaniasis [121]. Similarly, avian coccidiosis in poultry caused by *Eimeria* species (*E. tenella, E. acervulina, E. maxima*) can be alleviated using *Eimeria* parasite antigen-loaded DC exosomes which successfully alleviated clinical signs in poultry thereby reducing mortality rates [122].

With regards to bacterial infection, exosomes derived from the causative agent of bovine tuberculosis, *Mycobacterium bovis*-infected macrophages promote dendritic cell (DC) activation and the generation of antibacterial T cell response in vivo [47]. Similarly, *Mycobacterium tuberculosis* (human tuberculosis), induces the release of exosomes from infected macrophages with consequent recruitment of lymphocytes through the secretion of chemokine such as RANTES and MIP- 1α [123].

3.8. Therapeutic Potential of Exosomes in Lung Diseases

Bronchopulmonary dysplasia (BPD) is one of several conditions that could affect the lungs. BPD is a chronic disorder of the lungs in preterm infants [124]. Current therapeutic approaches using conventional therapy have only modest outcomes. The use of MDC-exosome therapeutics appears to be promising in clinical management, as shown in experimental models by alleviating neonatal lung injury [124]. Transplantation of various stem cell types—such as human amnion epithelial cells and endothelial colony-forming cells—appears to be promising with results from the preclinical models. MSC treatment of blunt hyperoxia-induced lung inflammation ameliorates vascular remodeling and improves exercise capacity with higher survival rates [46,125]. Furthermore, using allogeneic human umbilical cord blood-derived MSC in phase 1 clinical trials, it was observed that MSC administration brought about a reduction in inflammatory markers and lowered BPD severity [126]. The mechanism for now remains unknown. Furthermore, using bone marrow stromal cells (BMSCs) in a murine model with neonatal chronic lung disease, the BMSCs release immunomodulatory factors which ameliorated parenchymal and vascular injury of BPD in vivo [46]. Additionally, intratracheal distribution of BMSCs in a murine model with BPD, improved survival and exercise tolerance by attenuating alveolar and lung vascular injury as well as pulmonary hypertension in vivo, while in vitro, the BMSC-derived conditioned medium prevented O2-induced AEC2 apoptosis, boosted endothelial cord formation, and enhanced AEC2 wound healing [125].

3.9. Therapeutic Potential of Exosomes in Liver Diseases

Exosome-based therapeutic methods for the management of different kind of diseases of the liver, including liver tumor, is progressing. Exosomes obtained from adult human liver stem cells (HLSC) were shown to inhibit the growth and survival of HepG2 and primary hepatocellular carcinoma (HCC) in vitro via the use of antitumor miRNAs [127]. Furthermore, an in vivo experiment in a murine model showed accelerated morphological and functional gains of the liver despite 70% hepatectomy as well as proliferation and apoptosis resistance of rat and human hepatocytes in vitro [128]. Additionally, exosomes obtained from human umbilical cord Mesenchymal stem cells ameliorated fibrosis of the liver induced by carbon tetrachloride (CCl_4) [129]. It was demonstrated in vivo in mice that immunodeficient NOD/SCID mice with their liver engrafted with human hepatoma cells produce exosomes containing exogenous siRNA which shuttles between hepatic cells and subsequently produces CD81 siRNA [129].

4. Conclusions

Recently, exosomes are considered as significant mediators in intercellular communication. The exosomes' ability to transport DNA, proteins, non-coding RNAs, and mRNA, makes them an attractive target of research on the pathogenesis of several illnesses, including cardiovascular and autoimmune disease, cancer, obesity, etc. MSC and other cell type-derived exosomes play an important role in intercellular communication by using several neurotrophic factors, protein, and genetic materials to axons, and they can inhibit the issues resulting from stem cell transplantation. The involvement of exosomes in both biological and pathobiological functions makes them an interesting organelle, and hence it is interesting to study their potential biological applications in the treatment of diverse diseases and offer new possibilities in the amelioration and possible cure of some infectious and non-infectious diseases. Furthermore, results from both clinical and experimental studies on MSC-derived exosomes have shown their potential in the treatment of diverse diseases. However, future research must concentrate on improving the methods for the isolation of exosomes and other EVs to distinguish them appropriately, as well as their mechanism of action, before they are used in clinical studies. Success in this regard will promote their prompt application in the diagnosis, prevention, and treatment of diseases.

Author Contributions: Conceptualization, A.M.B., G.E.-S.B., and T.E.O.; methodology, A.M.B., G.E.-S.B., N.R.-P., and A.Z.-B.; software, A.M.B., and G.E.-S.B.; validation, A.M.B., G.E.-S.B., T.E.O., N.R.-P., H.F.H., M.Z., S.A., and A.Z.-B.; formal analysis G.E.-S.B., and T.E.O. investigation, A.M.B., G.E.-S.B., T.E.O., N.R.-P., H.F.H., M.Z., S.A., and A.Z.-B..; data curation, A.M.B., G.E.-S.B., M.A.G., and A.K.M.; writing—original draft preparation, A.M.B., T.E.O., and G.E.-S.B.; writing—review and editing, A.M.B., T.E.O., G.E.-S.B., N.R.-P., and A.Z.-B.; visualization, A.M.B., G.E.-S.B., N.R.-P., A.Z.-B., M.A.E.-E.; supervision, A.M.B., G.E.-S.B., and H.F.H.; funding acquisition, S.A., H.F.H., and M.Z. All authors have read and agreed to the published version of the manuscript.

Funding: This research received no external funding.

Acknowledgments: Graphical abstract and Figure 1 were made using biorender.com.

Conflicts of Interest: The authors declare no conflict of interest.

Abbreviations

EV:	extracellular vesicle
miRNA:	microRNA
mRNA:	messenger RNA
TEX:	Tumor-derived exosomes
PD-L1:	programmed death-ligand 1
IFN-γ:	interferon-γ
MVEs:	multivesicular endosomes
MVB:	multivesicular body
ILVs:	intraluminal vesicles
TEMs:	tetraspanin-enriched microdomains
HGF:	hepatocyte growth factor
MSCs:	mesenchymal stem cells
TGF-β1:	transforming growth factor-β1
TCF4:	human T-cell factor 4
IL-8:	interleukin-8
ECs:	endothelial cells
NF-κB:	nuclear factor-kappa B
CAD:	coronary artery disease
ox-LDL:	oxidized low-density lipoprotein
NO:	nitric oxide
KLF5:	krüppel-like factor 5
SMCs:	smooth muscle cells
PNI:	Peripheral nerve injury
SCs:	Schwann cells
NGF:	nerve growth factor
FGF-1:	fibroblast growth factor-1
GDNF:	glial cell-derived neurotrophic factor
IGF-1:	insulin-like growth factor-1
BDNF:	brain-derived neurotrophic factor
WD:	Wallerian degeneration

TGF-β3:	transforming growth factor-beta 3
TIMP1:	tissue inhibitor metalloproteinases 1
WAT:	white adipose tissue; IFN: interferon
BAFF:	B cell-activating factor
SGECs:	salivary gland epithelial cells
RNPs:	ribonucleoproteins
TLRs:	Toll-like receptors
STIM1:	Stromal interacting molecule 1
AQP5:	aquaporin 5
hUC-MSC-Exos:	human umbilical cord-derived MSC-Exos
IRBP:	interphotoreceptor retinoid-binding protein

References

1. Bonucci, E. Fine structure of early cartilage calcification. *J. Ultrastruct.* **1967**, *20*, 33–50. [CrossRef]
2. Anderson, H.C. Vesicles associated with calcification in the matrix of epiphyseal cartilage. *J. Cell Boil.* **1969**, *41*, 59–72. [CrossRef]
3. Trams, E.G.; Lauter, C.J.; Salem, J.N.; Heine, U. Exfoliation of membrane ecto-enzymes in the form of micro-vesicles. *Biochim. Biophys. Acta Biomembr.* **1981**, *645*, 63–70. [CrossRef]
4. Théry, C.; Zitvogel, L.; Amigorena, S. Exosomes: composition, biogenesis and function. *Nat. Rev. Immunol.* **2002**, *2*, 569–579. [CrossRef] [PubMed]
5. Théry, C.; Witwer, K.W.; Aikawa, E.; Alcaraz, M.J.; Anderson, J.D.; Andriantsitohaina, R.; Antoniou, A.; Arab, T.; Archer, F.; Atkin-Smith, G.K. Minimal information for studies of extracellular vesicles 2018 (MISEV2018): a position statement of the International Society for Extracellular Vesicles and update of the MISEV2014 guidelines. *J. Extracell. Vesicles.* **2018**, *7*, 1535750. [CrossRef]
6. Lawson, C.; Vicencio, J.M.; Yellon, D.M.; Davidson, S.M. Microvesicles and exosomes: new players in metabolic and cardiovascular disease. *J. Endocrin.* **2016**, *228*, R57–R71. [CrossRef] [PubMed]
7. Crenshaw, B.J.; Gu, L.; Sims, B.; Matthews, Q.L. Exosome biogenesis and biological function in response to viral infections. *Open Virol. J.* **2018**, *12*, 134. [CrossRef] [PubMed]
8. Gurunathan, S.; Kang, M.-H.; Jeyaraj, M.; Qasim, M.; Kim, J.-H. Review of the isolation, characterization, biological function, and multifarious therapeutic approaches of exosomes. *Cells* **2019**, *8*, 307. [CrossRef] [PubMed]
9. El-Esawi, M.A.; Alaraidh, I.A.; Alsahli, A.A.; Ali, H.M.; Alayafi, A.A.; Witczak, J.; Ahmad, M. Genetic variation and alleviation of salinity stress in barley (*Hordeum vulgare* L.). *Molecules* **2018**, *23*, 2488. [CrossRef]
10. El-Esawi, M.A.; Al-Ghamdi, A.A.; Ali, H.M.; Alayafi, A.A.; Witczak, J.; Ahmad, M. Analysis of genetic variation and enhancement of salt tolerance in French pea (*Pisum Sativum* L.). *Intern. J. Molec. Sci.* **2018**, *19*, 2433. [CrossRef]
11. Tung, K.H.; Ernstoff, M.S.; Allen, C.; La Shu, S. A Review of Exosomes and their Role in The Tumor Microenvironment and Host–Tumor "Macroenvironment". *J. Immunol. Sci.* **2019**, 3, 4. [CrossRef] [PubMed]
12. Whiteside, T.L. Exosomes and tumor-mediated immune suppression. *J. Clin. Investig.* **2016**, *126*, 1216–1223. [CrossRef] [PubMed]
13. Chen, G.; Huang, A.C.; Zhang, W.; Zhang, G.; Wu, M.; Xu, W.; Yu, Z.; Yang, J.; Wang, B.; Sun, H. Exosomal PD-L1 contributes to immunosuppression and is associated with anti-PD-1 response. *Nature* **2018**, *560*, 382–386. [CrossRef] [PubMed]
14. Gould, S.J.; Raposo, G. As we wait: coping with an imperfect nomenclature for extracellular vesicles. *J. Extracell. Vesicles* **2013**, *2*, 20389. [CrossRef]
15. Zheng, X.; Chen, F.; Zhang, J.; Zhang, Q.; Lin, J. Exosome analysis: a promising biomarker system with special attention to saliva. *J. Membr. Boil.* **2014**, *247*, 1129–1136. [CrossRef]
16. Pegtel, D.M.; Gould, S.J. Exosomes. *Annu. Rev. Biochem.* **2019**, *88*, 487–514. [CrossRef]
17. Kowal, J.; Tkach, M.; Théry, C. Biogenesis and secretion of exosomes. *Curr. Opin. Cell Boil.* **2014**, *29*, 116–125. [CrossRef]
18. Record, M.; Subra, C.; Silvente-Poirot, S.; Poirot, M. Exosomes as intercellular signalosomes and pharmacological effectors. *Biochem. Pharmacol.* **2011**, *81*, 1171–1182. [CrossRef]

19. Abels, E.R.; Breakefield, X.O. Introduction to extracellular vesicles: biogenesis, RNA cargo selection, content, release, and uptake. *Cell. Mol. Neurobiol.* **2016**, *36*, 301–312. [CrossRef]
20. Vlassov, A.V.; Magdaleno, S. Setterquist, R.; Conrad, R. Exosomes: current knowledge of their composition, biological functions, and diagnostic and therapeutic potentials. *Biochim. Biophys. Acta Gen. Subj.* **2012**, *1820*, 940–948. [CrossRef]
21. Hsu, C.; Morohashi, Y.; Yoshimura, S.-i.; Manrique-Hoyos, N.; Jung, S.; Lauterbach, M.A.; Bakhti, M.; Grønborg, M.; Möbius, W.; Rhee, J. Regulation of exosome secretion by Rab35 and its GTPase-activating proteins TBC1D10A–C. *J. Cell Biol.* **2010**, *189*, 223–232. [CrossRef] [PubMed]
22. Baietti, M.F.; Zhang, Z.; Mortier, E.; Melchior, A.; Degeest, G.; Geeraerts, A.; Ivarsson, Y.; Depoortere, F.; Coomans, C.; Vermeiren, E. Syndecan–syntenin–ALIX regulates the biogenesis of exosomes. *Nat. Cell Biol.* **2012**, *14*, 677–685. [CrossRef] [PubMed]
23. Zhang, Y.; Liu, Y.; Liu, H.; Tang, W.H. Exosomes: biogenesis, biologic function and clinical potential. *Cell Biosci.* **2019**, *9*, 19. [CrossRef]
24. Babst, M. MVB vesicle formation: ESCRT-dependent, ESCRT-independent and everything in between. *Curr. Opin. Cell Boil.* **2011**, *23*, 452–457. [CrossRef] [PubMed]
25. Perez-Hernandez, D.; Gutierrez-Vázquez, C.; Jorge, I.; López-Martín, S.; Ursa, A.; Sánchez-Madrid, F.; Vázquez, J.; Yáñez-Mó, M. The intracellular interactome of tetraspanin-enriched microdomains reveals their function as sorting machineries toward exosomes. *J.Biol.Chem.* **2013**, *288*, 11649–11661. [CrossRef]
26. Tancini, B.; Buratta, S.; Sagini, K.; Costanzi, E.; Delo, F.; Urbanelli, L.; Emiliani, C. Insight into the role of extracellular vesicles in lysosomal storage disorders. *Genes* **2019**, *10*, 510. [CrossRef]
27. Wang, Y.; He, G.; Guo, Y.; Tang, H.; Shi, Y.; Bian, X.; Zhu, M.; Kang, X.; Zhou, M.; Lyu, J.; et al. Exosomes from tendon stem cells promote injury tendon healing through balancing synthesis and degradation of the tendon extracellular matrix. *J. Cell Mol. Med.* **2019**, *23*, 5475–5485. [CrossRef]
28. Rong, Y.; Liu, W.; Wang, J.; Fan, J.; Luo, Y.; Li, L.; Kong, F.; Chen, J.; Tang, P.; Cai, W. Neural stem cell-derived small extracellular vesicles attenuate apoptosis and neuroinflammation after traumatic spinal cord injury by activating autophagy. *Cell Death Dis.* **2019**, *10*, 340. [CrossRef]
29. Xiong, L.; Sun, L.; Zhang, Y.; Jin, P.; Yan, J.; Liu, X. Exosomes from bone marrow mesenchymal stem cells can alleviate early brain injury following subarachnoid hemorrhage through miRNA129-5p-HMGB1 pathway. *Stem Cells Dev.* **2019**. [CrossRef]
30. Zhao, C.; Zhou, X.; Qiu, J.; Xin, D.; Li, T.; Chu, X.; Yuan, H.; Wang, H.; Wang, Z.; Wang, D. Exosomes Derived From Bone Marrow Mesenchymal Stem Cells Inhibit Complement Activation In Rats With Spinal Cord Injury. *Drug Des. Dev. Ther.* **2019**, *13*, 3693–3704. [CrossRef]
31. Lankford, K.L.; Arroyo, E.J.; Nazimek, K.; Bryniarski, K.; Askenase, P.W.; Kocsis, J.D. Intravenously delivered mesenchymal stem cell-derived exosomes target M2-type macrophages in the injured spinal cord. *PLoS ONE* **2018**, *13*, e0190358. [CrossRef] [PubMed]
32. Zhang, S.; Chu, W.; Lai, R.; Lim, S.; Hui, J.; Toh, W. Exosomes derived from human embryonic mesenchymal stem cells promote osteochondral regeneration. *Osteoarthr. Cartil.* **2016**, *24*, 2135–2140. [CrossRef] [PubMed]
33. Liu, Z.; Xu, Y.; Wan, Y.; Gao, J.; Chu, Y.; Li, J. Exosomes from adipose-derived mesenchymal stem cells prevent cardiomyocyte apoptosis induced by oxidative stress. *Cell Death Discov.* **2019**, *5*, 1–7. [CrossRef] [PubMed]
34. Sun, B.; Ma, Y.; Wang, F.; Hu, L.; Sun, Y. miR-644-5p carried by bone mesenchymal stem cell-derived exosomes targets regulation of p53 to inhibit ovarian granulosa cell apoptosis. *Stem Cell Res. Ther.* **2019**, *10*, 360. [CrossRef]
35. Xu, H.; Wang, Z.; Liu, L.; Zhang, B.; Li, B. Exosomes derived from adipose tissue, bone marrow, and umbilical cord blood for cardioprotection after myocardial infarction. *J. Cell. Biochem.* **2019**. [CrossRef]
36. Atay, S.; Banskota, S.; Crow, J.; Sethi, G.; Rink, L.; Godwin, A.K. Oncogenic KIT-containing exosomes increase gastrointestinal stromal tumor cell invasion. *Proc. Natl. Acad. Sci. USA* **2014**, *111*, 711–716. [CrossRef]
37. Thomi, G.; Surbek, D.; Haesler, V.; Joerger-Messerli, M.; Schoeberlein, A. Exosomes derived from umbilical cord mesenchymal stem cells reduce microglia-mediated neuroinflammation in perinatal brain injury. *Stem Cell Res. Ther.* **2019**, *10*, 105. [CrossRef]
38. Yan, L.; Wu, X. Exosomes produced from 3D cultures of umbilical cord mesenchymal stem cells in a hollow-fiber bioreactor show improved osteochondral regeneration activity. *Cell Boil. Toxicol.* **2019**. [CrossRef]

39. Yu, T.; Zhao, C.; Hou, S.; Zhou, W.; Wang, B.; Chen, Y. Exosomes secreted from miRNA-29b-modified mesenchymal stem cells repaired spinal cord injury in rats. *Braz. J. Med Boil. Res.* **2019**, *52*, e8735. [CrossRef]
40. Li, J.; Ju, Y.; Liu, S.; Fu, Y.; Zhao, S. Exosomes derived from lipopolysaccharide-preconditioned human dental pulp stem cells regulate Schwann cell migration and differentiation. *Connect. Tissue Res.* **2019**, 1–10. [CrossRef]
41. Bruno, S.; Tapparo, M.; Collino, F.; Chiabotto, G.; Deregibus, M.C.; Soares Lindoso, R.; Neri, F.; Kholia, S.; Giunti, S.; Wen, S. Renal regenerative potential of different extracellular vesicle populations derived from bone marrow mesenchymal stromal cells. *Tissue Eng. Part A* **2017**, *23*, 1262–1273. [CrossRef] [PubMed]
42. Xie, K.; Liu, L.; Chen, J.; Liu, F. Exosomes derived from human umbilical cord blood mesenchymal stem cells improve hepatic ischemia reperfusion injury via delivering miR-1246. *Cell Cycle.* **2019**, *18*, 3491–3501. [CrossRef] [PubMed]
43. El Harane, N.; Kervadec, A.; Bellamy, V.; Pidial, L.; Neametalla, H.J.; Perier, M.-C.; Lima Correa, B.; Thiébault, L.; Cagnard, N.; Duché, A. Acellular therapeutic approach for heart failure: in vitro production of extracellular vesicles from human cardiovascular progenitors. *Europ. Heart J.* **2018**, *39*, 1835–1847. [CrossRef] [PubMed]
44. Adamiak, M.; Cheng, G.; Bobis-Wozowicz, S.; Zhao, L.; Kedracka-Krok, S.; Samanta, A.; Karnas, E.; Xuan, Y.-T.; Skupien-Rabian, B.; Chen, X. Induced pluripotent stem cell (iPSC)–derived extracellular vesicles are safer and more effective for cardiac repair than iPSCs. *Circ. Res.* **2018**, *122*, 296–309. [CrossRef] [PubMed]
45. Bai, L.; Shao, H.; Wang, H.; Zhang, Z.; Su, C.; Dong, L.; Yu, B.; Chen, X.; Li, X.; Zhang, X. Effects of mesenchymal stem cell-derived exosomes on experimental autoimmune uveitis. *Sci.Rep.* **2017**, *7*, 1–11. [CrossRef]
46. Aslam, M.; Baveja, R.; Liang, O.D.; Fernandez-Gonzalez, A.; Lee, C.; Mitsialis, S.A.; Kourembanas, S. Bone marrow stromal cells attenuate lung injury in a murine model of neonatal chronic lung disease. *Amer. J. Resp. Crit. Care Med.* **2009**, *180*, 1122–1130. [CrossRef]
47. Giri, P.K.; Schorey, J.S. Exosomes derived from M. Bovis BCG infected macrophages activate antigen-specific CD4+ and CD8+ T cells in vitro and in vivo. *PLoS ONE* **2008**, *3*, e2461. [CrossRef]
48. Li, T.; Yan, Y.; Wang, B.; Qian, H.; Zhang, X.; Shen, L.; Wang, M.; Zhou, Y.; Zhu, W.; Li, W. Exosomes derived from human umbilical cord mesenchymal stem cells alleviate liver fibrosis. *Stem Cells Dev.* **2013**, *22*, 845–854. [CrossRef]
49. Das, C.K.; Jena, B.C.; Banerjee, I.; Das, S.; Parekh, A.; Bhutia, S.K.; Mandal, M. Exosome as a novel shuttle for delivery of therapeutics across biological barriers. *Mol. Pharm.* **2018**, *16*, 24–40. [CrossRef]
50. Ha, D.; Yang, N.; Nadithe, V. Exosomes as therapeutic drug carriers and delivery vehicles across biological membranes: current perspectives and future challenges. *Acta Pharm. Sin. B* **2016**, *6*, 287–296. [CrossRef]
51. Pullan, J.E.; Confeld, M.I.; Osborn, J.K.; Kim, J.; Sarkar, K.; Mallik, S. Exosomes as drug carriers for cancer therapy. *Mol. Pharmac.* **2019**, *16*, 1789–1798. [CrossRef] [PubMed]
52. Merino-González, C.; Zuñiga, F.A.; Escudero, C.; Ormazabal, V.; Reyes, C.; Nova-Lamperti, E.; Salomón, C.; Aguayo, C. Mesenchymal stem cell-derived extracellular vesicles promote angiogenesis: potencial clinical application. *Front. Physiol.* **2016**, *7*, 24. [CrossRef] [PubMed]
53. Wang, Y.; Yu, D.; Liu, Z.; Zhou, F.; Dai, J.; Wu, B.; Zhou, J.; Heng, B.C.; Zou, X.H.; Ouyang, H. Exosomes from embryonic mesenchymal stem cells alleviate osteoarthritis through balancing synthesis and degradation of cartilage extracellular matrix. *Stem Cell Res. Ther.* **2017**, *8*, 189. [CrossRef] [PubMed]
54. Hu, Y.; Rao, S.-S.; Wang, Z.-X.; Cao, J.; Tan, Y.-J.; Luo, J.; Li, H.-M.; Zhang, W.-S.; Chen, C.-Y.; Xie, H. Exosomes from human umbilical cord blood accelerate cutaneous wound healing through miR-21-3p-mediated promotion of angiogenesis and fibroblast function. *Theranostics* **2018**, *8*, 169. [CrossRef] [PubMed]
55. Zhao, H.; Shang, Q.; Pan, Z.; Bai, Y.; Li, Z.; Zhang, H.; Zhang, Q.; Guo, C.; Zhang, L.; Wang, Q. Exosomes from adipose-derived stem cells attenuate adipose inflammation and obesity through polarizing M2 macrophages and beiging in white adipose tissue. *Diabetes* **2018**, *67*, 235–247. [CrossRef] [PubMed]
56. Heldring, N.; Mäger, I.; Wood, M.J.; Le Blanc, K.; Andaloussi, S.E. Therapeutic potential of multipotent mesenchymal stromal cells and their extracellular vesicles. *Hum. Gene Ther.* **2015**, *26*, 506–517. [CrossRef]
57. Burke, J.; Hunter, M.; Kolhe, R.; Isales, C.; Hamrick, M.; Fulzele, S. Therapeutic potential of mesenchymal stem cell based therapy for osteoarthritis. *Clin. Transl. Med.* **2016**, *5*, 27. [CrossRef]

58. Ferreira, J.R.; Teixeira, G.Q.; Santos, S.G.; Barbosa, M.A.; Almeida-Porada, G.; Gonçalves, R.M. Mesenchymal stromal cell secretome: influencing therapeutic potential by cellular pre-conditioning. *Front. Immunol.* **2018**, *9*, 2837. [CrossRef]
59. Toh, W.S.; Lai, R.C.; Hui, J.H.P.; Lim, S.K. MSC exosome as a cell-free MSC therapy for cartilage regeneration: implications for osteoarthritis treatment. *Semin. Cell Dev. Boil.* **2017**, *67*, 56–64. [CrossRef]
60. Li, Z.; Wang, Y.; Xiao, K.; Xiang, S.; Li, Z.; Weng, X. Emerging role of exosomes in the joint diseases. *Cell. Physiol. Biochem.* **2018**, *47*, 2008–2017. [CrossRef]
61. Huang, L.; Ma, W.; Ma, Y.; Feng, D.; Chen, H.; Cai, B. Exosomes in mesenchymal stem cells, a new therapeutic strategy for cardiovascular diseases? *Int. J. Boil. Sci.* **2015**, *11*, 238. [CrossRef] [PubMed]
62. Lee, J.-K.; Park, S.-R.; Jung, B.-K.; Jeon, Y.-K.; Lee, Y.-S.; Kim, M.-K.; Kim, Y.-G.; Jang, J.-Y.; Kim, C.-W. Exosomes derived from mesenchymal stem cells suppress angiogenesis by down-regulating VEGF expression in breast cancer cells. *PloS ONE* **2013**, *8*, e84256. [CrossRef] [PubMed]
63. Ailawadi, S.; Wang, X.; Gu, H.; Fan, G.-C. Pathologic function and therapeutic potential of exosomes in cardiovascular disease. *Biochem. Biophys. Res. Commun.* **2015**, *1852*, 1–11. [CrossRef] [PubMed]
64. Chen, L.; Wang, Y.; Pan, Y.; Zhang, L.; Shen, C.; Qin, G.; Ashraf, M.; Weintraub, N.; Ma, G.; Tang, Y. Cardiac progenitor-derived exosomes protect ischemic myocardium from acute ischemia/reperfusion injury. *Biochem. Biophys. Res. Commun.* **2013**, *431*, 566–571. [CrossRef] [PubMed]
65. Zhang, H.; Xiang, M.; Meng, D.; Sun, N.; Chen, S. Inhibition of myocardial ischemia/reperfusion injury by exosomes secreted from mesenchymal stem cells. *Stem Cells Int.* **2016**, *2016*, 1–8. [CrossRef] [PubMed]
66. de Jong, O.G.; Verhaar, M.C.; Chen, Y.; Vader, P.; Gremmels, H.; Posthuma, G.; Schiffelers, R.M.; Gucek, M.; van Balkom, B.W. Cellular stress conditions are reflected in the protein and RNA content of endothelial cell-derived exosomes. *J. Extracell. Vesicles* **2012**, *1*, 18396. [CrossRef]
67. Hu, G.; Drescher, K.M.; Chen, X. Exosomal miRNAs: biological properties and therapeutic potential. *Front. Genet.* **2012**, *3*, 56. [CrossRef]
68. Gambim, M.H.; Do Carmo, A.D.O.; Marti, L.; Veríssimo-Filho, S.; Lopes, L.R.; Janiszewski, M. Platelet-derived exosomes induce endothelial cell apoptosis through peroxynitrite generation: experimental evidence for a novel mechanism of septic vascular dysfunction. *Crit. Care* **2007**, *11*, R107. [CrossRef]
69. Suzuki, E.; Fujita, D.; Takahashi, M.; Oba, S.; Nishimatsu, H. Stem cell-derived exosomes as a therapeutic tool for cardiovascular disease. *World J. Stem Cells* **2016**, *8*, 297. [CrossRef]
70. Bakogiannis, C.; Sachse, M.; Stamatelopoulos, K.; Stellos, K. Platelet-derived chemokines in inflammation and atherosclerosis. *Cytokine* **2019**, *122*, 154157. [CrossRef]
71. Hafiane, A.; Daskalopoulou, S.S. Extracellular vesicles characteristics and emerging roles in atherosclerotic cardiovascular disease. *Metabolism* **2018**, *85*, 213–222. [CrossRef] [PubMed]
72. Chen, L.; Yang, W.; Guo, Y.; Chen, W.; Zheng, P.; Zeng, J.; Tong, W. Exosomal lncRNA GAS5 regulates the apoptosis of macrophages and vascular endothelial cells in atherosclerosis. *PLoS ONE* **2017**, *12*, e0185406. [CrossRef] [PubMed]
73. Tsao, C.-R.; Liao, M.-F.; Wang, M.-h.; Cheng, C.-M.; Chen, C.-H. Mesenchymal stem cell derived exosomes: a new hope for the treatment of cardiovascular disease? *Acta Cardiol. Sin.* **2014**, *30*, 395. [PubMed]
74. Vella, L.J.; Sharples, R.A.; Nisbet, R.M.; Cappai, R.; Hill, A.F. The role of exosomes in the processing of proteins associated with neurodegenerative diseases. *Eur. Biophys. J.* **2008**, *37*, 323–332. [CrossRef] [PubMed]
75. Yuyama, K.; Sun, H.; Sakai, S.; Mitsutake, S.; Okada, M.; Tahara, H.; Furukawa, J.-i.; Fujitani, N.; Shinohara, Y.; Igarashi, Y. Decreased amyloid-β pathologies by intracerebral loading of glycosphingolipid-enriched exosomes in Alzheimer model mice. *J. Biol. Chem.* **2014**, *289*, 24488–24498. [CrossRef]
76. Wang, X.; Huang, W.; Liu, G.; Cai, W.; Millard, R.W.; Wang, Y.; Chang, J.; Peng, T.; Fan, G.-C. Cardiomyocytes mediate anti-angiogenesis in type 2 diabetic rats through the exosomal transfer of miR-320 into endothelial cells. *J. Mol. Cell. Cardiol.* **2014**, *74*, 139–150. [CrossRef]
77. Kalluri, R. The biology and function of exosomes in cancer. *J. Clin. Investig.* **2016**, *126*, 1208–1215. [CrossRef]
78. Wang, Z.; Wu, H.-j.; Fine, D.; Schmulen, J.; Hu, Y.; Godin, B.; Zhang, J.X.; Liu, X. Ciliated micropillars for the microfluidic-based isolation of nanoscale lipid vesicles. *Lab. Chip.* **2013**, *13*, 2879–2882. [CrossRef]
79. Mathivanan, S.; Fahner, C.J.; Reid, G.E.; Simpson, R.J. ExoCarta 2012: database of exosomal proteins, RNA and lipids. *Nucleic Acids Res.* **2011**, *40*, D1241–D1244. [CrossRef]
80. Chivet, M.; Hemming, F.; Fraboulet, S.; Sadoul, R. Emerging role of neuronal exosomes in the central nervous system. *Front. Physiol.* **2012**, *3*, 145. [CrossRef]

81. Yuyama, K.; Sun, H.; Usuki, S.; Sakai, S.; Hanamatsu, H.; Mioka, T.; Kimura, N.; Okada, M.; Tahara, H.; Furukawa, J.-i. A potential function for neuronal exosomes: Sequestering intracerebral amyloid-β peptide. *FEBS lett.* **2015**, *589*, 84–88. [CrossRef]
82. Frühbeis, C.; Fröhlich, D.; Krämer-Albers, E.-M. Emerging roles of exosomes in neuron–glia communication. *Front. Physiol.* **2012**, *3*, 119. [CrossRef] [PubMed]
83. An, K.; Klyubin, I.; Kim, Y.; Jung, J.H.; Mably, A.J.; T O'Dowd, S.; Lynch, T.; Kanmert, D.; Lemere, C.A.; Finan, G.M. Exosomes neutralize synaptic-plasticity-disrupting activity of Aβ assemblies in vivo. *Molec. Brain.* **2013**, *6*, 47. [CrossRef] [PubMed]
84. Pusic, A.D.; Kraig, R.P. Youth and environmental enrichment generate serum exosomes containing miR-219 that promote CNS myelination. *Glia* **2014**, *62*, 284–299. [CrossRef]
85. Hamlett, E.D.; Goetzl, E.J.; Ledreux, A.; Vasilevko, V.; Boger, H.A.; LaRosa, A.; Clark, D.; Carroll, S.L.; Carmona-Iragui, M.; Fortea, J. Neuronal exosomes reveal Alzheimer's disease biomarkers in Down syndrome. *Alzheimer's Dement.* **2017**, *13*, 541–549. [CrossRef] [PubMed]
86. Xin, H.; Katakowski, M.; Wang, F.; Qian, J.-Y.; Liu, X.S.; Ali, M.M.; Buller, B.; Zhang, Z.G.; Chopp, M. MicroRNA-17–92 cluster in exosomes enhance neuroplasticity and functional recovery after stroke in rats. *Stroke* **2017**, *48*, 747–753. [CrossRef]
87. Ching, R.C.; Kingham, P.J. The role of exosomes in peripheral nerve regeneration. *Neural Regen. Res.* **2015**, *10*, 743.
88. Dong, R.; Liu, Y.; Yang, Y.; Wang, H.; Xu, Y.; Zhang, Z. MSC-Derived Exosomes-Based Therapy for Peripheral Nerve Injury: A Novel Therapeutic Strategy. *BioMed Res. Int.* **2019**, *2019*, 6458237–6458312. [CrossRef]
89. Mahmood, A.; Lu, D.; Chopp, M. Intravenous administration of marrow stromal cells (MSCs) increases the expression of growth factors in rat brain after traumatic brain injury. *J. Neurotrauma.* **2004**, *21*, 33–39. [CrossRef]
90. Gaudet, A.D.; Popovich, P.G.; Ramer, M.S. Wallerian degeneration: gaining perspective on inflammatory events after peripheral nerve injury. *J. Neuroinflamm.* **2011**, *8*, 110. [CrossRef]
91. Glenn, T.D.; Talbot, W.S. Signals regulating myelination in peripheral nerves and the Schwann cell response to injury. *Curr. Opin. Neurol.* **2013**, *23*, 1041–1048. [CrossRef] [PubMed]
92. Benowitz, L.I.; Popovich, P.G. Inflammation and axon regeneration. *Curr. Opin. Neurol.* **2011**, *24*, 577–583. [CrossRef] [PubMed]
93. Fugleholm, K. The surgery of peripheral nerves (including tumors). In *Handbook of Clinical Neurology*; Elsevier: Amsterdam, The Netherlands, 2013; Volume 115, pp. 781–802.
94. De Albornoz, P.M.; Delgado, P.J.; Forriol, F.; Maffulli, N. Non-surgical therapies for peripheral nerve injury. *Br. Med. Bull.* **2011**, *100*, 73–100. [CrossRef]
95. Liu, X.; Li, Q.; Niu, X.; Hu, B.; Chen, S.; Song, W.; Ding, J.; Zhang, C.; Wang, Y. Exosomes secreted from human-induced pluripotent stem cell-derived mesenchymal stem cells prevent osteonecrosis of the femoral head by promoting angiogenesis. *Int. J. Boil. Sci.* **2017**, *13*, 232. [CrossRef]
96. Hu, G.-w.; Li, Q.; Niu, X.; Hu, B.; Liu, J.; Zhou, S.-m.; Guo, S.-c.; Lang, H.-l.; Zhang, C.-q.; Wang, Y. Exosomes secreted by human-induced pluripotent stem cell-derived mesenchymal stem cells attenuate limb ischemia by promoting angiogenesis in mice. *Stem Cell Res. Ther.* **2015**, *6*, 10. [CrossRef]
97. Lai, R.C.; Chen, T.S.; Lim, S.K. Mesenchymal stem cell exosome: a novel stem cell-based therapy for cardiovascular disease. *Regen. Med.* **2011**, *6*, 481–492. [CrossRef]
98. Ludwig, A.-K.; Giebel, B. Exosomes: small vesicles participating in intercellular communication. *Int. J. Biochem. Cell Boil.* **2012**, *44*, 11–15. [CrossRef]
99. Huang, C.-C.; Narayanan, R.; Alapati, S.; Ravindran, S. Exosomes as biomimetic tools for stem cell differentiation: applications in dental pulp tissue regeneration. *Biomaterials.* **2016**, *111*, 103–115. [CrossRef]
100. Qiu, H.; Liu, S.; Wu, K.; Zhao, R.; Cao, L.; Wang, H. Prospective application of exosomes derived from adipose-derived stem cells in skin wound healing: A review. *J. Cosmet. Dermatol.* **2019**. [CrossRef]
101. Zhang, W.; Bai, X.; Zhao, B.; Li, Y.; Zhang, Y.; Li, Z.; Wang, X.; Luo, L.; Han, F.; Zhang, J. Cell-free therapy based on adipose tissue stem cell-derived exosomes promotes wound healing via the PI3K/Akt signaling pathway. *Exp. Cell Res.* **2018**, *370*, 333–342. [CrossRef] [PubMed]
102. Nguyen, D.; Orgill, D.; Murphy, G. The pathophysiologic basis for wound healing and cutaneous regeneration. In *Biomaterials for Treating Skin Loss*; Elsevier: Amsterdam, The Netherlands, 2009; pp. 25–57.

103. Hong, P.; Yang, H.; Wu, Y.; Li, K.; Tang, Z. The functions and clinical application potential of exosomes derived from adipose mesenchymal stem cells: a comprehensive review. *Stem Cell Res. Ther.* **2019**, *10*, 242. [CrossRef] [PubMed]
104. Ma, T.; Fu, B.; Yang, X.; Xiao, Y.; Pan, M. Adipose mesenchymal stem cell-derived exosomes promote cell proliferation, migration, and inhibit cell apoptosis via Wnt/β-catenin signaling in cutaneous wound healing. *J. Cell. Biochem.* **2019**, *120*, 10847–10854. [CrossRef] [PubMed]
105. Dai, M.; Yu, M.; Zhang, Y.; Tian, W. Exosome-like vesicles derived from adipose tissue provide biochemical cues for adipose tissue regeneration. *Tissue Eng. Part A.* **2017**, *23*, 1221–1230. [CrossRef]
106. Thoene; Koh; Nishida-Aoki; Sims; Heijnen; Arraud; Brisson; Yoshioka, Y.; Im; Ishihara; et al. Abstract Book: ISEV2017. *J. Extracell. Vesicles.* **2017**, *6*, 77–227.
107. Wisse, B.E. The inflammatory syndrome the role of adipose tissue cytokines in metabolic disorders linked to obesity. *J. Am. Soc. Nephrol.* **2004**, *15*, 2792–2800. [CrossRef]
108. Zhang, Y.; Yu, M.; Tian, W. Physiological and pathological impact of exosomes of adipose tissue. *Cell Prolif.* **2016**, *49*, 3–13. [CrossRef]
109. Lazar, I.; Clement, E.; Dauvillier, S.; Milhas, D.; Ducoux-Petit, M.; LeGonidec, S.; Moro, C.; Soldan, V.; Dalle, S.; Balor, S. Adipocyte exosomes promote melanoma aggressiveness through fatty acid oxidation: a novel mechanism linking obesity and cancer. *Cancer Res.* **2016**, *76*, 4051–4057. [CrossRef]
110. Cai, Z.; Zhang, W.; Yang, F.; Yu, L.; Yu, Z.; Pan, J.; Wang, L.; Cao, X.; Wang, J. Immunosuppressive exosomes from TGF-β1 gene-modified dendritic cells attenuate Th17-mediated inflammatory autoimmune disease by inducing regulatory T cells. *Cell Res.* **2012**, *22*, 607–610. [CrossRef]
111. Brito-Zerón, P.; Theander, E.; Baldini, C.; Seror, R.; Retamozo, S.; Quartuccio, L.; Bootsma, H.; Bowman, S.J.; Dörner, T.; Gottenberg, J.-E. Early diagnosis of primary Sjögren's syndrome: EULAR-SS task force clinical recommendations. *Expert Rev. Clin. Immunol.* **2016**, *12*, 137–156. [CrossRef]
112. Yang, C.; Robbins, P.D. Immunosuppressive exosomes: a new approach for treating arthritis. *Int. J. Rheumatol.* **2012**, *2012*. [CrossRef] [PubMed]
113. Li, N.; Zhao, L.; Wei, Y.; Ea, V.L.; Nian, H.; Wei, R. Recent advances of exosomes in immune-mediated eye diseases. *Stem Cell Res. Ther.* **2019**, *10*, 278. [CrossRef]
114. Harrell, C.R.; Fellabaum, C.; Jovicic, N.; Djonov, V.; Arsenijevic, N.; Volarevic, V. Molecular mechanisms responsible for therapeutic potential of mesenchymal stem cell-derived secretome. *Cells* **2019**, *8*, 467. [CrossRef]
115. Tran, T.-H.; Mattheolabakis, G.; Aldawsari, H.; Amiji, M. Exosomes as nanocarriers for immunotherapy of cancer and inflammatory diseases. *Clin. Immun.* **2015**, *160*, 46–58. [CrossRef] [PubMed]
116. Hedlund, M.; Nagaeva, O.; Kargl, D.; Baranov, V.; Mincheva-Nilsson, L. Thermal-and oxidative stress causes enhanced release of NKG2D ligand-bearing immunosuppressive exosomes in leukemia/lymphoma T and B cells. *PLoS ONE* **2011**, *6*, e16899. [CrossRef]
117. Coakley, G.; Maizels, R.M.; Buck, A.H. Exosomes and other extracellular vesicles: the new communicators in parasite infections. *Trends Parasitol.* **2015**, *31*, 477–489. [CrossRef] [PubMed]
118. Couper, K.N.; Barnes, T.; Hafalla, J.C.; Combes, V.; Ryffel, B.; Secher, T.; Grau, G.E.; Riley, E.M.; de Souza, J.B. Parasite-derived plasma microparticles contribute significantly to malaria infection-induced inflammation through potent macrophage stimulation. *PLoS Pathog.* **2010**, *6*, e1000744. [CrossRef] [PubMed]
119. Campos, F.M.; Franklin, B.S.; Teixeira-Carvalho, A.; Agnaldo Filho, L.; de Paula, S.C.; Fontes, C.J.; Brito, C.F.; Carvalho, L.H. Augmented plasma microparticles during acute Plasmodium vivax infection. *Malaria J.* **2010**, *9*, 327. [CrossRef] [PubMed]
120. Hu, G.; Gong, A.-Y.; Roth, A.L.; Huang, B.Q.; Ward, H.D.; Zhu, G.; LaRusso, N.F.; Hanson, N.D.; Chen, X.-M. Release of luminal exosomes contributes to TLR4-mediated epithelial antimicrobial defense. *PLoS Pathog.* **2013**, *9*, e1003261. [CrossRef] [PubMed]
121. Schnitzer, J.K.; Berzel, S.; Fajardo-Moser, M.; Remer, K.A.; Moll, H. Fragments of antigen-loaded dendritic cells (DC) and DC-derived exosomes induce protective immunity against Leishmania major. *Vaccine* **2010**, *28*, 5785–5793. [CrossRef]
122. del Cacho, E.; Gallego, M.; Lee, S.H.; Lillehoj, H.S.; Quilez, J.; Lillehoj, E.P.; Sánchez-Acedo, C. Induction of protective immunity against Eimeria tenella, Eimeria maxima, and Eimeria acervulina infections using dendritic cell-derived exosomes. *Infect. Immun.* **2012**, *80*, 1909–1916. [CrossRef] [PubMed]

123. Singh, P.P.; Smith, V.L.; Karakousis, P.C.; Schorey, J.S. Exosomes isolated from mycobacteria-infected mice or cultured macrophages can recruit and activate immune cells in vitro and in vivo. *J. Immun.* **2012**, *189*, 777–785. [CrossRef] [PubMed]
124. Willis, G.R.; Mitsialis, S.A.; Kourembanas, S. "Good things come in small packages": application of exosome-based therapeutics in neonatal lung injury. *Ped. Res.* **2018**, *83*, 298–307. [CrossRef] [PubMed]
125. van Haaften, T.; Byrne, R.; Bonnet, S.; Rochefort, G.Y.; Akabutu, J.; Bouchentouf, M.; Rey-Parra, G.J.; Galipeau, J.; Haromy, A.; Eaton, F. Airway delivery of mesenchymal stem cells prevents arrested alveolar growth in neonatal lung injury in rats. *Am. J. Respir. Crit. Care Med.* **2009**, *180*, 1131–1142. [CrossRef]
126. Chang, Y.S.; Ahn, S.Y.; Yoo, H.S.; Sung, S.I.; Choi, S.J.; Oh, W.I.; Park, W.S. Mesenchymal stem cells for bronchopulmonary dysplasia: phase 1 dose-escalation clinical trial. *J. Pediatrics* **2014**, *164*, 966–972. e966. [CrossRef]
127. Fonsato, V.; Collino, F.; Herrera, M.B.; Cavallari, C.; Deregibus, M.C.; Cisterna, B.; Bruno, S.; Romagnoli, R.; Salizzoni, M.; Tetta, C. Human liver stem cell-derived microvesicles inhibit hepatoma growth in SCID mice by delivering antitumor microRNAs. *Stem Cells* **2012**, *30*, 1985–1998. [CrossRef]
128. Herrera, M.; Fonsato, V.; Gatti, S.; Deregibus, M.C.; Sordi, A.; Cantarella, D.; Calogero, R.; Bussolati, B.; Tetta, C.; Camussi, G. Human liver stem cell-derived microvesicles accelerate hepatic regeneration in hepatectomized rats. *J. Cell. Mol. Med.* **2010**, *14*, 1605–1618. [CrossRef]
129. Masyuk, A.I.; Masyuk, T.V.; LaRusso, N.F. Exosomes in the pathogenesis, diagnostics and therapeutics of liver diseases. *J. Hepat.* **2013**, *59*, 621–625. [CrossRef]

MDPI
St. Alban-Anlage 66
4052 Basel
Switzerland
Tel. +41 61 683 77 34
Fax +41 61 302 89 18
www.mdpi.com

Applied Sciences Editorial Office
E-mail: applsci@mdpi.com
www.mdpi.com/journal/applsci

www.ingramcontent.com/pod-product-compliance
Lightning Source LLC
Chambersburg PA
CBHW061626210926
51850CB00009B/2451
* 9 7 8 3 0 3 6 5 2 1 4 8 0 *